—— INTRODUCTORY ——
Digital Signal Processing
with
Computer Applications

INTRODUCTORY

Digital Signal Processing
with
Computer Applications

SECOND EDITION

PAUL A. LYNN

formerly
Department of Electrical Engineering
Imperial College of Science, Technology and Medicine
University of London
UK

WOLFGANG FUERST

Information Management Services
United Nations
New York
USA

JOHN WILEY & SONS
Chichester · New York · Weinheim · Brisbane · Singapore · Toronto

Telephone   (+44) 1243 779777

Email (for orders and customer service enquiries): cs-books@wiley.co.uk
Visit our Home Page on www.wileyeurope.com or www.wiley.com

First edition 1989, reprinted six times
Revised edition 1994, reprinted six times
Reprinted January 1999, April 2000, October 2004, November 2007

Other Wiley Editorial Offices

John Wiley & Sons Inc., 111 River Street, Hoboken, NJ 07030, USA

Jossey-Bass, 989 Market Street, San Francisco, CA 94103-1741, USA

Wiley-VCH Verlag GmbH, Boschstr. 12, D-69469 Weinheim, Germany

John Wiley & Sons Australia Ltd, 33 Park Road, Milton, Queensland 4064, Australia

John Wiley & Sons (Asia) Pte Ltd, 2 Clementi Loop #02-01, Jin Xing Distripark, Singapore 129809

John Wiley & Sons Canada Ltd, 22 Worcester Road, Etobicoke, Ontario, Canada M9W 1L1

Library of Congress Cataloging-in-Publication Data

Lynn, Paul A
     Introductory digital signal processing with computer applications
          Paul A. Lynn, Wolfgang Fuerst. — 2nd ed.
               p.   cm.
          Includes bibliographical references (p.   ) and index.
          ISBN 0-471-97631-8 (alk paper)
                    1. Signal processing — Digital techniques. 2. Signal processing —
     Digital techniques — Computer programs.   I. Fuerst, Wolfgang.
     II. Title
     TK5102.9.L95 1997
     621.382'2—dc21   97-22235
                                                                          CIP

British Library Cataloguing in Publication Data

A catalogue record for this book is available from the British Library

ISBN 978 0 471 97631 8 (P/B)

Typeset by MHL Typesetting Ltd, Coventry.

This book is printed on acid-free paper responsibly manufactured from sustainable forestry
in which at least two trees are planted for each one used for paper production.

Contents

*Sections marked with an asterisk are not essential for understanding subsequent material, and may be omitted on a first reading.

Preface to the First Edition

Digital Signal Processing (DSP) has already moved from being primarily a specialist research topic to one with practical applications in many disciplines. Whereas a few years ago DSP courses were only offered at postgraduate level, many undergraduate programs now include them. The major impact has so far been in electrical and electronic engineering. However, DSP techniques are also of great value in other branches of engineering and science which involve data acquisition — including aerospace, civil, mechanical and biomedical engineering, physics, meteorology, and information systems.

This book aims to serve the needs of an introductory, one-semester, course in DSP. We regard it primarily as 'DSP for non-DSP majors'. Our approach does not assume any expertise in electronics, computing, or data processing. The first few chapters should be quite accessible to undergraduate students midway through their courses. In addition, we have tried to give a reasonably practical account of DSP, pointing out some of the main problems and pitfalls, and showing how to interpret the results of signal processing.

A considerable number of computer programs are used to illustrate the text. They are listed in an appendix in both BASIC and PASCAL and are available separately on disk. We have included them because we believe that newcomers to DSP gain greatly by seeing the results of signal analysis and processing build up on a computer screen. Computer graphics are valuable for demonstrating key concepts and results, and also in design. The programs are simple in structure and ready to run on a wide variety of computers, from modest personal computers upwards. They have been developed on an industry-standard PC, with graphics based on a single graphics statement. It should therefore be a simple matter to adapt them to other machines. We should add that no claim is made for them as exercises in programming technique; they are provided purely as aids to understanding DSP.

Although we hope that students will have the opportunity to try the computer programs for themselves, it is not essential. Typical results and graphics plots from all the programs are illustrated and discussed in the main text.

In more detail, Chapter 1 provides a gentle introduction to the field of DSP, covering the basic notions of digital signals and linear time-invariant (LTI) systems. The approach is strictly digital, avoiding any mention of analog circuits or techniques. In this way we hope to make the book attractive across a wide range of disciplines.

Chapter 2 deals with time-domain analysis, including convolution, and Chapter 3 introduces digital Fourier techniques. Readers who are unfamiliar with classical (continuous-time) Fourier analysis will probably find the material in Appendix A2 useful at this stage. We restrict ourselves in Chapter 3 to digital equivalents of the classical Fourier Series and Fourier Transform — topics which will already be familiar to most readers. The very important Fourier representation known as the Discrete Fourier Transform (DFT), and its implementation by Fast Fourier Transform (FFT), is deliberately reserved for special treatment in Chapters 7 and 8.

Chapter 4, which introduces the z-Transform, concentrates on those aspects which are necessary to understand later chapters. An emphasis is placed on the z-plane pole-zero description of a signal or system, and its relationship with the Fourier Transform. We have used the unilateral version of the z-Transform, because it is adequate for our task and has much simpler convergence conditions than the bilateral version.

Chapters 5 and 6 describe some of the best-known techniques for designing nonrecursive and recursive digital filters respectively, and are supported by a number of filter design programs.

Chapter 7 gives an account of the DFT, the computational problems of implementing it directly, and the design of a range of FFT algorithms. We have concentrated on radix-2 algorithms, introducing the basic ideas by reference to a conventional decimation-in-time decomposition, and developing them further using the index-mapping approach. Some brief notes on higher radix algorithms are also included. Overall, our aim is to give a clear introduction to a number of FFT topics — including butterflies and twiddle factors — which often cause a certain amount of consternation!

Chapter 8 is essentially an applications chapter which shows how to use and interpret the results of FFT processing. It includes what we hope is a clear account of the rather difficult topic of spectral leakage.

Taking an overall view, Chapters 1 thru 4 present the basic theoretical background of linear DSP, providing the foundation for subsequent chapters. Chapters 5 and 6 are more or less independent of one another, although each relies heavily on earlier material. This also applies to Chapters 7 and 8. Therefore, having followed Chapters 1 thru 4 (preferably in sequence), the remaining chapters can be studied as required.

The text includes a number of Worked Examples, designed to illustrate and develop important ideas and design techniques. Problems are supplied at the end of each chapter, and selected answers are given at the end of the book. The problems are designed to test and consolidate work already done; we have avoided the temptation to use them as a means of introducing further theoretical material.

We have used a great many sources in writing this book. They include lecture notes, unrecorded conversations, results of personal investigations, and many textbooks and research papers. We have recorded our debt to the authors of a number of books listed in the Bibliography by mentioning them in the text, and suggesting them as valuable references for further study.

At a more personal level, it is a pleasure to thank Jeana Price for all her work on the manuscript, and the team at John Wiley & Sons for their enthusiasm throughout the project.

PAUL A. LYNN   WOLFGANG FUERST
4 Kensington Place   *15 Cathedral Court*
Clifton   *Hempstead*
Bristol BS8 3AH   *New York 11550*
UK   *USA*

Note to the Revised Edition

In this revised edition we have taken account of recent trends in the teaching and use of programming languages by converting all the book's computer programs into C. Our thanks are due to Barry Thomas of Thomas Technical Publications Ltd., Ashbourne (UK), and Ian Jack of John Wiley and Sons Ltd., Chichester, for their help with the translation.

The new C programs, as well as the PASCAL originals, are now listed in Appendix A1. We hope that this change, plus the inclusion of a free disk, will further encourage our readers to explore the principles and practice of DSP using personal computers.

A number of minor changes have been made to the main text to take account of the great increase in speed of personal computers over the past few years.

PAUL A. LYNN
WOLFGANG FUERST
May 1994

Preface to the Second Edition

Digital Signal Processing goes from strength to strength. The remarkable increase in speed and power of digital computers and special-purpose hardware over the last few years has ensured the continued growth of interest in this fascinating area, both as a subject for study and as a practical tool for problem-solving over a wide range of disciplines.

The major change in this new edition is the addition of two new chapters on Random Digital Signals and Random DSP. We hope that these chapters will enhance the book's value as an introduction to the fundamentals of DSP and its applications.

Random phenomena are widely regarded as difficult topics. Yet they have become so central to modern signal processing, communications, control, and data processing that they can hardly be omitted from even an introductory text on DSP. Actually, our own view is that such phenomena are perfectly accessible within the digital context, where the mathematics are simpler and the concepts easier to explain and illustrate than they are in analog terms. In addition, we make no apology for offering a strictly introductory approach. By concentrating exclusively on important aspects of Random DSP which fit naturally into the framework of earlier chapters of the book, we aim to help students' understanding of the time and frequency domain descriptions of signals and systems, and the constant interplay between them. It is our belief that a clear understanding of these central ideas, extended to random phenomena, will stand them in good stead for tackling more advanced topics which they may meet later.

The new chapters are supported by additional computer programs which, as before, we use to illustrate the main text. They have been added to the free disk which accompanies the book. It is remarkable how helpful computer-based illustrations are for understanding and appreciating random DSP: the rather daunting mathematics of many traditional textbooks is replaced by a clearer view of the processes involved, plus the chance to experiment by modifying the programs. Most students undoubtedly find this approach helpful.

Actually, in one respect the recent increase in speed of personal computers is something of a disadvantage. From a teaching point of view, it is very helpful to observe the steady build-up of digital signals or the effects of processing on the computer screen. When the first edition of this book was published, small computers were generally so slow that rather too much patience was required with some of our programs (especially those involving Fourier transformation).

Then, by the time of the revised edition in 1994, speeds had generally become close to ideal for 'watching DSP in action'. Now technology has moved on again, and personal computers can execute even the most demanding of the programs within the proverbial twinkling of an eye. It will not surprise us to learn in due course that some users of the book have deliberately introduced time-wasting routines in some of the programs to enhance their didactic value!

Apart from the two new chapters, little has changed in this edition. We have removed or adapted a few comments on the speed of computers and computer programs, inserted some comments to link the new chapters with the old, corrected a few minor errors in the text, and extended the bibliography.

It is a pleasure to thank the team at John Wiley & Sons for their continued enthusiasm and support for the book in its new edition.

PAUL A. LYNN   WOLFGANG FUERST
Sage's Mead   *15 Cathedral Court*
Butcombe   *Hempstead*
North Somerset BS18 6XF   *New York 11550*
UK   *USA*

Spring 1998

—— CHAPTER 1 ——
Introduction

1.1 THE SCOPE OF DIGITAL SIGNAL PROCESSING

1.1.1 BACKGROUND

The extraordinary growth of microelectronics and computing has had a major impact on digital signal processing (DSP). Powerful DSP techniques are now used to analyze and process signals and data arising in many areas of engineering and science, medicine, economics, and the social sciences. Most people trained in a numerate discipline can expect to encounter DSP at some stage in their careers, and there is every sign that the trend will continue.

DSP is concerned with the numerical manipulation of signals and data in sampled form. Using such elementary operations as digital storage and delay, addition, subtraction, and multiplication by constants, we can produce a wide variety of useful functions. For example, we may need to detect trends in data; to extract a wanted signal from unwanted 'noise'; or to assess the frequencies present in a signal.

Many readers of this book will have a background in electronic and electrical engineering. Others will come to DSP via the other branches of engineering, physics, computer science, or mathematics. And some of you will be interested in data processing applied to economics and the social sciences. We therefore assume no knowledge of electronics or computer hardware. Our approach is to introduce the basic theory and applications of DSP, illustrating important concepts and design techniques with the help of computer programs. The programs, listed in both C and PASCAL in Appendix A1, and provided on the free disk, are suitable for a wide range of personal computers — and, of course, more powerful machines. It is not essential that you run the programs for yourself, because their graphical outputs are reproduced at various points in the text. However, we hope that many of you will have the opportunity to do so.

The use of general-purpose computers is straightforward and convenient for illustrating DSP theory and applications. However if high-speed real-time signal processing is required, it may be essential to use special-purpose digital hardware. Intermediate between these two extremes come programmable microprocessors, possibly attached to a general-purpose host computer. Many manufacturers now produce DSP microprocessors, tailored to the requirements of signal processing. The use of such devices, and the design of dedicated DSP hardware, is a fast-growing field. So although we concentrate here on general-purpose computers, we must always remember that they are only a part of modern DSP.

Various terms are used to describe signals in the DSP environment. By *discrete-time signal*, we mean a signal which is defined only for a particular set of instants in time, or *sampling instants*. A good example would be the midday temperature at a certain place, measured on successive days. Discrete-time signals may be subdivided into two categories: *sampled-data signals*, which display a continuous range of amplitude values; and *digital signals*, in which the amplitude values are quantized in a series of finite steps. Signals stored or processed in a computer are digital, because each sample value is represented by a finite-length binary code. However, in practice the terms discrete-time signal and digital signal are often used interchangeably.

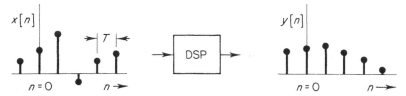

Figure 1.1 Input and output signals of a digital signal processor.

Two discrete-time signals are shown in Figure 1.1. Here $x[n]$ is the sampled input to a digital signal processor, and $y[n]$ is the sampled output. Successive input signal values ... $x[-1]$, $x[0]$, $x[1]$, $x[2]$ ... may be thought of as regularly-spaced samples of an underlying analog signal, and are separated from one another by the system sampling interval T. The integer variable n denotes the sample number, with $n=0$ corresponding to some convenient time origin or reference.

Note that x and y are not *necessarily* time functions. For example they may be functions of distance. However, for convenience we generally assume that n is a time variable.

In effect we may regard $x[n]$ and $y[n]$ as sequences of numerical data, which are binary-coded for the purposes of signal processing. Although we restrict ourselves in this book to one-dimensional signals, two-dimensional (or even multi-dimensional) DSP is of increasing practical importance. A good example is the processing of pictures and images, such as those obtained from X-ray brain-scanners or from satellite reconnaissance. Fortunately the techniques of one-dimensional DSP can be extended to such applications without too much difficulty.

Figure 1.2 illustrates a typical DSP scheme. It is often said that we live in an analog world, and most practical signals start off in analog form. Examples might be variations in voltage, temperature, pressure, or light intensity. If the signal is not inherently electrical, it is first converted to a proportional voltage fluctuation by a suitable transducer. The transducer provides the analog input to the signal processing chain. Very often, the first stage in the chain is an analog filter, designed to limit the frequency range of the signal prior to sampling. The reasons for this will become clear in Section 1.2.

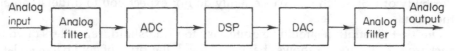

Figure 1.2 A typical DSP scheme.

The signal is next sampled and converted into a binary code by an analog-to-digital converter (ADC). After digital signal processing it may be changed back into analog form using a digital-to-analog converter (DAC). The final processing stage is often another analog filter, designed to remove sharp transitions from the DAC output.

There are a number of possible variations on the above theme. For example, after DSP the signal may be used to drive a computer display, or it may be transmitted in binary form to a remote terminal or location. Another important variation concerns data processing applications in such fields as economics and the social sciences. Here the numerical data representing the input signal may be available on a computer disk, or it may be entered on a keyboard by a computer operator. In such cases sampling and analog-to-digital conversion is an inherent part of the data entry process. Similarly, a computer print-out of processed sample values may take the place of digital-to-analog conversion.

To an electronic engineer, the DSP scheme in Figure 1.2 may seem rather a complicated way of proceeding. There are, after all, many well-established signal processing techniques based on analog electronic circuits and components. Why do we go to the trouble of converting signals into digital form, and then back again? There are several good reasons:

Signals and data of many types are increasingly stored in digital computers, and transmitted in digital form from place to place. In many cases it makes sense to process them digitally as well.

Digital processing is inherently stable and reliable. It also offers certain technical possibilities not available with analog methods.

Rapid advances in integrated circuit design and manufacture are producing ever more powerful DSP devices at decreasing cost.

In many case DSP is used to process a number of signals simultaneously. This may be done by *interlacing* samples obtained from the various signal channels — a technique known as *time-division-muliplexing.*

Of course, those of you who are interested in data processing applications will know that there is no realistic substitute for digital methods. The nature of your numerical data, and the way in which it is gathered, make computer storage and processing the obvious choice.

1.1.2 SOME PRACTICAL APPLICATIONS

We hope that you now have an idea of the general flavor of DSP. In this section we illustrate some practical applications drawn from different disciplines. Our aim is to provide motivation, and to show something of the power and flexibility of DSP methods. We certainly do not expect you to understand all the details at this stage.

Our first illustration should be quite easy to understand, regardless of your background. Figure 1.3 shows changes in the dollar price of gold between late 1979 and 1983. The heavy, rapidly fluctuating, curve represents the price recorded day by day. There are more than 1500 values altogether, so for convenience we have joined the samples together to give a continuous curve.

It is often helpful to smooth such 'raw data', to reduce rapid fluctuations and reveal underlying trends. A widely-used technique is to estimate a *moving average*. Each smoothed value is computed as the average of a number of preceding raw-data values, and the process is repeated sample-by-sample through the record. The figure shows the 200-day moving average (with individual values again joined together for convenience). 200-Day averages are popular with analysts of financial and stock market data for investigating long-term trends. Clearly, such processing gives a very pronounced smoothing action.

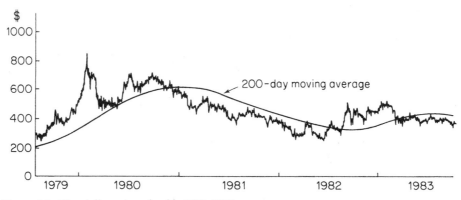

Figure 1.3 The dollar price of gold, 1979–1983.

From the DSP point of view, the daily gold price provides samples of an input signal $x[n]$. The smoothed values represent output samples $y[n]$ from the moving-average process. The 200-day average may be estimated using the algorithm:

$$y[n] = \frac{1}{200}\{x[n] + x[n-1] + x[n-2] + \cdots x[n-199]\}$$

$$= 0.005 \sum_{k=0}^{199} x[n-k] \tag{1.1}$$

It is a simple matter to store the raw data in a computer, and to write a program for calculating $y[n]$. Every time a new value of x becomes available, we can produce a new value of y.

Equation (1.1) is a *recurrence formula*, which is used over and over again to find successive values of y. It is also called a *difference equation*. It is *nonrecursive*, because each output sample is computed solely from input values. And the equation defines the operation of a simple *digital filter*, which in this case produces a smoothing action on the input data. Much of DSP is concerned with the design of digital filters, which can be tailored to produce a wide variety of processing functions.

You may have noticed that the above moving-average algorithm is not very efficient. The computation of each output value is almost exactly the same as the one before it, except that the most recent input sample is included, and the most distant one is discarded. Using this idea, we can generate a much more efficient algorithm. Equation (1.1) shows that the *next* smoothed output value must be:

$$y[n+1] = 0.005\{x[n+1] + x[n] + x[n-1] + \cdots x[n-198]\}$$
$$= y[n] + 0.005\ \{x[n+1] - x[n-199]\}$$

Since this is a recurrence formula which applies for *any* value of n, we may subtract 1 from each term in square brackets, giving:

$$y[n] = y[n-1] + 0.005\{x[n] - x[n-200]\} \tag{1.2}$$

Equation (1.2) confirms that we can estimate each output sample by *updating the previous output $y[n-1]$*. The equation defines a *recursive* version of the filter which is much more efficient, requiring far fewer additions/subtractions. Note that equations (1.1) and (1.2) are equivalent from the signal processing point of view.

Although Figure 1.3 refers to financial data, similar smoothing operations are widely used in DSP. For example, electronic engineers often need to remove unwanted interference, or 'noise', from a relatively slowly-varying signal. A smoothing filter used for this purpose is known as a *low-pass filter*. That is, it passes (transmits) low frequencies, representing slow fluctuations; but reduces high frequencies.

In medicine, the electrical activity of the heart can be recorded using electrodes placed on the chest. The resulting *electrocardiogram* (EKG) is often contaminated by rapid fluctuations due to electrical activity in nearby muscles. These, too, may be reduced by processing with a low-pass filter. We therefore see that the type of DSP action shown in Figure 1.3 has a wide range of practical uses. In fact there are many types of digital low-pass filter, and the simple moving-average design we have described would not be suitable for most applications.

Mention of the EKG leads to our next illustration of DSP. Figure 1.4(a) shows a typical EKG waveform, corresponding to a single heartbeat. In part (b) of the figure it is badly contaminated — not in this case by muscle 'noise', but by sinusoidal interference at mains supply frequency (60 Hz in the USA, 50 Hz in Europe). This is quite a common problem when recording biomedical signals, due to pick-up in electrode leads. It must obviously be reduced before the signal can have diagnostic value. Since biomedical signals are often stored in digital computers, we have the opportunity to reduce the interference with a digital filter.

Before describing the filter algorithm, we should explain the form of the sampled signals in the figure. They have been drawn by computer, and each sample value is represented by a thin vertical bar. In this case there are 640

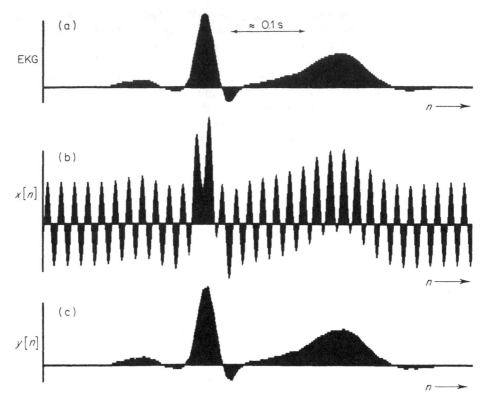

Figure 1.4 Removing mains-frequency interference from an electocardiogram (EKG) *(abscissa: 640 samples).*

sample values across the page, and the vertical bars are contiguous. This gives the plots their solid appearance. You will meet many more examples of such computer plots in this book, and will have the opportunity to try later programs for yourself on a personal computer. Note that in most cases there will be 320 sample values (or less) across the page, sometimes with gaps between adjacent vertical bars. The figure caption will make clear how many samples are plotted.

Let us now return to the problem of filtering the EKG. In this case we do not need a smoothing (low-pass) design, but one which rejects a narrow band of frequencies centered at mains supply frequency. This is called a *bandstop*, or *notch*, filter. The following recurrence formula may be used:

$$y[n] = 1.8523\, y[n-1] - 0.94833\, y[n-2] + x[n] - 1.9021\, x[n-1] + x[n-2]$$
$$(1.3)$$

Figure 1.4(c) shows the dramatic effect of this recursive filter on the contaminated signal of part (b). The interference has been greatly reduced, without significantly distorting the signal waveform.

At this stage we certainly do not expect you to understand how equation (1.3) has been derived. The illustration is intended to whet the appetite for some DSP theory in later chapters! However, we should just add that a digital filter is always designed for a particular sampling rate. If the interference is at 60 Hz, this algorithm is effective at a sampling frequency of 1200 samples per second (1.2 kHz); if it is at 50 Hz, the algorithm is effective at 1000 samples per second (1 kHz).

Our third introductory example is based on a computer program which you can try out for yourself. It shows the action of a simple *bandpass* filter — that is, a filter which passes a particular band of frequencies relatively strongly. By the time you have finished reading this section, you will therefore have met examples of low-pass, bandstop, and bandpass filters.

Bandpass filters are useful in many signal processing applications. A good example is electronic communications. Many communications systems — for example, radio, television, or data channels — send information from place to place within a well-defined frequency band. Signals can therefore be separated from one another at the receiving end on the basis of their different frequencies. Furthermore, interference or noise picked up during transmission often occupies a wide frequency range, and can be reduced with a suitable bandpass filter.

The recursive filter we have chosen is defined by:

$$y[n] = 1.5\, y[n-1] - 0.85\, y[n-2] + x[n] \tag{1.4}$$

This particular algorithm is designed to transmit a sampled sinusoidal signal relatively strongly when each cycle (or *period*) of the sinusoid contains about ten sample values. If there are more, or less, samples per period the output signal is reduced in amplitude. Computer Program no.1 in Appendix A1 produces a graphical output of the form shown in Figure 1.5. At the top is a sinusoidal input signal $x[n]$ of constant amplitude but variable frequency. It starts off with about twenty samples per period, decreasing to about six samples per period on the right-hand side. As it 'sweeps through' the frequency corresponding to ten samples per period, the output signal $y[n]$ displays an amplitude peak. By altering the multiplier coefficients in equation (1.4) we could easily achieve a more (or less) pronounced bandpass effect.

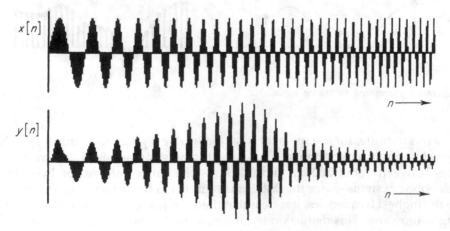

Figure 1.5 The action of a digital bandpass filter (*abscissa: 320 samples*).

If you refer to Appendix A1, you will see that Program no.1 has three main parts:

(a) It generates the 'swept-frequency' input signal. (You need not concern yourself with how this is done.)
(b) It uses an iterative loop to implement equation (1.4) over and over again for different values of n.
(c) It produces a graphical output of the type shown in Figure 1.5.

Part (b) is the most important from our point of view, since it shows just how easily a general-purpose computer can be programmed as a digital filter.

The four recurrence formulae discussed in this section should give you some idea of the possibilities offered by DSP. We hope that you are feeling sufficiently tantalized to want to tackle some theory! However, before we really start this in Chapter 2, we need to cover some further introductory aspects of digital signals and systems.

1.2   SAMPLING AND ANALOG-TO-DIGITAL CONVERSION

We have already mentioned the conversion from analog to digital signals, and back again, and have illustrated a DSP scheme in Figure 1.2. We should now look at the important processes of sampling and analog-to-digital conversion rather more carefully.

Suppose that an analog signal is to be represented by a set of equally-spaced samples. How often should it be sampled? An intuitive answer is suggested by Figure 1.6. Part (a) shows a signal sampled at a rate which is clearly too low to pick out the more rapid fluctutations. In part (b), however, sampling appears unnecessarily fast. A great many sample values would have to be stored, processed, or transmitted. Can we choose some intermediate sampling rate which is adequate, without being excessive?

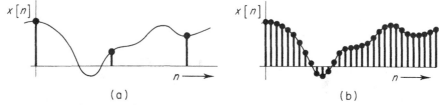

Figure 1.6 Sampling an analog signal.

Let us assume, for the moment, that the analog signal represents a speech waveform from a microphone. Then it will typically contain significant frequencies between about 300 Hz and 3 kHz — the range catered for by many telephone systems. Since the fastest fluctuations in the waveform correspond to the highest frequencies, it is clear that these frequencies will be lost if sampling is too slow. This deduction is confirmed by Shannon's famous *Sampling Theorem*, which may be stated as follows:

An analog signal containing components up to some maximum frequency f_1 Hz may be completely represented by regularly-spaced samples, provided the sampling rate is at least $2f_1$ samples per second.

Thus a speech waveform having f_1 = 3 kHz should be sampled at least 6000 times per second. Note that this corresponds to *two samples per period* of the highest frequency present. To take another example, we might need to sample a television signal having a maximum frequency of 5 MHz (5×10^6 Hz). The Sampling Theorem tells us that it should be sampled at least 10 million times per second.

If we use the minimum rate specified by the theorem, then the sampling interval T is clearly:

$$T = \frac{1}{2f_1} \tag{1.5}$$

Alternatively, if we have a digital system with sampling interval T, the maximum analog frequency which it can represent is:

$$f_1 = \frac{1}{2T} \text{ Hz}, \quad \text{or} \quad \omega_1 = 2\pi f_1 = \frac{\pi}{T} \text{ radians/sec} \tag{1.6}$$

At first sight it may seem surprising that a set of instantaneous samples can ever give a *complete* representation of an analog signal. It implies that we could recover the original signal perfectly from its sample values if we wished. This is indeed the case, and means that sampling need not involve any loss of information. We now present, in simplified form, the reasoning behind this extremely important idea.

Figure 1.7(a) represents the frequency distribution, or *spectrum*, of an analog speech signal. Its precise shape is unimportant; what matters is that there are no significant components above some maximum frequency f_1 — here taken as 3 kHz. (We have drawn the spectral *magnitude* $|H(f)|$, ignoring the *phases* of the various components. Once again, this is adequate for the present discussion.) Note that we have shown the spectrum as an *even* function, extending to negative frequencies. This widely-used representation results from expressing each frequency component as the sum of two exponentials. Thus a component $A \cos(\omega t)$ of amplitude A and frequency ω radians per second may be written as:

$$A \cos (\omega t) = \frac{A}{2} \exp (j\omega t) + \frac{A}{2} \exp (j\{-\omega\}t) \tag{1.7}$$

and represented by *two* spectral components of amplitude $A/2$ at frequencies $\pm \omega$.

It may be shown that sampling the analog signal causes the original spectrum to repeat around multiples of the sampling frequency. This is illustrated in part (b) of the figure for a sampling frequency of 8000 samples per second (rather higher than required by the Sampling Theorem). The original spectrum repeats around 8 kHz, 16 kHz, and so on. We may think of the spectral repetitions, which in theory extend to infinitely high frequencies, as a consequence

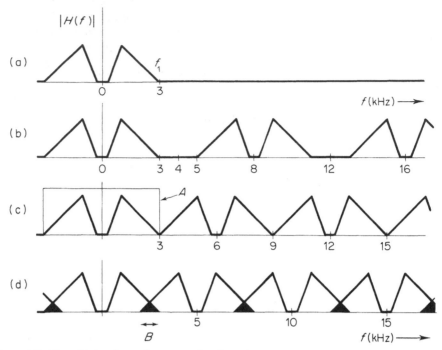

Figure 1.7   The effects of sampling on a signal spectrum.

of representing a smooth analog waveform by a set of narrow samples, or *impulses*.

In principle we could recover the spectrum shown in part (a) — and hence the original analog signal — with a low-pass filter. Such a *reconstituting filter* should transmit components up to 3 kHz, but reject all the spectral repetitions caused by sampling. Since the first of the repetitions in part (b) of the figure begins at 5 kHz, the filter should eliminate components at 5 kHz and above.

What happens if we reduce the sampling rate to the minimum value demanded by the Sampling Theorem — in this example, 6 kHz? Figure 1.7(c) shows the resulting spectrum. The repetitions now occur around 6 kHz, 12 kHz, and so on. In the region of 3 kHz they touch, but do not overlap, so it is still just theoretically possible to recover the original spectrum by low-pass filtering. The required filter characteristic is labeled 'A' in the figure. It must have an infinitely sharp cut-off at 3 kHz, and is referred to as the 'ideal' reconstituting filter.

If we reduce the sampling frequency further — say to 5 kHz — we get overlap between successive spectral repetitions. The effect, known as *aliasing*, is illustrated in part (d) of the figure. The overlap in the region of 3 kHz (labeled 'B') corrupts the original spectrum, which cannot now be recovered. Information has been lost through inadequate sampling.

We have discussed the problem in relation to a speech signal with $f_1 = 3$ kHz. But the same basic ideas apply to a very wide variety of signals, data, and sampling rates.

We therefore see that the Sampling Theorem defines the minimum sampling rate which avoids spectral overlap, or aliasing. In practice we choose a somewhat higher sampling rate — for example, speech signals having $f_1 = $ 3 kHz are commonly sampled at 8 kHz. The main reason is that we cannot make an ideal reconstituting filter with an infinitely sharp cut-off. By sampling rather faster than the minimum rate we introduce a *guard band* between adjacent spectral repetitions (for example, the range 3–5 kHz in part (b) of the figure). The reconstituting filter can now have a more gentle cut-off. Also we have a small safety margin, in case the maximum frequency f_1 has been underestimated.

While on the subject of sampling, we should introduce a little more terminology. The maximum frequency f_1 contained in an analog signal is widely referred to as the *Nyquist frequency*. The minimum sampling rate ($2f_1$ samples per second) at which it can theoretically be recovered is known as the *Nyquist rate*. And the so-called *folding frequency*, which equals half the sampling frequency actually used, is the highest frequency which can be adequately represented according to the Sampling Theorem. When the theorem is just obeyed, the Nyquist and folding frequencies are equal.

Referring back to Figure 1.2, it is now clear why our general DSP scheme includes an analog filter on the input side. Known as an *anti-aliasing filter*, it has low-pass characteristics and is designed to ensure that the maximum frequency $f_1 = 1/2T$ is not exceeded.

Let us now consider the sampling process in a little more detail. In most electronic DSP applications, sampling is performed by an analog-to-digital converter (ADC), which also transforms the stream of samples into a binary code. It is simple to show that a binary code N bits long allows 2^N separate numbers, or signal values, to be represented. Thus if $N = 8$ we may encode $2^8 = 256$ discrete values; if $N = 16$ we may encode $2^{16} = 65\,536$ values, and so on.

Analog signals normally take on a continuous range of amplitudes, so when they are sampled and binary-coded small amplitude errors are bound to be introduced. This is illustrated by Figure 1.8. For simplicity, we have assumed

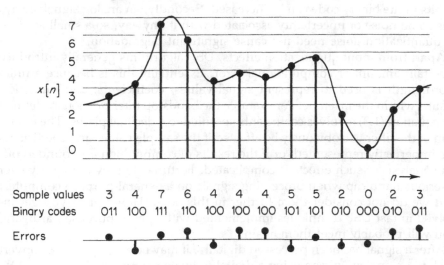

Sample values	3	4	7	6	4	4	4	5	5	2	0	2	3
Binary codes	011	100	111	110	100	100	100	101	101	010	000	010	011

Errors

Figure 1.8 Converting an analog signal into a binary code.

a 3-bit code, which can represent just $2^3 = 8$ separate sample values. The total amplitude range is therefore divided into eight *quantization levels* or 'slots', and each sample value is allocated the appropriate binary code. The sequence of coded values now represents the original signal, and may be processed digitally.

The maximum error introduced into each sample value by this process is plus or minus half a quantization level. With a 3-bit code this equals $\pm\frac{1}{16}$ of the total available amplitude range. In most practical applications such large errors would be unacceptable, and a longer binary code would be used. Note that the magnitudes of successive errors, also shown in the figure, are often more or less random and independent of one another. We can therefore regard them as unwanted *quantization noise*, introduced by the coding process. It cannot subsequently be removed.

Note that we are assuming the quantization to be uniform: that is, the various quantization levels are assumed equally spaced. Nonuniform quantization is sometimes encountered — for example, in coding schemes for digital communication systems, and in the roundoff of floating point numbers. However it is more difficult to describe and visualize, and is not covered in this book.

Quantization also occurs when a signal is coded 'manually' rather than automatically. For example, you may wish to record some midday temperature values, and enter them into a computer via a keyboard. You make a decision about the accuracy of the data, and quantize the numerical values accordingly. Thus a true temperature value of 18.27 °C (if you can read it that accurately!) might be entered as 18.3 °C, or just as 18 °C — and so on. Once again, the sample values have been quantized, and some quantization noise has been introduced. Of course, in this case the samples are converted into a binary code inside the computer. And if you attempt to enter an extremely accurate value, say 18.2 737 948 65, the amount of quantization will probably be determined by the binary wordlength of the computer itself.

Analog-to-digital conversion therefore always degrades a signal to some extent. This may appear to be a serious drawback of digital processing. However we must remember that if the quantization errors are too large, then the number of bits in the binary code can be increased. Secondly, an analog signal always has some noise or uncertainty associated with it anyway, so a small amount of quantization noise need not cause significant degradation.

Apart from input quantization effects, DSP algorithms generally introduce a certain amount of computational error. Essentially this is because a finite wordlength is used to represent signal values, coefficients, and numerical results within the machine. For example the bandstop filter previously defined by equation (1.3) involves coefficients specified to 4 decimal places. The operation of the filter will obviously be affected if the sample values and coefficients are imperfectly represented, or if the results of computation are rounded off. The analysis of such effects is complicated. Fortunately they are hardly ever a serious handicap when processing signals on a general-purpose computer, and we do not consider them further in this book. However, if you get involved in fast DSP techniques implemented with special-purpose hardware, you will probably meet them again!

After a signal has been processed digitally, it may be appropriate to convert back to an analog voltage using a digital-to-analog converter (DAC). A DAC

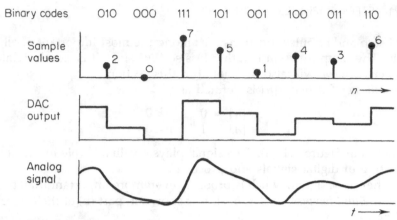

Figure 1.9 Digital-to-analog conversion.

usually produces an output of the *zero-order-hold* type, also known as *sample-and-hold*. This is illustrated in Figure 1.9. Each signal sample value, corresponding to the binary code delivered to the DAC input, is 'held' during the following sampling interval. The resulting staircase waveform is adequate for many practical applications (including the computer plots illustrated in this book). If a 'true' analog output is required, a further smoothing filter must be employed. The fully reconstituted signal is then as shown at the bottom of the figure.

1.3 BASIC TYPES OF DIGITAL SIGNAL

Our next task is to describe some basic types of digital signal. You may feel that these signals, with their rather simple waveforms and straightforward mathematical descriptions, are of limited value. After all, Figures 1.3 and 1.4 have shown examples of real-life signals which are very complicated in form and structure, so we begin by summarizing the main reasons why basic signals are so valuable:

Complicated, real-life, signals may generally be considered as the summation of a number of simpler, basic, signals.

Many useful DSP algorithms are *linear*. The response of a linear algorithm, or processor, to a number of signals applied simultaneously equals the summation of its responses to each signal applied separately. Thus if we can define the response of a linear processor to basic signals, we can predict its response to more complicated ones.

Basic signals are simple to describe and generate, and are widely used as *test signals* for investigating the properties of linear processors and systems.

We might note in passing that this book concentrates exclusively on linear DSP techniques. Nonlinear ones are certainly important, and have found many practical applications. But their theoretical treatment is much more difficult, and cannot easily be covered in an introductory text.

1.3.1 STEPS, IMPULSES, AND RAMPS

Digital step and impulse functions are among the most important of all basic
signals. The ramp function is rather less so, but since it is closely related to
the other two, we will also describe it in this section.

The *unit step function u[n]* is defined as:

$$u[n] = 0, \qquad n < 0$$
$$u[n] = 1 \qquad n \geq 0 \tag{1.8}$$

It is shown in Figure 1.10(a). This signal plays a valuable role in the analysis
and testing of digital signals and processors.

A further basic signal, which is probably even more important than the unit
step, is the *unit impulse function δ[n]*. It is shown in part (b) of the figure, and
is defined as:

$$\delta[n] = 0, \qquad n \neq 0$$
$$\delta[n] = 1, \qquad n = 0 \tag{1.9}$$

Also known as the *unit sample*, $\delta[n]$ consists of an isolated unit-valued sample
at $n = 0$, surrounded on both sides by zeros. We shall see later that this signal
is of the greatest value in DSP theory and practice.

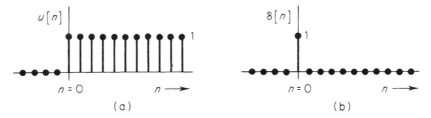

Figure 1.10 Basic digital signals: (a) the unit step function, and (b) the unit impulse
function.

$u[n]$ and $\delta[n]$ are closely related to one another. $u[n]$ is said to be the *running
sum* of $\delta[n]$. This is simple to visualize. If we start at the left-hand side of Figure
1.10(b), and move to the right summing the sample values of $\delta[n]$ as we go,
we generate the unit step function $u[n]$. The relationship may be written formally as:

$$u[n] = \sum_{m=-\infty}^{n} \delta[m] \tag{1.10}$$

Conversely, we can easily generate $\delta[n]$ from $u[n]$. The recurrence formula:

$$\delta[n] = u[n] - u[n-1] \tag{1.11}$$

holds good for all integer values of n. Therefore $\delta[n]$ is referred to as the *first-
order difference* of $u[n]$.

Ramp functions also deserve a mention here. A digital ramp signal rises,
or falls, linearly with the variable n. It is sometimes used as a test signal, or
for approximating a practical signal by a set of straight line sections. The *unit*

ramp function $r[n]$ is defined as:

$$r[n] = n\,u[n] \tag{1.12}$$

Since $u[n]$ is zero for $n<0$, so also is the ramp function. Note that $r[n]$ is a running sum of $u[n]$; alternatively, $u[n]$ is a first-order difference of $r[n]$. The unit-ramp signal is shown in Figure 1.11.

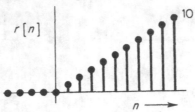

Figure 1.11 The unit ramp function.

A number of worked examples are included in this book, to help you consolidate your understanding of the main text. In the first of these, we look at the effects of scaling, shifting, and time-reversing steps, impulses, and ramps; and we also show how other simple digital signals may be built up by summation of such basic functions.

Example 1.1 Find expressions for the various signals shown in Figure 1.12.

Solution

(a) This is a unit step function which has been scaled (weighted) by a factor of -2; it starts at $n=-4$, rather than $n=0$; and it is time-reversed.
Now:
Scaling by -2 gives the function $-2u[n]$.
Time-shifting so that it starts at $n=4$ gives the function $-2u[n=-4]$.
Reversal gives the function $-2u[-n-4]$. Hence the required function is:

$$x[n] = -2u[-n-4]$$

(b) This digital 'rectangular pulse' may be considered as the summation, or superposition, of a unit step starting at $n=-3$, and an equal but opposite unit step starting at $n=5$. Hence:

$$x[n] = u[n+3] - u[n-5]$$

(c) The signal is a weighted unit impulse of value 8, time-shifted to occur at $n=6$. Hence:

$$x[n] = 8\delta[n-6]$$

(d) This signal may be considered as the superposition of two ramp functions. One starts at $n=-6$ and has a slope of 2. Its upward trend may be stopped by adding another ramp starting at $n = -2$, with a slope of -2. Hence:

$$x[n] = 2r[n+6] - 2r[n+2]$$

Note: The answers we have given are not the only possible ones.

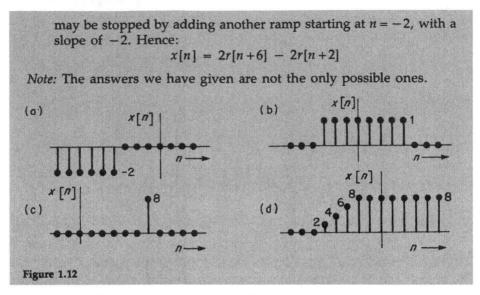

Figure 1.12

You have now met the basic ideas and mathematical notation of steps, impulses, and ramps. Some further properties of these signals — particularly of impulses — will be developed in later chapters.

1.3.2 EXPONENTIALS, SINES, AND COSINES

Exponential signals — and their close relations, sine and cosine waves — are extremely important in many branches of engineering and applied science. They are widely used in analysis, and often occur in the natural world and in technology. In the DSP context, sampled exponentials are central to a powerful set of *frequency-domain techniques* used to analyze and process signals, and to design systems. These form the main material for Chapters 3 and 4. It is therefore important to become familiar with such signals. Those of you who have previously worked with their analog (continuous-time) counterparts will notice many similarities, and a few differences.

We start by considering a function of the form:

$$x[n] = A \exp(\beta n) \tag{1.13}$$

where A and β are constants, and exp denotes the exponential function. Although we restrict ourselves to real values of A, we will allow β to be real, imaginary, or complex.

Suppose that β is real. Figure 1.13 shows the forms of signal when it is negative, zero, and positive. When $\beta < 0$, we get a *decaying exponential*. In theory it falls towards zero as n increases, and rises without limit for negative n. (We have, of course, shown only a few sample values of each function.) When $\beta = 0$, all sample values are equal. The signal is just a sampled 'steady level', often referred to in electronic engineering as a *DC level*. If $\beta > 0$, we get a *rising exponential* which grows without limit as n increases. In all three cases, the sample value at $n = 0$ equals the constant A.

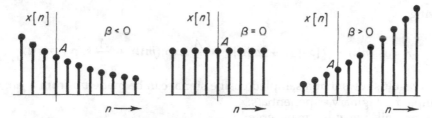

Figure 1.13 Basic digital signals: real exponentials.

Such real exponentials are theoretically *eternal* — they continue forever in both directions. In practice, however, we often work with signals which are zero prior to some reference instant, normally taken as $n = 0$. An exponential of this type may be written as:

$$x[n] = A \exp(\beta n), \qquad n \geq 0$$
$$= 0, \qquad\qquad n < 0 \qquad\qquad (1.14)$$

Alternatively, we can make use of the fact that the unit step function $u[n]$ is zero for $n < 0$, and unity for $n \geq 0$. Therefore if we multiply, or *modulate*, an eternal exponential by $u[n]$ we specify the function:

$$x[n] = A \exp(\beta n) \, u[n] \qquad\qquad (1.15)$$

This signal is effectively 'switched on' at $n = 0$ by the unit step function. Equations (1.14) and (1.15) are equivalent.

Another interesting property of sampled real exponentials is that successive sample values form a simple geometric progression — each value equals that of its neighbor, multiplied by a constant. We may show this by rewriting equation (1.13) as:

$$x[n] = A \exp(\beta n) = AB^n, \qquad \text{where } B = \exp(\beta)$$

Hence

$$x[n+1] = AB^{n+1} = Bx[n] \qquad\qquad (1.16)$$

If we next allow the constant β in equation (1.13) to be purely imaginary, we can generate sine and cosine signals. Suppose $\beta = j\Omega$, where $j = \sqrt{-1}$ and Ω is a real constant. The resulting signal $x_1[n]$ is given by:

$$x_1[n] = A \exp(jn\Omega) = A \cos(n\Omega) + jA \sin(n\Omega) \qquad\qquad (1.17)$$

It contains a cosinusoidal real part and a sinusoidal imaginary part. This is a little awkward, since (in this book at least!) we deal mainly with real signals. Note, however, that if we put $\beta = -j\Omega$, we define another signal:

$$x_2[n] = A \exp(-jn\Omega) = A \cos(n\Omega) + jA \sin(-n\Omega) \qquad\qquad (1.18)$$

Now cosines are *even* functions, thus $\cos(-n\Omega) = \cos(n\Omega)$; and sines are *odd* functions, thus $\sin(-n\Omega) = -\sin(n\Omega)$. Hence by adding $x_1[n]$ and $x_2[n]$ together we form the signal:

$$x_1[n] + x_2[n] = 2A \cos(n\Omega) \qquad\qquad (1.19)$$

giving:

$$A \cos(n\Omega) = \tfrac{1}{2}\{x_1[n] + x_2[n]\} = \frac{A}{2} \exp(jn\Omega) + \frac{A}{2} \exp(-jn\Omega) \tag{1.20}$$

This result shows that a sampled cosine signal can be made up from a *pair* of sampled imaginary exponentials.

In a similar way we may write:

$$x_1[n] - x_2[n] = 2jA \sin(n\Omega)$$

giving:

$$A \sin(n\Omega) = \frac{A}{2j} \exp(jn\Omega) - \frac{A}{2j} \exp(-jn\Omega) \tag{1.21}$$

Therefore a real, sampled, sine signal may also be made up from a pair of imaginary exponentials. We now see why exponentials, sines, and cosines are so closely related.

What are the differences between sampled sines and cosines, and their analog counterparts? You probably know that analog sine and cosine signals are oscillatory, and *periodic*. That is to say, they repeat indefinitely along the time axis. For example, the signal $A \sin(\omega t)$ is wave-like, and repeats every $2\pi/\omega$ seconds. Its peak value, or amplitude, is A; and its frequency is ω radians per second, or $\omega/2\pi$ Hz.

Sampled sines and cosines, however, are not necessarily periodic. Although their sample values lie on a periodic *envelope*, the numerical values may not form a repetitive sequence. We illustrate this in Figure 1.14. Part (a) shows a discrete-time sinusoid, and part (b) a cosinusoid, both of which *are* periodic. But part (c), although its samples lie along a sinusoidal envelope, does not have repeating numerical values. Indeed, it is hard to tell merely by inspection whether it is a sampled sinusoid or not.

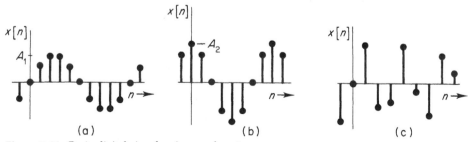

Figure 1.14 Basic digital signals: sines and cosines.

Intuitively, it is clear that exact repetition will only occur if the sampling interval bears some simple relationship to the repetition time, or period, of the underlying analog waveform. We may quantify this idea by returning to equation (1.17). For the moment let us assume that $x[n]$ *is* truly periodic, and repeats every N sample values. This means that:

$$x[n] = A \exp(jn\Omega) = A \exp(j\{n+N\}\Omega)$$
$$= A \exp(jn\Omega) \exp(jN\Omega) \tag{1.22}$$

and hence that:

$$\exp{(jN\Omega)} = 1 \tag{1.23}$$

The last equation only holds good if $N\Omega$ is a multiple of 2π. In other words there must be an integer m such that:

$$N\Omega = 2\pi m, \quad \text{or} \quad \frac{\Omega}{2\pi} = \frac{m}{N} \tag{1.24}$$

Hence $x[n]$ is only periodic if $\Omega/2\pi$ is a rational number (the ratio of two integers). Otherwise its sample values do not repeat. We have derived this result for an imaginary exponential, but it applies equally well to sines and cosines.

A second difference between analog and digital sinusoidal signals concerns the question of time and frequency scales. When we write a digital signal as $x[n] = A \sin{(n\Omega)}$, n is assumed to be a dimensionless integer variable. Therefore the constant Ω must be measured in radians rather than radians per second. Strictly speaking it is not a frequency — although we generally describe it as such. However our formulation for $x[n]$ allows us to define instead the number of *samples in each cycle, or period*, of the sinusoid. Since one complete period corresponds to:

$$n\Omega = 2\pi, \quad \text{or} \quad n = 2\pi/\Omega \tag{1.25}$$

there must be $2\pi/\Omega$ samples per period, regardless of the time or frequency scales of the signal. For example, referring to Figure 1.14(a) we see that the sinusoid has exactly ten samples per period. Therefore $2\pi/\Omega = 10$, or $\Omega = \pi/5$. The signal is given by $x[n] = A_1 \sin{(n\pi/5)}$. Similarly, the cosinusoid in part (b) of the figure has eight samples per cycle, and is given by $x[n] = A_2 \cos{(n\pi/4)}$.

In some cases, however, we really do need to work in terms of time and frequency. At the beginning of section 1.2 we introduced the Sampling Theorem, which states that an analog signal with frequency components up to some maximum frequency f_1 Hz may be represented by samples taken at a rate of at least $2f_1$ samples per second. Clearly, when considering this theorem we *are* interested in time and frequency scales. A good way of introducing them is to work in terms of the sampling interval T. The sampling instants are given by:

$$t = nT, \quad n = \ldots -2, -1, 0, 1, 2, 3 \ldots \tag{1.26}$$

and the sampling frequency is:

$$f_s = \frac{1}{T} \tag{1.27}$$

We now write our digital signal as:

$$x[n] = \sin{(n\Omega)} = \sin{(n\omega T)} \tag{1.28}$$

Since nT represents time in seconds, ω must be an angular frequency in radians per second. If f is the corresponding frequency in hertz, then $\omega = 2\pi f$ and equation (1.28) may be written as:

$$x[n] = \sin\left(\frac{n2\pi f}{f_s}\right) \tag{1.29}$$

We can now generate the sample series $x[n]$ if we know the frequency f of the underlying analog sinusoid, and the sampling frequency f_s. Our sampled signal has been redefined in terms of frequency in hertz, and time in seconds.

It may be helpful to mention the practical importance of sines and cosines once again. Such signals arise widely in engineering and science, and in the natural world: the mains electricity supply; the signal generators used in electronic laboratories; oscillations in electronic circuits containing capacitance and inductance; the vibrations of tuning forks and mechanical structures; and various types of wave motion. Many of these signals can be stored and processed in digital computers. But from our point of view, it is also very important that other, *completely different,* signals may be considered as the summation of a number of sines and cosines of appropriate amplitudes and frequencies. Seen in this light, sines and cosines can even be useful for DSP applications in economics and the social sciences!

We end our survey of exponential and sinusoidal signals by letting the constant β in equation (1.13) become complex. You can probably imagine what will happen. We have seen that *real* values of β give signals which rise or fall exponentially. *Imaginary* values produce sines and cosines. Therefore *complex* values may be expected to produce oscillatory signals with rising or falling amplitudes.

Consider the signal:

$$x[n] = A \exp(\beta n) = A \exp(\{\beta_0 + j\Omega\}n) \tag{1.30}$$

where β_0 is a real part, and $j\Omega$ the imaginary part, of β. Hence:

$$x[n] = A \exp(\beta_0 n) \exp(j\Omega n) \tag{1.31}$$

The frequency term $\exp(j\Omega n)$ is multiplied, or modulated, by the rising (or falling) amplitude term $A \exp(\beta_0 n)$. Of course, we again have the difficulty that $\exp(j\Omega n)$ is not a real function of n. We therefore need to consider a pair of imaginary exponentials which produce a real sine or cosine. The argument is just as before, leading to functions of the form:

$$x[n] = A \exp(\beta_0 n) \sin(n\Omega)$$

or

$$x[n] = A \exp(\beta_0 n) \cos(n\Omega) \tag{1.32}$$

Typical examples are given in parts (c) and (d) of the worked example below.

These signals are also important in practice. Sinusoidal oscillations with decaying amplitudes are often observed in circuits and filters — both analog and digital. They also arise in vibrating mechanical systems. Exponentially increasing oscillations tend to occur in *unstable* systems — including digital processors which have been badly designed! We shall have more to say about this later.

If you require further familiarization with sampled exponentials and sinusoids, you may like to try Computer Program no.2 in Appendix A1. This simple program draws signals of the general form $x[n] = \exp(\beta_0 n)\cos(n\Omega)$ on the screen, allowing you to control β_0 and Ω. Note that if $\Omega = 0$, $x[n]$ is a real exponential. If $\beta_0 = 0$, it is a cosine. Whereas if neither is zero, $x[n]$ is an exponentially rising (or falling) cosine. (Suitable values to start with are $\beta_0 = 0.01$ and $\Omega = 0.2$.)

Example 1.2 Sketch carefully, and label, the following signals:

(a) $x[n] = \exp(0.2n)$

(b) $x[n] = \cos\left(\dfrac{\pi n}{4}\right)$

(c) $x[n] = \exp\left(\dfrac{n}{15}\right)\sin\left(\dfrac{\pi n}{6}\right)$

(d) $x[n] = \exp\left(\dfrac{-n}{5}\right)\cos(n)\,u[n]$

Solution: The signals are shown in Figure 1.15, and should be self-explanatory.

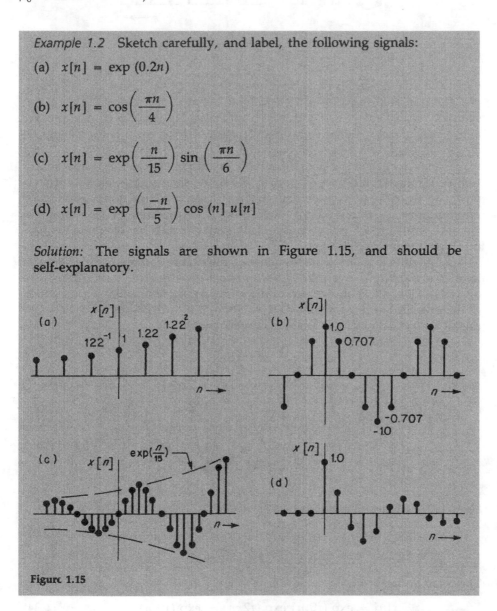

Figure 1.15

1.4 AMBIGUITY IN DIGITAL SIGNALS

A digital signal $x[n]$ generally represents some underlying analog function. However, if we start with a given set of sample values, it is clear that a huge

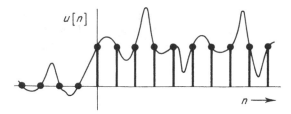

Figure 1.16 A unit step function, and one of the many analog signals which can be drawn through its sample points.

variety of analog signals *could* be drawn through them. Consider, for example, the unit step function $u[n]$. Although we may think of it as being the sampled equivalent of an analog step function, Figure 1.16 shows another analog waveform with the same sample values. It is clear that digital signals are, in an important sense, ambiguous.

You may feel that the above example is rather exaggerated, and that a sampled signal should have essentially the same 'shape' as its analog counterpart. However, we must remember that many sampled functions — including random noise — simply do not have an obvious underlying shape. Nor can we necessarily expect the systems and computers which handle them to interpret our intentions correctly.

The difficulty is made worse when we realize that sampled sinusoids also involve ambiguity — even though we restrict ouselves to a particular shape of signal. Figure 1.17 shows two of the many analog sinusoidal waves which fit the given set of sample values. Here, then, is a further difference between digital sinusoids and their analog counterparts.

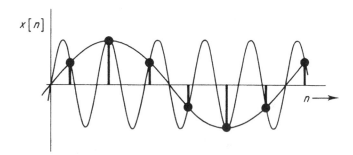

Figure 1.17 Ambiguity in digital sinusoids.

We can clarify the situation by considering the complex exponential signal:

$$x[n] = \exp\left(j\{\Omega + 2\pi\}n\right) \tag{1.33}$$

This may be written as:

$$x[n] = \exp\left(j\Omega\, n\right) \exp\left(j2\pi n\right) = \exp\left(j\Omega\, n\right) \tag{1.34}$$

Therefore a sampled exponential of frequency Ω is *identical* to those of frequency $\Omega \pm 2\pi$, $\Omega \pm 4\pi$, and so on. Although we have shown this for exponentials, the same is true for sines and cosines. As we increase the frequency of a digital sinusoid from zero, its rate of oscillation initially *increases*. Beyond $\Omega = \pi$ it starts to *decrease* again. At $\Omega = 2\pi$, the rate of oscillation is again zero — and so on, indefinitely. Ambiguity arises because a digital sinusoid can represent any one of an infinite set of underlying frequencies, separated from one another by 2π.

Program no.3 in Appendix A1 illustrates this important point. It generates and plots a digital sinusoid whose frequency Ω increases in ten equal steps across the screen. (For convenience we use a cosine rather than a sine.) The signal is given by:

$$x[n] = \cos(n\Omega) = \cos\left(\frac{n2\pi m}{8}\right) \tag{1.35}$$

with the integer m increasing from 0 to 9 inclusive.

The plot is reproduced in Figure 1.18. We have also marked the values of Ω against it. On the left-hand side $\Omega = 0$, and on the right $\Omega = 2.25\pi$. As expected, the rate of oscillation increases as far as $\Omega = \pi$, then decreases again. At $\Omega = 2\pi$, the signal is the same as at $\Omega = 0$. At $\Omega = 2.25\pi$ it is the same as at 0.25π, and so on. You may think of the underlying analog frequency as continuously increasing from left to right.

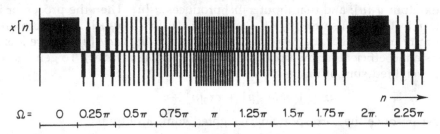

Figure 1.18 A digital cosinusoid whose frequency is increased in ten equal increments (*abscissa: 320 samples*).

The ambiguity problem is closely related to our discussion of the Sampling Theorem and aliasing in Section 1.2. To obey the theorem we must make the sampling frequency at least twice the highest frequency present in the underlying analog waveform ($f_s \geq 2f_1$). This is equivalent to taking at least two samples per period of the highest frequency. Now two samples per period corresponds to $\Omega = \pi$ (see equation (1.25)). Hence there is no ambiguity over which frequency is intended within the range $\Omega = 0$ to $\Omega = \pi$. It is only when we go outside this range — as in Figure 1.18 — that we get into difficulty. In effect the Sampling Theorem resolves ambiguity by restricting the allowable frequency range of the underlying analog signal.

1.5 DIGITAL PROCESSORS

It is now time to consider some basic aspects of digital signal processors. By 'processor' we mean any system which carries out a DSP function. It may be a general-purpose computer programmed for a particular DSP task, a microprocessor, dedicated digital hardware — or a combination of these. Since this book's emphasis is on the use of general-purpose computers, we will generally think of a processor as a computer plus relevant software. Incidentally, we make no particular distinction between the terms 'processor' and 'system', using them interchangeably.

1.5.1   LINEAR TIME-INVARIANT (LTI) SYSTEMS

This book covers only *linear* DSP. Nonlinear processing is certainly important, but its theoretical treatment is much more difficult. Fortunately, linear processing includes a wide range of DSP algorithms of great practical interest.

A linear system, or processor, may be defined as one which obeys the *Principle of Superposition*. The principle may be stated as follows:

> If an input consisting of the sum of a number of signals is applied to a linear system, then the output is the sum, or *superposition*, of the system's responses to each signal considered separately.

Let us suppose than an input $x_1[n]$, applied to a digital processor, produces the output $y_1[n]$; and that input $x_2[n]$ produces $y_2[n]$. Then the processor is linear if its response to $\{x_1[n] + x_2[n]\}$ is $\{y_1[n] + y_2[n]\}$. Furthermore, linearity implies that the response to an input $ax_1[n]$ is $ay_1[n]$, where a is a constant coefficient, or multiplier (also called a *weighting factor*). To generalize, the weighted sum of inputs:

$$ax_1[n] + bx_2[n] + cx_3[n] + \ldots \tag{1.36}$$

must produce the corresponding weighted sum of outputs:

$$ay_1[n] + by_2[n] + cy_3[n] + \ldots \tag{1.37}$$

You may feel that the property of linearity appears obvious, or even trivial. So we should point out straightaway that many quite simple processes are *not* linear. For example, suppose we have a system which *squares* each sample value applied to its input. Thus for two different inputs applied separately, we have:

$$y_1[n] = (x_1[n])^2 \quad \text{and} \quad y_2[n] = (x_2[n])^2 \tag{1.38}$$

When the same two inputs are summed, and applied simultaneously, the output is:

$$\begin{aligned} y_3[n] &= (x_1[n] + x_2[n])^2 \\ &= (x_1[n])^2 + (x_2[n])^2 + 2x_1[n]x_2[n] \end{aligned} \tag{1.39}$$

This is clearly *not* the sum of its responses to $x_1[n]$ and $x_2[n]$ applied separately.

A major property of linear systems, which is closely related to the Principle of Superposition, is known as *frequency-preservation*. It means that if we apply an input signal containing certain frequencies to a linear system, the output can contain only the same frequencies, and no others. The property depends upon the fact that a sampled sinusoid, applied to any linear processor, always produces a similar form of output, *at the same frequency as the input*. If we apply a complicated signal containing many sinusoidal frequencies, the Principle of Superposition tells us that the output must be the sum of the outputs due to each input frequency, considered separately. The output therefore contains only those frequencies present in the input.

We can easily demonstrate that this property does not hold for the squaring system mentioned above. Suppose a signal $\sin(n\Omega)$ is applied to its input. Its output is therefore:

$$y[n] = (x[n])^2 = \sin^2(n\Omega)$$
$$= \tfrac{1}{2}\{1 + \cos(2n\Omega)\} = \tfrac{1}{2} + \tfrac{1}{2}\cos(n2\Omega) \qquad (1.40)$$

There are two separate, additive, frequency components in the output: one at zero frequency, of amplitude $\tfrac{1}{2}$; the other at frequency 2Ω, also of amplitude $\tfrac{1}{2}$. There is no component in the output at the frequency Ω.

We next describe *time-invariance*. A time-invariant system is one whose properties do not vary with time. The only effect of a time-shift in an input signal to the system is a corresponding time-shift in its output. The majority of technological systems and processes are of this type — including the DSP algorithms we cover in this book. Thus we concern ourselves exclusively with *linear time-invariant (LTI)* systems.

We can illustrate the above discussion with a simple example. Consider a cash register, of the type used at the check-out of a supermarket. It is a digital processor, which adds up the prices of individual items (the input signal) to produce a total (the output signal). We may think of it as a form of *digital integrator*, or *accumulator*, which estimates the running sum:

$$y[n] = x[n] + x[n-1] + x[n-2] + \ldots \qquad (1.41)$$

Of course, no sensible cash register works nonrecursively — repeating the complete calculation every time a new item is included! It keeps a running total, which is updated. This is done using the equivalent recursive algorithm:

$$y[n] = y[n-1] + x[n] \qquad (1.42)$$

when a customer checks out, the cash register is reset to zero, and the summation starts again.

It is fairly obvious that such a device is an LTI system. Its linearity may be deduced as follows. Suppose one customer checks out and the cash register estimates a total payment of y_1. The next customer's total is y_2. Now suppose the two customers are friends, and they decide to pool their goods before checking out. The grand total will be $(y_1 + y_2)$, which is the superposition of the two previous totals calculated separately. The cash register is also time-invariant, because its performance is (we hope!) unaffected by the time of day.

The cash register clearly involves storage/delay and addition of numerical values. More generally, digital LTI processors involve the following types of operation on input and/or output samples.

storage/delay
addition/subtraction
multiplication by constants

As another example, consider the digital notch filter previously illustrated in Figure 1.4. Equation (1.3) gave its recurrence formula, or difference equation, as:

$$y[n] = 1.8523\ y[n-1] - 0.94833\ y[n-2] + x[n]$$
$$- 1.9021\ x[n-1] + x[n-2] \qquad (1.43)$$

Although we are not yet in a position to explain how the filter works, the equation shows that it requires only the above types of operation. We can safely conclude that the filter is a LTI processor.

This is a convenient moment to show how digital LTI systems can be represented in block diagram form. Figure 1.19(a) illustrates the nonrecursive version of the cash register equation — see equation (1.41). Input samples are stored and delayed successively by an amount equal to the sampling interval T. This produces the values $x[n-1]$, $x[n-2]$, $x[n-3]$ ... Tapping points lead to an adder unit, whose output is $y[n]$. The figure implies that input samples are equally spaced in time. Although this is unlikely to be true of a cash register, regular sampling is used in most types of digital processor.

Part (b) of the figure shows the recursive version of the cash register equation — see equation (1.42). The current output $y[n]$ is *fed back* via a single delay unit T to produce the previous output $y[n-1]$, and this is updated by the current input $x[n]$ in an adder unit.

In part (c) we show a block diagram for the digital notch filter defined by

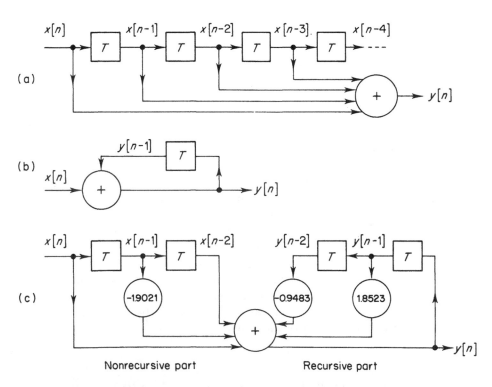

Figure 1.19 System block diagrams of digital signal processors.

equation (1.43). The current output $y[n]$ depends on three input sample values, weighted and added in the nonrecursive part of the filter; and on two previous outputs, provided by the recursive part. As before, the recursive part involves feedback.

When a general-purpose computer is used to implement a DSP algorithm, it *behaves as if* it were such a set of digital building blocks — delay/storage units, multipliers, and adders. This is achieved by the software. On the other hand when a DSP system is realized in special-purpose hardware, its architecture is likely to resemble such a block diagram much more closely. We should, however, bear in mind that sample values are represented in binary form, and that each block must process multi-bit codes.

You should by now have a good idea of what a digital LTI system is, and how it may be represented — either as a recurrence formula, or as a block diagram. In the final part of this chapter we cover some additional ideas about digital processors, and introduce some more terminology.

1.5.2 OTHER SYSTEM PROPERTIES

We have already mentioned that LTI systems obey the Principle of Superposition, and have the property of frequency preservation. They possess other important properties, including *association* and *commutation*. The associative property means that we may analyze a complicated LTI system by breaking it down into a number of simpler subsystems. Alternatively, we can synthesize an overall system — perhaps a very complicated one — by designing a number of independent subsystems. We know that, when we put them all together, overall performance can be predicted by straightforward rules of association. These rules will be explained in later chapters.

The commutative property of LTI systems means that if subsystems are arranged in series, or *cascade*, then they may be rearranged in any order without affecting overall performance. This can sometimes offer advantages — for example, it may reduce complexity.

Although the associative and commutative properties of LTI systems may appear rather obvious, they are not generally shared by nonlinear systems.

There are four other properties of systems which are often mentioned in the DSP literature. They are: *causality*; *stability*; *invertibility*; and *memory*. A processor may or may not possess them, independently of whether it is LTI or not. We now briefly summarize each in turn.

In a *causal* system, the output signal depends only on present and/or previous values of the input. You might assume that practical signal processors are always causal, because they cannot anticipate the future! However, if we record a signal or data and subsequently process it by computer, the software need not be causal. Of course it is reasonable to object that the complete recording/processing system *is* causal. But at this point we may become entangled in semantics!

A *stable* system is one which produces a finite, or *bounded*, output in response to a bounded input. This means that if the system is disturbed from its resting state by an input signal which does not grow without limit, or *diverge*, then the output must not diverge either. In Section 1.3.2 we considered some exponential signals which grow without limit (Figures 1.13 and 1.15). Clearly, if a system produced such a signal in response to a bounded input (such as an impulse), it would be unstable. In practice the output of an unstable system must sooner or later cease to diverge, due to some *limiting effect* (such as numerical 'overflow' in a computer). We shall see later that instability is closely associated with the idea of feedback.

We now discuss *invertibility*. If a digital processor with input $x[n]$ gives an output $y[n]$, then its *inverse* would produce $x[n]$ if fed with $y[n]$. Most practical systems are invertible. An exception in the 'squaring device' mentioned earlier, for which $y[n] = (x[n])^2$. Knowledge of an output value $y[n]$ does not allow us to find $x[n]$ unambiguously, since it could be positive or negative.

A processor possesses *memory* if its present output $y[n]$ depends upon one or more previous input values $x[n-1]$, $x[n-2]$ ... In other words it must contain storage/delay elements, such as those shown in Figure 1.19. Interesting DSP systems invariably possess memory.

We can summarize the above discussion by saying that the main interest of this book is in LTI systems which are also causal, stable, invertible, and possess memory.

Example 1.3 $x[n]$ and $y[n]$ are the input and output signals of a DSP system. Determine which of the following properties are possessed by systems defined by the recurrence formulae (a) to (d) below:

linearity	time-invariance
causality	stability
invertibility	memory

(a) $y[n] = 3x[n] - 4x[n-1]$
(b) $y[n] = 2y[n-1] + x[n+2]$
(c) $y[n] = nx[n]$
(d) $y[n] = \cos(x[n])$

Solution
(a) The output is a weighted sum of present and previous inputs. It is bounded if the input is bounded. A given input sequence is unambiguously related to the output sequence. Therefore the system possesses all six properties mentioned above.
(b) The present output depends on a future input, so the system is not causal. If the input signal ceases, the output goes on rising without limit, since each output value is twice the previous one. Therefore the system is unstable. But it possesses the other four properties mentioned above, namely: linearity, time-invariance, invertibility and memory.
(c) The output depends on the present input only, so the system has no memory. Since it also depends on the independent variable, the

> system is time-variant. But it possesses the properties of linearity, causality, stability, and invertibility.
>
> (d) A cosine function is periodic, so many different values of $x[n]$ would produce the same value of $y[n]$. Hence the system is not invertible. Neither is it linear, because if (for example) we double $x[n]$, we do not double $y[n]$. Also, since $y[n]$ depends only on $x[n]$, the system has no memory. However, it is time-invariant, causal, and stable.

PROBLEMS

SECTION 1.2

Q1.1 An analog television signal is to be coded by an ADC for transmission via a digital communication system. The signal contains significant frequencies up to 6 MHz. The quantization error in any one sample value, introduced by coding, must be less than 1 per cent of the total amplitude range of the ADC. Uniform quantization is to be used.

Assuming the Sampling Theorem is obeyed, what is the minimum rate of transmission of binary pulses, expressed in bits per second?

Q1.2 The analog television signal in Q1.1 fluctuates in amplitude, sometimes filling only 20 per cent of the amplitude range of the ADC. If the amount of quantization noise, *relative to the signal*, is to be no worse than before, what minimum transmission rate is required?

Q1.3 Repeat problem Q1.1, but for an electrocardiogram (EKG) signal containing significant components up to 200 Hz. The quantization error introduced by coding should again be less than 1 per cent of the ADC's amplitude range.

SECTION 1.3.1

Q1.4 Figure Q1.4 shows a digital signal $x[n]$. Sketch and label carefully the following signals:
(a) $x[n-2]$
(b) $x[3-n]$

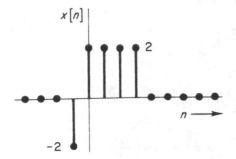

Figure Q1.4

(c) $x[n-1]\,u[n]$

(d) $x[n-1]\,\delta[n]$

(e) $x[1-n]\,\delta[n-2]$

Q1.5 If $x_1[n]$ is an even signal, such that $x_1[n] = x_1[-n]$; and $x_2[n]$ is an odd signal, such that $x_2[n] = -x_2[-n]$; then show that:

(a) $(x_1[n]\,x_2[n])$ is an odd signal

(b) $(x_2[n])^2$ is an even signal

(c) $\displaystyle\sum_{n=-\infty}^{\infty} x_2[n] = 0$

Q1.6 Find mathematical expressions for the various step, impulse, and ramp signals illustrated in Figure Q1.6.

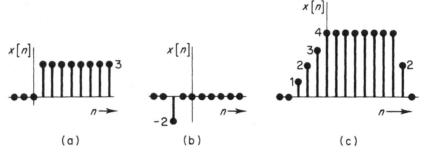

(a)   (b)   (c)

Figure Q1.6

Q1.7 Sketch, and label carefully, the following signals:

(a) $-u[n-2]$

(b) $u[n+1] + \delta[n]$

(c) $2u[n+2] - u[3-n]$

(d) $r[n] - 2r[n-3]$

SECTION 1.3.2

Q1.8 Determine which of the following signals is strictly periodic. If a signal is periodic, find its period.

(a) $x[n] = \sin\left(\dfrac{\pi n}{9}\right)$

(b) $x[n] = \sin(n\pi^2)$

(c) $x[n] = \cos\left(\dfrac{\pi n^2}{15}\right)$

(d) $x[n] = \sin\left(\dfrac{\pi n}{5} + \pi\right) + \cos\left(\dfrac{\pi n}{10} - \pi\right)$

Q1.9 Sketch carefully portions of the following signals:

(a) $x[n] = 3 \exp(-n/2)$
(b) $x[n] = \exp(n/5)$

(c) $x[n] = \sin\left(\dfrac{\pi n}{6}\right) u[n]$

(d) $x[n] = \exp(-n/4) \cos\left(\dfrac{\pi n}{2}\right) u[n]$

SECTION 1.5

Q1.10 $x[n]$ and $y[n]$ are the input and output signals of a digital processor. Determine which of the following properties are exhibited by each of the systems defined below: linearity, time-invariance, causality, stability, and memory.

(a) $y[n] = x[5-n]$
(b) $y[n] = x[n] \, x[n-3]$
(c) $y[n] = x[n] + x[n-1] + 3x[n-2]$
(d) $y[n] = 1.5 \, y[n-1] + x[n-1]$
(e) $y[n] = nx[n]$

Q1.11 Which of the following systems are invertible? If a system is not invertible, find two values of the input which produce the same output.

(a) $y[n] = x[2-n]$
(b) $y[n] = x[n] \, x[n-1]$
(c) $y[n] = \sin(nx[n])$

Q1.12 Figure Q1.12 shows a block diagram of a digital processor. Assuming $y[n] = 0$ for $n < 0$, sketch the output from the processor when the input is (a) the unit impulse function $\delta[n]$, and (b) the unit step function $u[n]$. Assume the constant multiplier α is between 0 and 1. What would happen if $\alpha > 1$, or $\alpha < -1$?

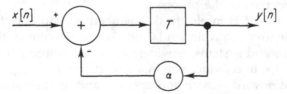

Figure Q1.12

Q1.13 Draw a block diagram for the digital bandpass filter defined by equation (1.4) in the main text.

Q1.14 Draw a block diagram for a digital processor with the following recurrence formula. Distinguish clearly between its nonrecursive and recursive parts.

$$y[n] = 1.625 \, y[n-1] - 0.934 \, y[n-2] + 0.5 \, x[n] - 0.1 \, x[n-2]$$

-------- CHAPTER 2 --------
Time-Domain Analysis

2.1 INTRODUCTION

In this chapter we develop the basic techniques for describing digital signals and processors in the time domain. Foremost among them is *convolution*, which allows us to find the output signal from any LTI processor in response to any input signal. It also gives valuable insights into the ways in which trains of signal samples are modified by linear processing. A sound understanding of digital convolution is essential to anyone interested in DSP.

It is fair to add that time-domain methods, including convolution, are not a great help in designing new types of processor. With a few notable exceptions, these methods are more useful for analysis than synthesis. The main reason is that design specifications are often based on performance in the frequency domain. In other words they concentrate on the response to the various frequencies present in an input signal. We shall develop such ideas in later chapters.

As far as the time domain is concerned, we can find the output signal $y[n]$ from an LTI processor by *convolving* its input signal with a second time function representing the processor itself. This second function is the processor's response to the unit impulse function $\delta[n]$ — in other words, its *impulse response.* The convolution takes place entirely in the time domain, without any need to consider the frequency components of the input signal, or the frequency-dependent properties of the processor.

If your background is in electronic engineering, you may feel somewhat dismayed by the word 'convolution'! Continuous-time convolution, as applied to analog signals and systems, is widely regarded as a rather difficult topic. However, we hope to convince you that discrete-time, or digital, convolution is quite straightforward — both conceptually and mathematically.

Convolution is a form of superposition. It relies upon the fact that any input signal can be built up by summation of a number of weighted, shifted, impulses. Since an LTI processor obeys the Principle of Superposition, the output signal must equal the summation of its responses to all such impulses, considered separately.

Before we can describe the convolution process in detail, it is therefore necessary to develop two subsidiary themes. First, we must be able to characterize any input signal as a set of impulse functions. And secondly, we need to consider how a digital LTI system responds to an individual impulse.

2.2 DESCRIBING DIGITAL SIGNALS WITH IMPULSE FUNCTIONS

A portion of a digital signal $x[n]$ is shown in part (a) of Figure 2.1. It is clear that $x[n]$ may be considered as the superposition, or summation, of the more basic impulse signals shown below in parts (b) to (f). Each of these is a unit impulse which has been *weighted* by the appropriate value of $x[n]$, and *shifted* by a number of sampling intervals. Thus the signal shown in part (b) may be written as $x[-2]\,\delta[n+2]$, and so on. We can define the complete signal $x[n]$ as:

$$x[n] = \cdots + x[-2]\,\delta[n+2] + x[-1]\,\delta[n+1] + x[0]\,\delta[n]$$
$$+ x[1]\,\delta[n-1] + x[2]\,\delta[n-2] + \cdots \qquad (2.1)$$

or

$$x[n] = \sum_{k=-\infty}^{\infty} x[k]\,\delta[n-k] \qquad (2.2)$$

Since the integer k takes on all values between $\pm\infty$, $x[n]$ is a completely general digital signal. Of course, if we know that $x[n] = 0$ for $n < 0$, we can limit k to the range 0 to ∞, and so on.

Note that, for any value of n, only one term on the right-hand side of equation (2.2) is nonzero. This demonstrates the important *sifting property* of the

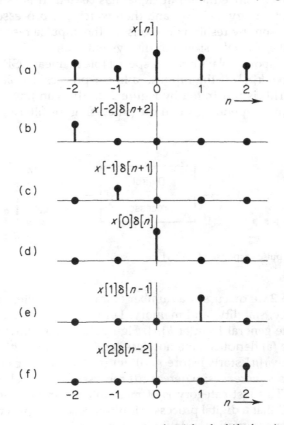

Figure 2.1 Representing a digital signal as a set of weighted, shifted, unit impulses.

unit impulse function. For example if we choose $n = 4$, equation (2.2) becomes:

$$x[4] = \sum_{k=-\infty}^{\infty} x[k]\,\delta[4-k] \qquad (2.3)$$

Now $\delta[4-k]$ is only nonzero for $k = 4$, when its value is unity. The product $x[k]\,\delta[4-k]$ is likewise nonzero only for $k = 4$, when it has the value $x[4]$. The impulse function has therefore *sifted out*, or selected, the value of $x[n]$ which coincides with it. We shall meet this valuable property again later.

2.3 DESCRIBING DIGITAL LTI PROCESSORS

2.3.1 THE IMPULSE RESPONSE

Our next task is to examine the nature of an LTI processor's response to an individual impulse. Before getting down to detail, it is helpful to make some general comments. First, we note that an impulse is the ultimate *time-limited* signal. The unit impulse $\delta[n]$ is finite at $n = 0$, but zero elsewhere. If we deliver such a signal to the input of a digital processor, the 'excitation' is confined to the instant $n = 0$. Any output signal observed after $n = 0$ must be *characteristic of the processor itself*, since the input signal has ceased. It is as if we deliver a sudden 'burst of energy' at $n = 0$, and then watch the processor settle on its own. For this reason the resulting response — the impulse response — is often referred to as the *natural response* of the system.

The impulse response plays a rather special role in linear DSP, and is usually given the symbol $h[n]$. If the input to a linear processor is $\delta[n]$, the output must be $h[n]$. This is illustrated by Figure 2.2. $h[n]$ can have a wide variety of different forms, depending upon the processing, or filtering, action of the system.

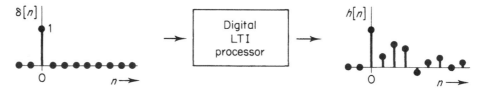

Figure 2.2 The impulse response of an LTI processor.

In Section 1.5.2 we discussed a number of important system properties, including causality, stability, and memory. Let us consider how these properties relate to the general form of $h[n]$. Figure 2.3 shows four representative examples. Case (a) denotes a memoryless (and rather boring!) system; (b) is noncausal, since $h[n]$ starts before $n = 0$; (c) is unstable, because $h[n]$ grows without limit as n increases; whereas (d) represents a causal, stable, system with memory. This last category is of most interest in DSP applications.

You will recall that a digital processor is often described by a recurrence formula, or difference equation, relating its input and output signals. It is quite

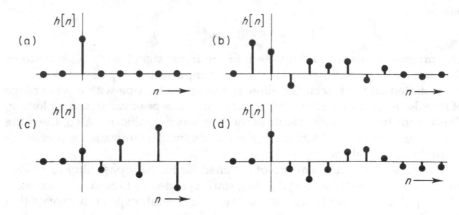

Figure 2.3 Various forms of impulse response.

easy to find the impulse response from such an equation. For example, consider the bandpass filter discussed in Section 1.1.2 and illustrated by Figure 1.5. Its recurrence formula (see equation (1.4)) is:

$$y[n] = 1.5\, y[n-1] -0.85\, y[n-2] + x[n] \tag{2.4}$$

If we deliver a unit impulse to the filter input, then $x[n]$ becomes $\delta[n]$. The output signal must be $h[n]$. Hence the following recurrence formula must also hold:

$$h[n] = 1.5\, h[n-1] -0.85\, h[n-2] + \delta[n] \tag{2.5}$$

Bearing in mind that $\delta[n]$ contributes the value 1.0 when $n=0$, but is zero elsewhere, it is simple to evaluate $h[n]$ term-by-term. Assuming the system is causal, then $h[n]=0$ for $n<0$, and we obtain:

$$h[0] = 1.5\, h[-1] -0.85\, h[-2] + \delta[0] = 0-0+1 = 1$$
$$h[1] = 1.5\, h[0] -0.85\, h[-1] + \delta[1] = 1.5 -0+0 = 1.5$$
$$h[2] = 1.5\, h[1] -0.85\, h[0] + \delta[2] = 2.25 -0.85 + 0 = 1.4$$

and so on. Note that, in this example, the input impulse only contributes at $n=0$. Thereafter, the values of $h[n]$ depend entirely on previous values, being generated recursively.

Of course it is rather tedious to estimate $h[n]$ 'by hand' — or even with a calculator. A computer program makes it very much easier. Program no.4 in Appendix A1 is designed for this purpose. It estimates and plots the impulse response of any LTI filter or processor with up to three recursive, and three nonrecursive, terms in its recurrence formula. It is therefore suitable for formulae of the general form:

$$y[n] = a_1 y[n-1] + a_2 y[n-2] + a_3 y[n-3]$$
$$+ b_0 x[n] + b_1 x[n-1] + b_2 x[n-2] \tag{2.6}$$

The program requests the values of coefficients a_1, a_2, a_3 and b_0, b_1, b_2. If they are not required in a particular case, they should be entered as zeros. For the bandpass filter defined by equation (2.4), we therefore enter the coefficient values:

$$1.5, \ -0.85, \ 0 \quad \text{and} \quad 1, \ 0, \ 0$$

The program loads a unit impulse into the input signal array X, and stores the impulse response in array H. The input impulse and impulse response are both plotted on the screen. To allow the program to cope with a wide range of impulse responses, all are normalized to the same peak value before plotting. This means that the vertical scale of the plot is not significant. Also, the finite vertical resolution of the screen display causes small quantization errors in the plotted sample values.

Figure 2.4 shows the form of plot obtained for the bandpass filter discussed above. The filter is (fortunately!) stable, since $h[n]$ decays to zero as n increases. The impulse response is also oscillatory, and careful inspection reveals that it oscillates at about ten samples per period. This is the frequency which is 'most natural' to it. We now see why the filter responds enthusiastically to a sampled sinusoid at about ten samples per period — as shown previously by Figure 1.5. Other frequencies are transmitted with reduced amplitudes, because they are not so close to the filter's natural frequency.

In this intuitive way, we realize that there must be a close relationship between a processor's impulse response $h[n]$, and its response to various sinusoidal frequencies. The time-domain and frequency-domain descriptions are interrelated. We will develop this important idea in Chapters 3 and 4.

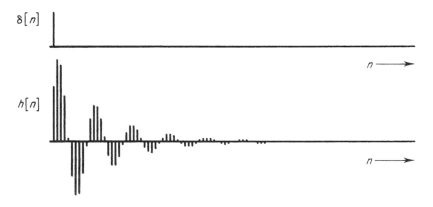

Figure 2.4 Computed impulse response of a bandpass filter *(abscissa: 100 samples).*

Example 2.1 Find the first four sample values of the impulse response $h[n]$ for each of the following digital processors:

(a) The system illustrated in Figure 2.5.
(b) The cash register previously defined by equations (1.41) and (1.42).

Solution
(a) By inspection we see that the recurrence formula for this system is:

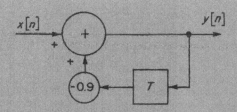

Figure 2.5

$$y[n] = -0.9\,y[n-1] + x[n]$$

The impulse response is therefore given by:

$$h[n] = -0.9\,h[n-1] + \delta[n]$$

The system is clearly causal, so that $h[n]=0$ for $n<0$. Hence:

$$h[0] = -0.9h[-1] + \delta[0] = 0 + 1 = 1$$
$$h[1] = -0.9h[0] = -0.9$$
$$h[2] = -0.9h[1] = 0.81$$
$$h[3] = -0.9h[2] = -0.729$$

Each sample value is -0.9 times the previous one, and $h[n]$ therefore follows a decaying, real, exponential envelope, with alternate samples inverted.

(b) The nonrecursive version of the cash register equation is:

$$y[n] = x[n] + x[n-1] + x[n-2) + \cdots$$

Hence:

$$h[n] = \delta[n] + \delta[n-1] + \delta[n-2] + \cdots$$

Once again we have a causal system. The first two values of $h[n]$ are:

$$h[0] = \delta[0] + \delta[-1] + \delta[-2] + \cdots = 1$$
$$h[1] = \delta[1] + \delta[0] + \delta[-1] + \cdots = 1$$

Similarly $h[2] = h[3] = 1$, and so on. The impulse response therefore equals the unit step function $u[n]$. This is correct since the cash register is required to perform a running sum.

The recursive version of the cash register, equation (1.42), is:

$$y[n] = y[n-1] + x[n]$$

giving:

$$h[n] = h[n-1] + \delta[n]$$

we again obtain:

$$h[0] = h[-1] + \delta[0] = 0 + 1 = 1$$
$$h[1] = h[0] + \delta[1] = 1 + 0 = 1$$

and so on. The form of this impulse response can also be inferred quite easily from the block diagrams of Figure 1.19(a) and (b), assuming a unit impulse is delivered to the input in each case.

> We hope that these simple examples have given useful practice in finding $h[n]$ for different processors — either from the block diagram or from the recurrence formula. We could, of course, have used Computer Program no.4 to do the work for us. You may like to try this; but beware of being accused of laziness!

2.3.2 THE STEP RESPONSE

Just as the unit step function $u[n]$ is the running sum of the unit impulse $\delta[n]$, so the step response of a LTI processor is the running sum of its impulse response. Therefore if we denote the step response by $s[n]$, we have:

$$s[n] = \sum_{m=-\infty}^{n} h[m] \qquad (2.7)$$

Alternatively, $h[n]$ is the first-order difference of $s[n]$:

$$h[n] = s[n] - s[n-1] \qquad (2.8)$$

These relationships stem from the commutative property of LTI systems and processes, mentioned in Section 1.5.2. The property tells us that the processing chain:

$$\delta[n] \;\rightarrow\; \text{running sum} \;\rightarrow\; u[n] \;\rightarrow\; \text{LTI system} \;\rightarrow\; s[n]$$

must be equivalent to:

$$\delta[n] \;\rightarrow\; \text{LTI system} \;\rightarrow\; h[n] \;\rightarrow\; \text{running sum} \;\rightarrow\; s[n]$$

A step response gives essentially the same information as an impulse response, but in slightly different form. Step responses are useful for several reasons. They are important in their own right because step signals occur quite often in practice: for example, as test signals, or in the form of rectangular pulses or pulse sequences. Also the response of a system to a sudden disturbance is quite often assessed in terms of its step response, particularly in the field of automatic control. Finally, the process of convolution can be defined in terms of step signals and step responses — rather than the more common approach based on impulses and impulse responses which we use in this book.

A little earlier we used Computer Program no.4 in Appendix A1 to plot the impulse response of a bandpass filter (Figure 2.4). A minor modification allows it to plot the step response instead — all we need do is load a unit step signal into array X, rather than a unit impulse.

The result is illustrated by Figure 2.6. The unit step, shown at the top, starts at the left of the figure and continues indefinitely to the right. The step response $s[n]$ initially displays the characteristic oscillation of this particular bandpass filter. It eventually settles to a constant value which reflects the filter's response to an input 'DC level'. An advantage is that it clearly shows the DC, or zero frequency, response of the system in this way. (Note, however, that the relative vertical scales of the plots are not correct.)

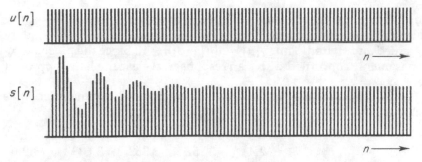

Figure 2.6 Computed step response of a bandpass filter *(abscissa: 100 samples)*.

Example 2.2 Find, and sketch, the first few sample values of the impulse and step responses of the system shown in Figure 2.7(a). Also determine the final value of $s[n]$ as $n \to \infty$. Use your results to find the response to the 'rectangular pulse' input shown in part (b) of the figure.

Solution By inspection we see that the recurrence formula is:

$$y[n] = 0.8 \, y[n-1] + x[n]$$

Its impulse response is therefore given by:

$$h[n] = 0.8 \, h[n-1] + \delta[n]$$

Evaluating $h[n]$ term-by-term, we readily obtain:

$$h[0] = 1 \qquad\qquad h[1] = 0.8$$
$$h[2] = 0.8^2 = 0.64 \qquad h[3] = 0.8^3 = 0.512$$

and so on. $h[n]$ is illustrated in part (c) of the figure.

The step response equals the running sum of $h[n]$. Hence its first few values are:

$$s[0] = h[0] = 1 \qquad s[1] = h[0] + h[1] = 1.8$$

$$s[2] = h[0] + h[1] + h[2] = s[1] + h[2] = 2.44$$

$$s[3] = s[2] + h[3] = 2.952$$

$$s[4] = s[3] + h[4] = 3.3616 \qquad\qquad \text{and so on.}$$

The final value of $s[n]$ is given by:

$$s[\infty] = 1 + 0.8 + 0.8^2 + 0.8^3 + \cdots = \frac{1}{(1 - 0.8)} = 5.0$$

The step response is shown in part (d) of the figure.

We may find the response to the input signal in part (b) of the figure by one of two methods. Either, noting that $x[n]$ consists of four consecutive unit impulses, we may superpose four versions of $h[n]$, duly shifted. Alternatively, we may consider $x[n]$ as the summation of a unit step starting at $n = 0$, and an equal but opposite step starting at $n = 4$. The output is therefore found by superposing two step responses, duly weighted and shifted. Thus $y[n] = s[n] - s[n-4]$. Since the present section is mainly concerned with step responses, we use the latter approach here.

We may tabulate the first few values, correct to 3 decimal places, as follows:

n	0	1	2	3	4	5	6
$s[n]$	1	1.8	2.44	2.952	3.362	3.689	3.951
$-s[n-4]$					-1	-1.8	-2.44
$y[n]$	1	1.8	2.44	2.952	2.362	1.889	1.511

The output signal is shown in part (e) of the figure.

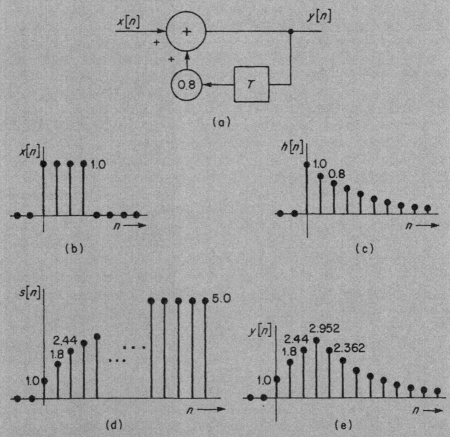

Figure 2.7

The form of $y[n]$ demonstrates some interesting points. Since $x[n]$ starts off like the unit step function $u[n]$, the output is initially the same as $s[n]$. However, after only four finite sample values the input returns to zero. So the output stops rising, and displays an exponentially-falling transient.

It is simple to check that each value in the 'tail' of $y[n]$ is 0.8 times the previous value. In other words the transient has the same form as the system's impulse response $h[n]$. This must be so, because the system is 'on its own'. More generally, our example illustrates the spreading of a 'rectangular pulse' signal along the time axis, when it is processed by a LTI system containing memory. The spreading effect is closely related to the form and duration of $h[n]$.

2.4 DIGITAL CONVOLUTION

We have seen how to describe digital signals as sets of impulse functions, and how to characterize LTI processors by their impulse or step responses. Example 2.2 has shown, in a particular case, how to estimate a system's response to a nontrivial input. However we clearly need a general, computer-based, method for doing this — a method which will work for any LTI system and any form of input signal. It is known as *digital convolution*.

2.4.1 THE CONVOLUTION SUM

We can introduce you to digital convolution with a simple example. Consider an input signal with just four arbitrary, nonzero, sample values, as shown in Figure 2.8(a). We also have a digital processor with the impulse response $h[n]$ illustrated in part (b) of the figure. In the rest of the figure, part (c), $x[n]$ is decomposed into a set of weighted, shifted, impulses. Each of these generates its own version of the system's impulse response. Note carefully how each version is *weighted* by the value of $x[n]$ which causes it, and *shifted* to begin at the correct instant. The output signal $y[n]$ is found by superposition of all these individual responses. It is shown at the bottom of the figure. Convolution of the input signal $x[n]$ and the impulse response $h[n]$ has been accomplished.

You are recommended to study this figure carefully, making sure you understand exactly how $y[n]$ has been formed. It is one of the most important figures in the book!

In most practical cases there would be many more finite sample values in $x[n]$, and probably also more in $h[n]$, and we would have to superpose many more individual responses to find the output signal. In the most general case, we may write:

$$y[n] = \cdots + x[-2]\,h[n+2] + x[-1]\,h[n+1] + x[0]\,h[n]$$
$$+ x[1]\,h[n-1] + x[2]\,h[n-2] + \cdots \tag{2.9}$$

or:

$$y[n] = \sum_{k=-\infty}^{\infty} x[k]\,h[n-k] \tag{2.10}$$

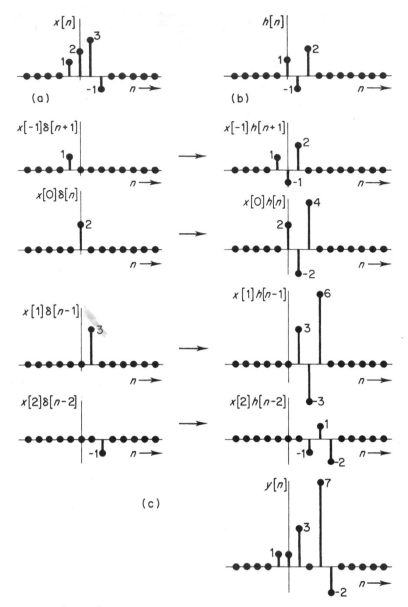

Figure 2.8 Digital convolution.

The last equation is very important, and is known as the *convolution sum*. Note its similarity with the earlier equation (2.2), which expressed any input signal $x[n]$ as the summation of a set of weighted, shifted impulses. What we are now doing is to express an output signal $y[n]$ as an equivalent set of weighted, shifted impulse responses. So the term $\delta[n-k]$ in equation (2.2) is replaced by $h[n-k]$ in equation (2.10).

Equation (2.10) has a very convenient and useful graphical interpretation, illustrated in Figure 2.9 for the particular input signal and impulse response

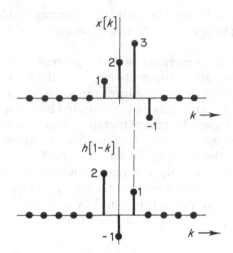

Figure 2.9 Graphical interpretation of convolution.

we have chosen. Instead of considering each constituent impulse of $x[n]$, the complete input signal is drawn. Next, the impulse response is laid out backwards (or *time-reversed*) beneath it, starting from the instant for which we wish to calculated the value of $y[n]$. In the figure, the impulse response is laid out backwards starting beneath $x[1]$, so the output value we will calculate is $y[1]$. We now multiply together all coincident sample values in the input signal and reversed impulse response, and sum all the products so formed. In this case we obtain:

$$y[1] = (3)(1) + (2)(-1) + (1)(2) = 3$$

The *next* output value $y[2]$ is found by shifting the reversed impulse response along to the right by one sampling interval, followed again by cross-multiplication and summation. Thus

$$y[2] = (-1)(1) + (3)(-1) + (2)(2) = 0$$

The same process must be repeated for each output signal value. We should note that Figure 2.9 is just a graphical representation of equation (2.10) for the particular case when $n=1$, that is:

$$y[1] = \sum_{k=-\infty}^{\infty} x[k]\, h[1-k] \tag{2.11}$$

The two discrete-time functions in the figure are, indeed, $x[k]$ and $h[1-k]$, k being a 'dummy' variable which disappears when we perform the cross-multiplication and summation. Of course, in this case we need only consider values of k between -1 and $+2$, because the product is zero outside these limits.

In normal English usage, the word convolution means a kind of 'coiling or twisting'. In the graphical technique just described one of the functions is first time-reversed, then shifted along in a series of steps. At each step it is multiplied by the other function, and all finite products are added together. You may or may not feel that this sequence of events is tantamount to a 'coiling or twisting' — perhaps a 'rolling together' of the two functions would be a more accurate description?

This is a good point to mention that the process of convolution is often denoted by an asterisk, and is commutative. Thus the convolution of $x[n]$ with $h[n]$ is written as $x[n]*h[n]$ and produces the same result as the convolution $h[n]*x[n]$. This may be quite easily demonstrated by reference to Figure 2.10, which is identical to Figure 2.9 except that the integer variable k is defined differently. It is now measured 'back into the past' from the instant for which we wish to calculate the output. Thus $k=0$ coincides with the origin of the reversed impulse response, and k increases towards the *left-hand* side of the diagram. The reversed impulse response is now given by $h[k]$, and the input signal by $x[1-k]$. The output $y[1]$ is found by multiplying the two functions together, and adding all finite products, as before

$$y[1] = \sum_{k=-\infty}^{\infty} h[k]x[1-k] \qquad (2.12)$$

and in general we may write the output as

$$y[n] = \sum_{k=-\infty}^{\infty} h[k]x[n-k] \qquad (2.13)$$

The last result is the one we shall generally use in this text. It is identical to equation (2.10) except that x and h are interchanged. (One consequence is that the graphical procedure shown in Figures 2.9 and 2.10 could as well have been carried out by reversing the input signal beneath the impulse response, rather than vice versa. However, to avoid confusion in this text, we will standardize

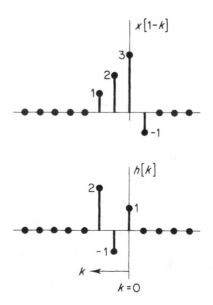

Figure 2.10 Alternative graphical interpretation of convolution.

by always reversing the impulse response.) Once again, the limits of k may generally be reduced in practice.

Example 2.3 Find the output sample sequence from a digital processor in the following cases:

(a) The input signal is as in Figure 2.10, and the impulse response is given by:

$h[n]=0, \quad n<0 \text{ and } n>2$
$h[0]=2; \quad h[1]=1; \quad h[2]=-1$

(b) The input signal is the sample sequence:

... 0, 0, 1, 1, 1, 1, 1, 1, 0, 0 ...

and the impulse response is given by:

$h[n]=0, \quad n<0 \text{ and } n>1$
$h[0]=1; \quad h[1]=-1$

Solution In simple cases such as these, it is hardly necessary to draw figures; we can treat the input signal and impulse response as number sequences.

(a) We have:

input signal: 1, 2, 3, −1
reversed impulse response: −1, 1, 2

We now 'slide' the reversed impulse response past the input signal, cross-multiplying and summing terms at each step. Successive output values are:

$y[-1] = (1 \times 2) = 2$
$y[0] = (2 \times 2) + (1 \times 1) = 5$
$y[1] = (3 \times 2) + (2 \times 1) + (-1 \times 1) = 7$
$y[2] = (-1 \times 2) + (3 \times 1) + (2 \times -1) = -1$
$y[3] = (-1 \times 1)) (3 \times -1) + -4$
$y[4] = (-1 \times -1) = 1$
$y[5] = 0, y[6] = 0, \quad$ and so on

Hence the output signal is the sequence:

... 0, 0, 2, 5, 7, −1, −4, 1, 0, 0 ...

The first finite value occurs at $n = -1$, since the input signal begins at this instant and $h[n]$ begins at $n=0$. Note that the reduced limits for k in equation (2.13) are automatically taken care of by this procedure.

(b) We have:
input signal: 1, 1, 1, 1, 1, 1
reversed impulse response: −1, 1

Hence successive output signal values are:

1, 0, 0, 0, 0, 0, −1

Since $h[n]$ begins at $n = 0$, the first finite output coincides with the first finite input value. This particular processor estimates the first-order difference of its input signal.

The graphical interpretation of convolution can provide some valuable insights into the behavior of practical LTI systems. However, the simple forms of input signal and impulse response we have used so far are not really suitable for developing such insights. We need a more realistic input signal, and an impulse response which represents a more obviously useful system.

Back in Section 1.1.2, we introduced various practical applications of DSP. One related to the smoothing of financial or stock market data using a moving-average low-pass filter (see Figure 1.3). We can use a filter of this general type to demonstrate further important aspects of convolution. The particular application we now consider is the smoothing of temperature data, such as might be recorded at a meteorological station.

Figure 2.11(a) represents variations in midday temperature at a certain place over a 5-week period. Remember that such a signal will generally contain a mixture of sinusoidal frequencies — the lowest ones representing slow fluctuations or long-term trends, the higher ones representing day-to-day variations or more rapid fluctuations in the weather. A meteorologist or weather forecaster may well wish to smooth the signal, suppressing the day-to-day effects but preserving the long-term trends. This is not a critical processing task, so a simple moving-average filter should be quite adequate.

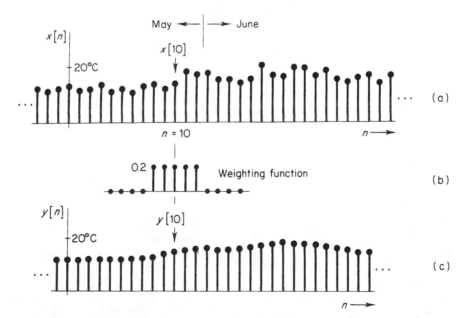

Figure 2.11 Smoothing temperature data with a 5-point moving-average filter.

Suppose we take a 5-point moving average. In other words, each smoothed sample value $y[n]$ is found as the average of $x[n]$ and two values to either side of it:

$$y[n] = 0.2\{x[n+2] + x[n+1] + x[n] + x[n-1] + x[n-2]\} \qquad (2.14)$$

This averaging process is illustrated graphically in part (b) of the figure, for the particular case $n = 10$. The input signal is multiplied by the *weighting function* shown, and all finite products are summed to give the smoothed value $y[10]$. The weighting function is then moved along step-by-step, and the process is repeated to give the complete output $y[n]$.

It is not difficult to see why $y[n]$ is a smoothed version of $x[n]$. Since any one input sample contributes only 20 per cent of any one output, rapid input fluctuations tend to be 'ironed out'. Clearly, this effect would be more pronounced if we averaged over more input samples (for example, Figure 1.3 shows the effect of a 200-point moving average). Of course, if we averaged over *too* many, we would 'iron out' the required slow trends as well.

You have probably noticed that the process just described is identical to time-domain convolution. The weighting function used to calculate the moving average of $x[n]$ behaves just like an impulse response. And the nonrecursive recurrence formula, equation (2.14), is simply an expression of the convolution sum for the particular filter we are considering.

This important conclusion means that we could smooth $x[n]$ automatically, by passing it into a digital LTI processor with an impulse response $h[n]$ identical to a time-reversed version of the weighting function shown in Figure 2.11. (Actually in this case time-reversal makes no difference, because the function is symmetrical, with five equal sample values.)

Note also that this particular filter is noncausal. Each value of $y[n]$ depends on *future* inputs $x[n+1]$ and $x[n+2]$. This is possible because we are dealing with a *pre-recorded* signal. We know all its values in advance. Of course, if a signal is to be processed as it is generated, the impulse response must be causal. To achieve this in the present example, we could shift the weighting function two sampling intervals to the left. The output signal would simply be delayed by the same amount (output sample $y[10]$ being the value previously calculated for $y[8]$, and so on).

You probably agree that it is time to make the digital computer do some work for us! Program no.5 in Appendix A1 carries out a digital convolution for any form of input signal, and any form of impulse response. Input and output signals are normalized to the same peak value before plotting on the screen. (This means that the vertical scales of the plots have no particular significance.) The input signal is loaded into array X, starting at location 60. The impulse response $h[n]$ is assumed causal, and should have less than 60 terms. The program loads $h[n]$ into array H, then implements the convolution sum directly.

The program can be used to explore various aspects of digital convolution. Our first demonstration again concerns a moving-average filter. And in the next section we will use it to investigate transients in LTI processors.

Figure 2.12 shows the screen plot for an input signal containing two distinct sinusoidal frequency components:

$$x[n] = \sin\left(\frac{2\pi n}{60}\right) + \sin\left(\frac{2\pi n}{10}\right), \qquad 60 \le n \le 320 \qquad (2.15)$$

processed by the 10-term moving-average filter:

$$\begin{aligned} h[n] &= 0.1, \qquad 0 \le n \le 9 \\ &= 0 \text{ elsewhere} \end{aligned} \qquad (2.16)$$

(The program listings in Appendix A1 include these choices for $x[n]$ and $h[n]$.) The form of $y[n]$ clearly illustrates the low-pass action of this type of filter. The lower of the two frequencies in $x[n]$, with sixty samples per period, is transmitted. The higher one, with ten samples per period, is not.

In fact we have chosen a rather special case, in which the higher frequency is completely suppressed. The reason is quite straightforward. The filter averages over ten input sample values, so it eliminates a frequency component with exactly ten samples per period. More generally, we would expect such a filter to reduce high frequencies — without eliminating them completely.

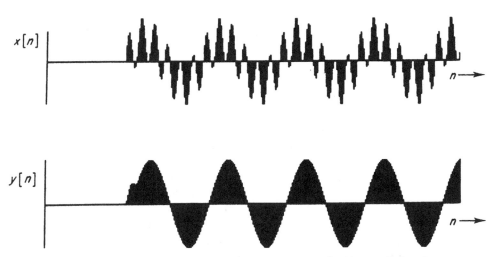

Figure 2.12 A moving-average filter applied to an input signal with two distinct frequency components *(abscissa: 320 samples)*.

You have probably noticed one more thing about Figure 2.12: the rather curious *transient* which occurs at the start of $y[n]$. This 'start-up' effect is due to the sudden application of the input signal. Such transients are an important aspect of LTI processing, and we shall discuss them properly in the following section.

So far we have treated digital convolution as a nonrecursive operation, in which $x[n]$ and $h[n]$ are convolved to produce $y[n]$. A nonrecursive recurrence formula, such as equation (2.14), is equivalent to the convolution sum, equation (2.13). The coefficients by which input samples are multiplied are simply equal to successive terms in $h[n]$. However such nonrecursive convolution is rather slow — especially when $x[n]$ and/or $h[n]$ contain many sample values. An equivalent recursive process — assuming we can find one — is generally much faster.

Whereas a nonrecursive recurrence formula defines an *explicit* convolution, a recursive formula defines an *implicit* one. It, too, implements the convolution sum — but in a way which may be far from obvious. As an example, equation (2.14) could be implemented recursively as follows:

$$y[n] = y[n-1] + 0.2\{x[n+2] - x[n-3]\} \qquad (2.17)$$

This processor has the same impulse response and gives the same performance. But its multiplier coefficients are not obviously related to $h[n]$. Of course we could always generate $h[n]$ from such a formula term-by-term, and use it to implement a nonrecursive version of the same processor. However, in most cases we prefer to use the recursive form (if available), because it involves less computation.

A good example of the computational economy of a recursive processor is the bandpass filter defined by equation (2.4). Its impulse response has been illustrated in Figure 2.4. The recursive formula:

$$y[n] = 1.5y[n-1] - 0.85y[n-2] + x[n] \qquad (2.18)$$

has only three terms. However the impulse response theoretically continues for ever, and even if we ignore the small terms in its 'tail', a nonrecursive version of the same filter would involve about fifty multiplications on input samples alone.

We have so far presented convolution as a method of finding the output signal from an LTI processor, given its input signal and impulse response. This operation is indeed one of the most valuable and widely used in DSP. However, convolution has rather wider implications than this, and we end this section by summarizing some of them.

Earlier in the section we noted that convolution is *commutative*, and is often denoted by an asterisk. Given two digital signals or functions $x_1[n]$ and $x_2[n]$, we may write:

$$x_1[n]*x_2[n] = x_2[n]*x_1[n] \qquad (2.19)$$

In the graphical interpretation of convolution, one function is reversed beneath the other, followed by cross-multiplication and summation. The commutative property implies that it makes no difference which of the two functions is reversed. It also means that the roles of the two functions are interchangeable. An input signal $x[n]$ applied to a processor with impulse response $h[n]$ produces the same output as a signal $h[n]$ applied to a processor with impulse response $x[n]$. In this sense, signals and processors are analogous.

Two further basic aspects of convolution are its *associative* and *distributive* properties. We may summarize the associative property by the following expression:

$$x[n]*\{h_1[n]*h_2[n]\} = \{x[n]*h_1[n]*h_2[n]\} \qquad (2.20)$$

It implies that a cascaded combination of two or more LTI systems can be condensed into a single LTI system. The overall impulse response is found by *convolving the individual responses*. For example, suppose we have two individual processors with impulse response terms:

$$1, 1, 1, 1, 0, 0, 0 \ldots$$

and: 1, −2, 1, 0, 0, 0 . . .

Let us assume that the first term corresponds to $n = 0$ in each case, and that we now cascade the two processors to give a single, overall, system. You may like to check that the overall impulse response, found by convolution, has the values:

$$1, \ -1, \ 1, \ 1, \ 0, \ 2, \ 0, \ 0, \ 0 \ . . .$$

with the first term again corresponding to $n = 0$.

Whereas the associative property has important implications for cascaded LTI systems, the *distributive* property has important implications for LTI systems in *parallel*. We may summarize it as follows:

$$x[n] * \{h_1[n] + h_2[n]\} = \{x[n] * h_1[n]\} + \{x[n] * h_2[n]\} \qquad (2.21)$$

This means that two (or more) parallel systems are equivalent to a single system whose impulse response equals the *sum* of the individual impulse responses. You may feel that this property is rather more obvious than the associative or commutative ones, so we will not illustrate it with an example. However, you will find a problem on it at the end of the chapter.

It is finally worth noting that convolution is not *only* a time-domain procedure used in signal and system analysis. It can be thought of as a mathematical operation to be carried out on two functions — regardless of what they represent. Taking this broader view, we need not be surprised that convolution is important, for example, in the theory of probability and random processes. Please bear in mind, therefore, that the story of convolution presented here, although sufficient for our purposes, is not complete.

2.4.2 TRANSIENTS IN LTI PROCESSORS

No real-life signal continues forever. Even if it did, we could only observe or record it over a finite interval. Furthermore, we cannot store and process infinitely long signals in a digital computer. Hence practical DSP is concerned with signals which are effectively 'switched on' at one instant, and 'switched off' again later.

When an input signal is first switched on, and applied to a digital processor containing memory, it generates a *start-up transient*. When the signal is switched off again, a *stop transient* is produced. We should not regard these transients as an unnecessary, avoidable, nuisance. Nuisance they may be; but they are an inevitable consequence of working with processors containing memory.

Start-up transients are visible in several previous figures in this book. If you are very observant, you may have noticed that the initial portion of $y[n]$ in Figure 1.5 has a rather strange shape. A much more obvious example is the step response of Figure 2.6. The filter is responding to a steady input applied at $n = 0$ (a unit step), but its eventual response is only seen after the oscillatory transient has died away. A third example — already commented on — occurs in Figure 2.12.

Transients are important for several reasons:

A start-up transient may mask the initial portion of an output signal, preventing us from seeing the desired response.

We often assume that the output signal of a digital processor is initially zero. But this will only be true if it has 'settled' following any previous input. In other words, any stop transient must have died away.

Transients are closely related to the 'natural' response of a system, and to our earlier work on impulse responses. They give further valuable insights into the behaviour of linear processors.

We now demonstrate some typical transients using a digital computer. We will look more carefully at their theoretical aspects during our discussion of difference equations in the next section.

Figure 2.13(a) shows an input signal $x[n]$ consisting of five periods of a 'rectangular pulse' waveform. Each period contains 40 samples — 20 of value +1, and 20 of value −1. The signal is switched on at $n = 60$, and switched off again at $n = 260$. This allows us to observe the start-up and stop transients produced by linear processing.

We have again chosen a simple moving-average smoothing filter as the pro-

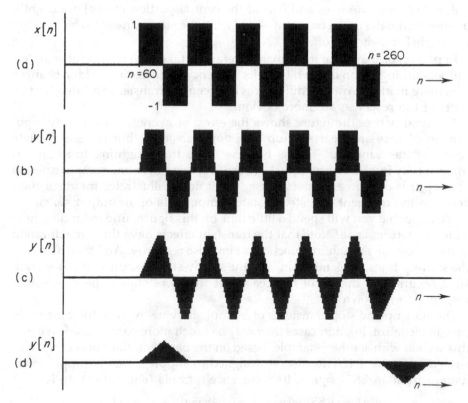

Figure 2.13 Transient and steady-state performance of three moving-average filters, in response to a 'rectangular pulse' input signal *(abscissa: 320 samples)*.

cessor, because its action is easily visualized. Later we will consider a more complicated example.

Parts (b), (c) and (d) of the figure show the effects of using 5-point, 15-point, and 40-point moving-average filters respectively. We have again used Computer Program No.5 to produce the outputs $y[n]$, by changing the input signal, and loading each impulse response in turn into array H. The plots have been assembled into a single figure, to aid comparison. One other point deserves mention. We have normalized $x[n]$ and the three versions of $y[n]$ so as to preserve their *relative* vertical scales.

Let us start with part (b) of the figure, which shows the effect of a 5-point moving average. It is helpful to consider a convolution as we did in Figure 2.11. A weighting function, equivalent to a reversed version of $h[n]$, is cross-multiplied with the input signal, and all products are summed to give a single value of $y[n]$. The weighting function is moved along one step at a time, and the process is repeated. The filter is assumed causal, and has the recurrence formula:

$$y[n] = 0.2\{x[n] + x[n-1] + x[n-2] + x[n-3] + x[n-4]\} \qquad (2.22)$$

When the weighting function first 'moves into' the input signal, it generates a start-up transient. If you look carefully at the initial part of $y[n]$, you will see five individual steps, corresponding to the five weighting function terms. Thereafter the filter displays its *steady-state response* to the input rectangular pulse. Note that the rises and falls of the central portion of $y[n]$ are slightly different from the initial transient — and from the stop transient which occurs when $x[n]$ is switched off.

In part (c) of the figure the moving average is taken over 15 input samples. The smoothing action of the filter — its tendency to 'iron out' sudden changes — is more marked. And the differences between the transient and steady-state parts of the response are more obvious.

The final part of the figure shows the effect of averaging over forty input samples. There are clear start-up and stop transients; but the steady-state response has vanished. This is because, once the weighting function has 'moved fully into' the input signal, we are averaging over one complete period of $x[n]$. It is of course, a special case. But it neatly illustrates the distinction between the transient and steady-state components of an output signal.

We hope that you will spend a little time on this figure, understanding how its main features arise. Note that the transient effects have the same duration as the processor's weighting function, or impulse response. And that, although the start-up transient is mixed in with the steady-state response, the stop transient occurs after the input has ceased. It therefore shows the processor's response 'on its own'.

The start-up and stop transients of a simple moving-average filter are fairly easy to visualize. In other cases they may be much more complicated. We end this section with another example, based on the bandstop filter previously illustrated in Figure 1.4. This filter is designed to suppress mains-supply interference from an EKG signal. Its recurrence formula (equation (1.3)) is:

$$y[n] = 1.8523y[n-1] - 0.94833y[n-2] + x[n]$$
$$- 1.9021x[n-1] + x[n-2] \qquad (2.23)$$

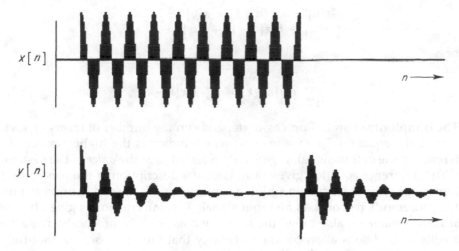

Figure 2.14 Start-up and stop transients of a bandstop filter *(abscissa: 320 samples)*.

Figure 2.14 shows the filter's response to a switched sinusoid at mains supply frequency. This is, of course, the frequency which the filter is required to reject. There is a long-lived start-up transient, which takes the form of an exponentially decaying oscillation. As the transient decays to zero, we notice that the steady-state filter output is indeed very small. When the input sinusoid is switched off, there is a long-lived stop transient. Such effects could have considerable practical importance. For example, we can hardly expect the filter to process an EKG signal effectively if it is in the middle of a start-up transient.

The figure has been produced by Computer Program No. 6 in Appendix A1, which implements the convolution recursively using equation (2.23).

We have tried to give you a general appreciation of transients in LTI processors. In the next section we explore the theoretical relationships between transient and steady-state components of a response, by looking more carefully at the properties of difference equations.

2.5 DIFFERENCE EQUATIONS

You have already met a number of recursive and nonrecursive recurrence formulae. They describe the operation of digital processors in the time-domain, and are also referred to as *difference equations*.

Equation (2.6) gave a general form of difference equation with three recursive and three nonrecursive terms:

$$y[n] = a_1 y[n-1] + a_2 y[n-2] + a_3 y[n-3]$$
$$+ b_0 x[n] + b_1 x[n-1] + b_2 x[n-2] \qquad (2.24)$$

To make the equation even more general, we allow an arbitrary number of terms and recast in the form:

$$a_0 y[n] + a_1 y[n-1] + a_2 y[n-2] + \cdots$$
$$= b_0 x[n] + b_1 x[n-1] + b_2 x[n-2] + \cdots$$

or

$$\sum_{k=0}^{N} a_k y[n-k] = \sum_{k=0}^{M} b_k x[n-k] \tag{2.25}$$

The complexity of an LTI processor depends on the number of terms on each side of the equation. The value of N, which indicates the highest-order difference of the output signal, is generally referred to as the *order* of the system.

The difference equation gives a fundamental description of the processor it represents. However, certain additional information is needed in order to find the processor's response to an input signal. This information is given by the *auxiliary conditions*, also called the *boundary conditions*. In practical terms, the auxiliary conditions allow for the possibility that the processor has not fully 'come to rest', or settled, following some previous input.

As a simple example, consider a system with the difference equation:

$$y[n] - 0.8y[n-1] = x[n] \quad \text{or} \quad y[n] = 0.8y[n-1] + x[n] \tag{2.26}$$

Suppose we have an input signal which is zero perior to $n=0$, and has finite sample values for $n=0, 1, 2 \ldots$ Clearly, we can only determine $y[0]$ if we have a value for $y[-1]$, and $y[-1]$ could be nonzero if the system was still responding to a previous input. Fortunately, we are generally interested in systems which have come to rest. We have assumed this in all previous figures and computer simulations.

How do we proceed if the auxiliary conditions are not zero? The most obvious way is to insert the nonzero value of $y[-1]$, and use the difference equation to estimate subsequent output values term-by-term. This has the effect of mixing, or superposing, any residual transient with the response to the new input signal. However, an alternative approach is to separate the two components of the total response. These are known as the *homogeneous* and *particular* solutions respectively. Their separation gives us valuable insights into the distinction between transient and steady-state responses.

Broadly speaking, the *homogeneous* solution accounts for any nonzero auxiliary conditions, and for transients caused by switching an input signal on and off. The *particular* solution represents the steady-state response of the system to a continuing input signal. The superposition of the two components gives us the complete output. It will probably not surprise you to know that the homogeneous component has a lot to do with the system's impulse response, and therefore with its behavior when left 'on its own'.

It is worth repeating that time-domain convolution, whether performed recursively or nonrecursively, makes no clear distinction between the homogeneous and particular components of an output. It computes the complete output signal 'all at once'. Indeed it is often difficult to disentangle the homogeneous and particular components (see for example Figure 2.12, and Figure 2.13(b) and (c)). This is not necessarily a disadvantage; but if we do disentangle them, we can learn quite a lot about the behavior of LTI processors.

We now illustrate the above ideas in detail, using a second-order system with the difference equation:

$$y[n] - y[n-1] + 0.5y[n-2] = x[n] \qquad (2.27)$$

There is no particular significance in this choice of equation — except that it gives simple numerical values in our calculations! Let us assume a sinusoidal input signal, switched on at $n = 0$:

$$x[n] = \left\{ \sin\left(\frac{2\pi n}{6} + \frac{\pi}{6}\right) \right\} u[n] \qquad (2.28)$$

This signal is shown in Figure 2.15(a). We will proceed in three stages. First,

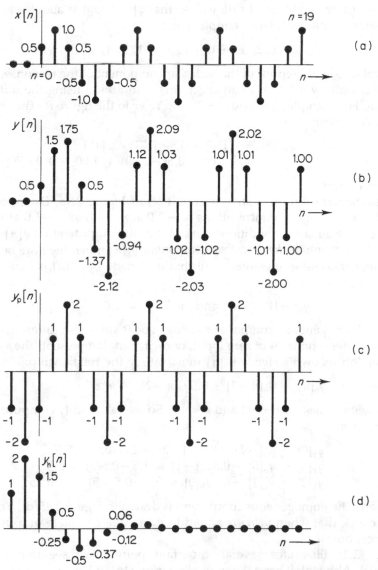

Figure 2.15 The response of a system to a switched sinusoidal input, together with the particular and homogeneous components.

we will find the system's complete output signal using the difference equation. Next, we will infer the particular component of the output. Finally, we will show that a homogeneous component is also present, which allows the auxiliary conditions to be satisfied. Since we are assuming that the system has 'come to rest', the auxiliary conditions relate to the fact that $x[n]$ is not 'eternal', but is switched on at $n = 0$.

Using the difference equation to calculate the output term-by-term (and assuming $y[-1] = y[-2] = 0$), we get the result shown in part (b) of the figure. Sample values are given to two decimal figures. It is clear that, as expected, there is a start-up transient due to the sudden application of the input. It is mixed together with the ongoing response of the system to the sinusoid. However, on the right-hand side we see that $y[n]$ has just about reached its steady-state, with successive sample values.

$$\dots \ 1, \ 2, \ 1, \ -1, \ -2, \ -1, \ 1, \ 2, \ 1, \ \dots$$

These values should represent the *particular* component of the response, which we will denote by $y_p[n]$. We can check this assumption using the difference equation. For example, if we put $n = 19$ and refer to the figure for the required values, we obtain:

$$\begin{aligned} y_p[19] &= y_p[18] - 0.5y_p[17] + x[19] \\ &= -1.00 - 0.5(-2.00) + 1.00 = 1.00 \end{aligned} \qquad (2.29)$$

which is correct.

The particular solution is shown in part (c) of Figure 2.15. Extended back beyond $n = 0$, it makes contributions of -2.0 at $n = -1$, and -1.0 at $n = -2$. However, the auxiliary conditions *demand* that the output signal $y[n]$ is zero at both these points. A *homogeneous* component $y_h[n]$ must therefore be added in. It must cancel the unwanted contributions made by $y_p[n]$, so we require that:

$$y_h[-1] = 2.0 \quad \text{and} \quad y_h[-2] = 1.0 \qquad (2.30)$$

Now the homogeneous component always obeys the same difference equation, but under conditions of zero input, or excitation. It represents the system's behavior 'on its own'. Hence $y_h[n]$ must satisfy the relationship:

$$y_h[n] = y_h[n-1] - 0.5y_h[n-2], \qquad n \geq 0 \qquad (2.31)$$

We know its values for $n = -1$ and $n = -2$. So we can readily compute subsequent values:

$$\begin{aligned} y_h[0] &= y_h[-1] - 0.5y_h[-2] = 2 - 0.5 = 1.5 \\ y_h[1] &= y_h[0] - 0.5y_h[-1] = 1.5 - 0.5(2) = 0.5 \\ y_h[2] &= y_h[1] - 0.5y_h[0] = 0.5 - 0.5(1.5) = -0.25 \end{aligned}$$

and so on. The homogeneous component is drawn in Figure 2.15(d). You may like to check that when $y_h[n]$ is added to $y_p[n]$, the complete output signal $y[n]$ is produced.

Figure 2.15 illustrates several important points. We see that $y_p[n]$ is sinusoidal. Although it has a different *phase* from $x[n]$, it has the same *frequency*. This must be so, because we are dealing with an LTI system. On the other hand, $y_h[n]$ displays a decaying oscillation at a *different* frequency. Remember that a homogeneous solution is characteristic of the system, not of the input, so there is no reason why it *should* oscillate at the input frequency. In fact $y_h[n]$ always takes the same general form as the impulse response of the system.

Although derivation of the output signal in terms of its particular and homogeneous components is rather long-winded, compared with a normal convolution, we hope you will agree that it gives valuable insights into the transient and steady-state behavior of an LTI processor.

You may feel that our discussion so far has been rather theoretical. To give some further experience of transient and steady-state analysis in the time domain, we end the chapter with an example having a more practical flavor.

Example 2.4 A computer system is used to monitor and control various automatic processes in the paint-shop of a sheet-metal factory. One of its tasks is to measure and record the temperature of an oven using a thermocouple. The system is illustrated in Figure 2.16(a). The output signal from the amplifier is sampled every 10 seconds, using an ADC. The scaling of the signal is such that the sample values in the computer represent the oven temperature in °C, divided by 10. Thus a temperature of 200 °C is represented by the value 20, and so on. The temperature data are entered into a storage array X as they are generated. The signal is then filtered by a first-order low-pass filter to reduce unwanted high-frequency noise. The filtered output is stored in array Y. On command from an operator, the contents of array Y may be output via a DAC to a display.

The filter's difference equation is:

$$y[n] - 0.5y[n-1] = 0.5x[n]$$

If the oven temperature is 40 °C when sampling begins, and is rising at a constant rate of 1 °C per second:

(a) Find the impulse response of the filter, and use it to estimate the first five values of the output signal $y[n]$ loaded into array Y.
(b) Check the five values of $y[n]$ already found by using the filter's difference equation.
(c) Infer the particular component $y_p[n]$ of the output, and hence find the homogeneous component $y_h[n]$.
(d) Find which initial oven temperature would cause $y_h[n]$ to be zero for all n, assuming the oven temperature rises at 1 °C per second.

Solution
(a) The impulse response may be found by making $x[n]$ a unit impulse. Thus:

$$y[n] = 0.5y[n-1] + 0.5x[n]$$

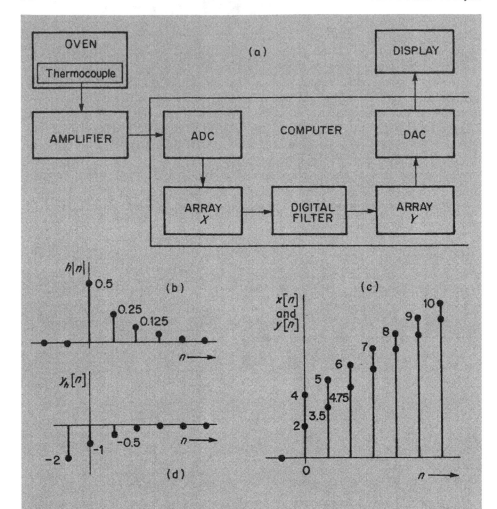

becomes

$$h[n] = 0.5h[n-1] + 0.5\delta[n]$$

This yields:

$$h[0] = 0.5; \quad h[1] = 0.25; \quad h[2] = 0.125$$

and so on. $h[n]$ is shown in part (b) of the figure.

Taking the scaling factor of 10 into account, the input signal $x[n]$ is as shown in part (c) of the figure. Nonrecursive convolution of $x[n]$ with $h[n]$ gives the following first five filter output values:

$$y[0] = 0.5(4) = 2.0$$
$$y[1] = 0.5(5) + 0.25(4) = 3.5$$
$$y[2] = 0.5(6) + 0.25(5) + 0.125(4) = 4.75$$
$$y[3] = 0.5(7) + 0.25(6) + 0.125(5) + 0.0625(4) = 5.875$$
$$y[4] + 0.5(8) + 0.25(7) + 0.125(6) + 0.0625(5) + 0.03125(4)$$
$$= 6.9375$$

$y[n]$ is also shown in part (c) of the figure.

(b) The recursive recurrence formula of the filter is:

$$y[n] = 0.5y[n-1] + 0.5x[n]$$

We may use it to compute successive values of $y[n]$, assuming that the first value in storage array Y is $y[-1]$ and equals zero. Thus:

$y[0] = 0.5(0) + 0.5(4) = 2$
$y[1] = 0.5(2.0) + 0.5(5) = 3.5$
$y[2] = 0.5(3.5) + 0.5(6) = 4.75$
$y[3] = 0.5(4.75) + 0.5(7) = 5.875$
$y[4] = 0.5(5.875) + 0.5(8) = 6.9375$
$y[5] = 0.5(6.9375) + 0.5(9) = 7.9688$
$y[6] = 0.5(7.9688) + 0.5(10) = 8.9844$
$y[7] = 0.5(8.9844) + 0.5(11) = 9.9922$
$y[8] = 0.5(9.9922) + 0.5(12) = 10.9961$
$y[9] = 0.5(10.9961) + 0.5(13) = 11.9981$ etc.

It is therefore clear that $y[n]$ is tending to follow $x[n]$, but with a steady-state error of -1.0, equivalent to $-10\ °C$. We see that the first five values of $y[n]$ agree with the values estimated nonrecursively in part (a) above.

(c) The particular component of the solution, $y_p[n]$, represents the filter's steady-state response to the ramp input signal. The above results show that it must be given by:

$$y_p[n] = x[n] - 1.00, \qquad n \geq -1$$

Extended back to $n = -1$, it would have the value $y_p[-1] = 2$. However $y[-1]$ must be zero to satisfy the initial conditions. We therefore require a homogeneous component $y_h[n]$ such that $y_h[-1] = -2$. Now $y_h[n]$ must obey the relationship:

$$y_h[n] - 0.5\, y_h[n-1] = 0, \qquad n \geq 0$$

Hence:

$$y_h[0] = 0.5(-2) = -1.00$$
$$y_h[1] = 0.5(-1.00) = -0.50$$
$$y_h[2] = 0.5(-0.50) = -0.25, \quad \text{and so on.}$$

$y_h[n]$ is shown in part (d) of the figure. This homogeneous component is a scaled (and inverted) version of the impulse response $h[n]$.

(d) If the particular component $y_p[n]$ had *zero* value at $n = -1$, there would be no need for a homogeneous component. The initial conditions would already be satisfied. The result of part (c) above shows that this would occur if $x[0] = 2$; that is, if the initial oven temperature was 20 °C. We may check this using the filter's recurrence formula with input signal values:

$$2, 3, 4, 5, 6, 7 \ldots$$

Again assuming $y[-1] = 0$, we obtain:

$$y[0] = 0.5(0) + 0.5(2) = 1.0$$
$$y[1] = 0.5(1.0) + 0.5(3) = 2.0$$
$$y[2] = 0.5(2.0) + 0.5(4) = 3.0, \quad \text{and so on.}$$

As expected there is no start-up transient, and no homogeneous component.

PROBLEMS

SECTION 2.2

Q2.1 Describe the signals shown in Figure Q2.1 using sets of weighted, shifted, unit impulse functions.

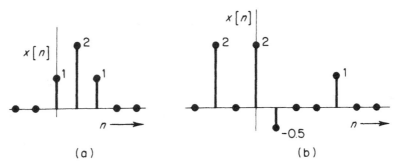

(a)   (b)

Figure Q2.1

SECTION 2.3.1

Q2.2 Which of the following impulse responses describe causal, stable, LTI processors? Give reasons for your answers.

(a) $h[n] = 3\delta[n-2] + \delta[n-4]$
(b) $h[n] = u[n-3] - u[n+5]$
(c) $h[n] = \cos(n\pi/8), \ -1 < n < 20; \ = 0$ elsewhere
(d) $h[n] = \exp(-0.1n) \, u[n]$
(e) $h[n] = \sin(n) \exp(n) \, u[n]$

Q2.3 Sketch the impulse responses of Q2.2.

Q2.4 Sketch the first ten terms of the impulse responses of digital filters described by the following recurrence formulae:

(a) $y[n] = x[n] + x[n-4] + x[n-8]$

(b) $y[n] = \sum_{k=0}^{6} n \, x[n-k]$

(c) $y[n] = y[n-1] + x[n] - x[n-8]$
(d) $y[n] = y[n-1] - 0.5 \, y[n-2] + x[n]$

Q2.5 Use Computer Program no.4 in Appendix A1 to investigate the effect of changing the recursive coefficients of the bandpass filter given in equation (2.4), paying particular attention to its stability.

Q2.6 Estimate the first ten impulse response terms (to 4 decimal places) of the bandstop filter defined by equation (1.3), and illustrated in Figure 1.4 of the main text.

SECTION 2.3.2

Q2.7 Sketch the step responses corresponding to the impulse responses of Q2.2(a) and (b).

Q2.8 Sketch the step responses of the digital filters defined in Q2.4(a) and (c).

Q2.9 Sketch the step responses of the LTI processors defined by the following recurrence formulae:

$$\text{(a)} \quad y[n] = 0.5\, y[n-1] + x[n]$$
$$\text{(b)} \quad y[n] = -0.5\, y[n-1] + x[n]$$

In each case find the value reached by $s[n]$ as $n \to \infty$, and hence infer the response of the processor to a unit-height, sampled, DC-level.

SECTION 2.4.1

Q2.10 (a) Use the graphical interpretation of convolution to find the output $y[n]$ for the input $x[n]$ and impulse response $h[n]$ shown in part (a) of Figure Q2.10. Sketch $y[n]$ carefully.

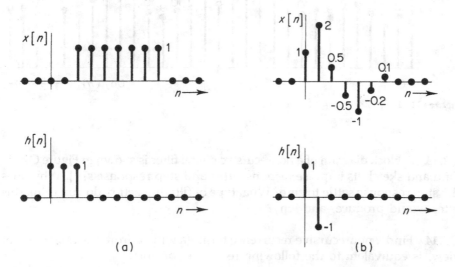

(a)   (b)

Figure Q2.10

(b) Repeat the above for $x[n]$ and $h[n]$ as shown in part (b) of the figure.

(c) Find an expression for the convolution $y[n] = x[n] * h[n]$, given that:

$$x[n] = \alpha^n u[n]$$
$$h[n] = \beta^n u[n]$$
and
$$\alpha \neq \beta$$

Q2.11 If the solution to part (a) of Q2.10 is expressed in the form of a convolution sum:

$$y[n] = \sum_{k=-\infty}^{\infty} h[k]x[n-k]$$

What actual range of the integer variable k is required, given the time-limited nature of $x[n]$ and $h[n]$?

Q2.12 A 7-term moving-average filter defined by the difference equation.

$$y[n] = \frac{1}{7} \sum_{k=-3}^{3} x[n-k]$$

is to be used to smooth the signal shown in Figure Q2.12. This represents a daily wind-speed record through the month of November at Hempstead, Long Island, NY. Write a computer program to implement the filter, and to accept the raw data values given in the figure. What are the filtered values for $y[0]$, $y[6]$, and $y[20]$? Sketch $y[n]$ (or plot it on the computer screen) and make sure you can identify the transients. Can you think of any way of reducing them and, if so, do you think it would have much practical value?

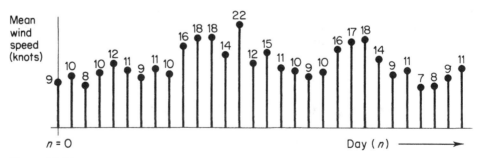

Figure Q2.12

Q2.13 A block diagram of a nonrecursive digital filter is shown in Figure Q2.13. Find and sketch its impulse response $h[n]$ and step response $s[n]$. Why does the step response settle to zero? What type of filtering action do you think this filter would produce, and why?

Q2.14 Find a nonrecursive recurrence formula which, from the DSP point of view, is equivalent to the following recursive formula:

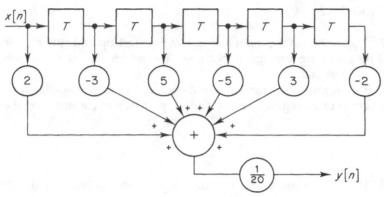

Figure Q2.13

(a) $y[n] = y[n-1] + x[n] - x[n-7]$
(b) $y[n] = 0.9y[n-1] + x[n]$

What is the relative computational economy of the recursive and nonrecursive versions of (a)? Why could the nonrecursive form of (b) not be exactly implemented in practice?

Q2.15 Find the impulse response of an overall system formed by cascading two LTI processors with the impulse responses:

$$h_1[n] = \frac{1}{n}, \quad 0 < n < 4; \quad = 0 \text{ elsewhere}$$

$$h_2[n] = n, \quad 0 < n < 4; \quad = 0 \text{ elsewhere}$$

Your answer may be expressed as a series of sample values.

Q2.16 Two digital LTI processors, with the recurrence formulae given in Figure Q2.16, are connected in parallel to form an overall system. Use the distributive property of convolution to find the impulse response $h[n]$ of the overall system for $0 \le n \le 8$, and sketch it over this range.

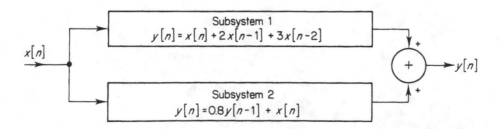

Figure Q2.16

SECTION 2.4.2

Q2.17 Figure 2.13 in the main text shows the start-up and stop transients produced by a moving-average smoothing filter when processing a 'rectangular
pulse' input signal.

   Modify Computer Program no.5 in Appendix A1 to give similar screen plots.
Try the effect of averaging over 30, and 50, points, and comment on the results.

SECTION 2.5

Q2.18 It may be shown that a recursive processor defined by the difference
equation:

$$y[n] = 0.8y[n-1] + 0.2x[n]$$

possesses low-pass filtering properties. Separate the particular and
homogeneous components of its response to the unit step function $u[n]$. Can
you explain their form?

Q2.19 A digital filter has the difference equation:

$$y[n] + y[n-1] + 0.5y[n-2] = x[n]$$

Its input signal is given by:

$$x[n] = \frac{2}{\sqrt{3}} \sin\left(\frac{2n\pi}{3} + \frac{\pi}{3}\right) u[n]$$

Use the difference equation to compute output signal values $y[0]$ to $y[12]$
inclusive, given the auxiliary conditions $y[-2] = y[-1] = 0$.

   Infer the form of the particular solution $y_p[n]$ and, by extending it back to
$n = -1$ and $n = -2$, infer the homogeneous solution $y_h[n]$. Tabulate the values
of $y[n]$, $y_p[n]$ and $y_h[n]$ to 2 decimal places over the range $-2 \le n \le 9$, and
check that $y[n] = y_p[n] + y_h[n]$.

CHAPTER 3

Frequency-Domain Analysis:
the Discrete Fourier Series and
the Fourier Transform

3.1 INTRODUCTION

Jean Baptiste Joseph, Baron de Fourier, was thirty years old when he took part in Napoleon's Egyptian campaign of 1798. He was made governor of Lower Egypt, and contributed many scientific papers to the Egyptian Institute founded by Napoleon. He subsequently returned to France, and became Prefect of Grenoble.

Fourier submitted his ideas on the solution of heat flow problems using trigonometric series to the Institut de France in 1807. The work was considered controversial, and publication of his monumental book on heat had to wait another fifteen years. In it he showed that periodic signals can be represented as weighted sums of harmonically-related sinusoids. He also showed that nonrepetitive, or aperiodic, signals can be considered as the integral of sinusoids which are not harmonically related. These two key ideas form the basis of the famous *Fourier Series* and *Fourier Transform* respectively. They have had a profound influence on many branches of engineering and applied science, including electronics and signal processing.

Fourier analysis has been applied to continuous-time (analog) phenomena for almost two hundred years. Recent dramatic developments in digital electronics and computing have stimulated great interest in corresponding discrete-time (digital) techniques. Digital computers and special-purpose hardware are now widely used for analyzing the frequency components of signals, and the frequency-domain performance of systems.

Before we explain the principles of digital Fourier analysis, and its role in DSP, it is worth considering why the general approach is so useful. After all, we have seen in the previous chapter that time-domain convolution gives a powerful method of analyzing signal flow through linear processors. Why do we also need a frequency-domain approach? There are perhaps three main reasons:

Sinusoidal and exponential signals occur in the natural world, and in the world of technology. Even when a signal is not of this type, it can be analyzed into component frequencies. The response of an LTI processor to each such component is quite simple: it can only alter the amplitude and phase, not the frequency. The overall output signal can then be found by superposition.

If an input signal is described by its *frequency spectrum*, and an LTI processor by its *frequency response*, then the output signal spectrum is found by *multiplication*. This is generally simpler to perform, and to visualize, than the equivalent time-domain convolution.

The design of DSP algorithms and systems often starts with a frequency-domain specification. In other words, it specifies which frequency ranges in an input signal are to be enhanced, and which suppressed. (The low-pass, bandstop, and bandpass filters illustrated in Section 1.1.2 are good examples.)

Many of you will have come to DSP via electronic or electrical engineering, physics or mathematics, and will be familiar with continuous-time Fourier analysis. Other readers may find the account given in Appendix A2 useful. We can summarize its main features, as applied to signals and systems, as follows:

A practical signal may always be analyzed into, or synthesized from, a set of sine and cosine components with appropriate amplitudes and frequencies.

If the signal is an *even* function (symmetrical about the time origin), it contains only cosines. If it is an *odd* function (antisymmetrical about the time origin), it contains only sines.

Approximation of the signal by a limited number of frequency components gives a 'best fit' in the least-squares sense.

If the signal is strictly periodic, its frequency components are harmonically related. The spectrum has a number of discrete spectral lines, and is called a *line spectrum*. It is described mathematically by a *Fourier Series*.

The trigonometric form of the Fourier Series may be recast in an *exponential form*, by expressing each sine and cosine as a pair of imaginary exponentials.

When a signal is non-repetitive (aperiodic), it can be expressed as the infinite sum (integral) of sinusoids, or exponentials, which are not harmonically related. The corresponding spectrum is *continuous* and is described mathematically by the *Fourier Transform*.

Fourier Transformation of a signal gives us its spectrum. A complementary process, *Inverse Fourier Transformation*, allows us to regenerate the signal.

Just as a signal can be described in the frequency domain by its spectrum, so an LTI system can be described by its *frequency response*. This indicates how each sinusoidal (or exponential) component of an input signal is modified in amplitude and phase as it passes through the system. The product of frequency response and input signal spectrum gives the spectrum of the output signal.

Our aim in this chapter is to show that there is a parallel set of Fourier techniques which apply to digital signals. In particular, we discuss two Fourier representations — a discrete-time version of the Fourier Series, which applies to strictly periodic digital signals; and a discrete-time version of the Fourier Transform, relevant to aperiodic signals and LTI processors. You will see that there are many similarities with the continuous-time Fourier Series and Transform, as well as a few important differences.

We should also mention that there is a third type of Fourier representation, known as the *Discrete Fourier Transform (DFT)*, which is of key significance for the computer analysis of digital signals and systems. The DFT is widely implemented using so-called *Fast Fourier Transform (FFT)* algorithms. Not surprisingly, these techniques are closely related to the work of this chapter. However they are of such central importance to DSP that they deserve special treatment. We will discuss them in Chapter 7.

One final point should be made. At this stage of our discussion we will generally assume that signals are real functions, because they are easier to visualize. Thus each sample value is assumed to be defined by a single real number, without an imaginary part. However, we shall have more to say about complex signals in Chapter 7, during our work on the DFT.

3.2 THE DISCRETE FOURIER SERIES

3.2.1 SPECTRA OF PERIODIC DIGITAL SIGNALS

A periodic digital signal can be represented by a Fourier Series. Like its analog counterpart, it has a line spectrum. The features of such a line spectrum form a good starting point for our discussion of digital Fourier analysis. Rather than getting too involved in mathematical derivations and details, we will start with a definition, and illustrate it with the help of a computer program.

Let us take an arbitrary periodic digital signal such as the one shown in Figure 3.1(a). The coefficients of its line spectrum indicate the 'amount' of various frequencies contained in the signal. They may be found using the equation:

$$a_k = \frac{1}{N} \sum_{n=0}^{N-1} x[n] \exp(-j2\pi kn/N) \tag{3.1}$$

a_k represents the kth spectral component, or *harmonic*, and N is the number of sample values in each period of the signal. Equation (3.1) is known as the *analysis equation* of the discrete Fourier Series. Conversely, if we know the coefficients a_k, we may regenerate $x[n]$ using the *synthesis equation*:

$$x[n] = \sum_{k=0}^{N-1} a_k \exp(j2\pi kn/N) \tag{3.2}$$

(Note that, in some texts, the $1/N$ multiplier appears in the synthesis equation, rather than in the analysis equation. However, this is not an important

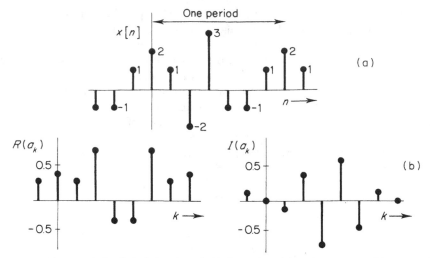

Figure 3.1 (a) A periodic digital signal, and (b) the real and imaginary parts of its spectral coefficients.

difference, since it is only a scaling factor. One advantage of the present definition is that it leads naturally to the Fourier Transform equations which we develop in the next section.)

We see that the analysis and synthesis equations are very similar in form. The process of transforming from the time domain to the frequency domain is essentially the same as that of inverse transformation — going from the frequency domain back to the time domain.

The signal we have chosen contains just seven sample values per period, so $N = 7$. Equation (3.1) becomes:

$$a_k = \frac{1}{7} \sum_{n=0}^{6} x[n] \exp(-j2\pi kn/7) \tag{3.3}$$

Computer Program no.7 in Appendix A1 estimates the various coefficients a_k, given the values of $x[n]$ as data. The real and imaginary parts of the coefficients are calculated separately and held in different arrays (AR and AI). This

	Spectral coefficient a_k	
k	Real part	Imaginary part
0	0.4 285 715	0
1	0.3 018 007	−0.1 086 581
2	0.7 864 066	0.3 847 772
3	−0.3 024 935	−0.6 687 913
4	−0.3 024 928	0.6 687 927
5	0.7 864 058	−0.3 847 782
6	0.3 018 006	0.1 086 581

is done by writing the exponential as $\cos(2\pi kn/7) - j\sin(2\pi kn/7)$. If you try this program, you will obtain the values shown in the table for the various spectral components.

If we ignore the small arithmetic errors in the last one or two decimal places, it is clear that the values display a 'mirror-image' pattern. For example, the real parts of a_1 and a_6 are equal, so are those of a_2 and a_5. The imaginary parts show a similar pattern, but with a change of sign. This always happens when $x[n]$ is a real function of n.

Now suppose that the computer program is used to estimate additional coefficients, outside the range a_0 to a_6. You may like to try this. If you do, you will find that they form part of a repetitive, periodic, sequence. For example, the next seven coefficients, a_7 to a_{13} inclusive, are identical to the set a_0 to a_6. Futhermore, the repetition extends to negative values of k, such that the real parts of a_k form an even function of k, and the imaginary parts form an odd function. This is because the exponential term in equation (3.3) is, like $x[n]$ itself, strictly periodic. The real and imaginary parts of a_k for the signal we have chosen are illustrated in part (b) of the figure (using the symbol $\Re$ to denote real part, and $\Im$ to denote imaginary part).

A periodic digital signal with N samples per period may therefore be completely specified in the frequency domain by a set of N consecutive harmonics. Half this number of harmonics is adequate if $x[n]$ is real, because of the mirror-image pattern we have already noted. For example, coefficients a_0 to a_3 can be used to define the spectrum of the signal in Figure 3.1(a). There is an intuitively appealing reason for this. Our signal has seven independently-adjustable sample values, and is said to have seven *degrees of freedom* in the time domain. It is therefore reasonable that it should have seven degrees of freedom in the frequency domain. Each harmonic contributes two of these (independent real and imaginary parts). The zero-frequency coefficient a_0, which always has zero imaginary part, contributes one. So we reach the expected total of seven.

Our discussion shows that periodic digital signals — unlike their analog counterparts — have spectra which repeat indefinitely along the frequency axis (we shall see later that this is also true of *aperiodic* digital signals). The feature is closely tied up with the ambiguity of sampled sinusoids, discussed in Section 1.4. We showed that a digital sinusoid (or exponential) of frequency Ω is identical to others of frequency $\Omega+2\pi$, $\Omega+4\pi$, and so on. The point was illustrated by Figure 1.17. Now Fourier analysis of a signal is equivalent to finding out 'how much' of each frequency is present. The question is ambiguous with a digital signal, because a whole set of spectral representations are possible. Digital Fourier analysis acknowledges this by producing spectra which repeat indefinitely along the frequency axis. However, this need not cause confusion. As long as we have obeyed the Sampling Theorem, it is only the *first* of these repetitions which reflects frequencies in the underlying analog signal. The rest are simply a consequence of sampling.

The digital signal in Figure 3.1(a) has been chosen arbitrarily, and it is hard to explain the values of its spectral coefficients. We can illustrate the relationships between a signal and its spectrum more clearly, by choosing a signal with a few known frequency components.

Example 3.1 Sketch the periodic digital signal:

$$x[n] = 1 + \sin\left(\frac{\pi n}{4}\right) + 2\cos\left(\frac{\pi n}{2}\right)$$

Find its Fourier Series coefficients a_k, and sketch their real and imaginary parts.

Solution $x[n]$ contains a digital sinusoid with eight samples per period, and a cosinusoid with four samples per period. There is also a zero-frequency, or DC, term. We may tabulate the values of $x[n]$ over one complete period ($n = 0$ to 7 inclusive) as follows:

n	$\sin\left(\dfrac{\pi n}{4}\right)$	$2\cos\left(\dfrac{\pi n}{2}\right)$	$x[n]$
0	0	2	3
1	0.707	0	1.707
2	1	−2	0
3	0.707	0	1.707
4	0	2	3
5	−0.707	0	0.293
6	−1	−2	−2
7	−0.707	0	0.293

The signal is drawn in Figure 3.2(a). We could modify Computer Program no.7 to accommodate eight samples per period, and use it to find the Fourier Series coefficients a_k. However, since $x[n]$ is expressed as a sum of sines and cosines, we can easily expand it as a complex exponential series. Thus:

$$x[n] = 1 + \sin\left(\frac{\pi n}{4}\right) + 2\cos\left(\frac{\pi n}{2}\right)$$

$$= 1 + \frac{1}{2j}\{\exp(j\pi n/4) - \exp(-j\pi n/4)\}$$

$$+ \{\exp(j\pi n/2) + \exp(-j\pi n/2)\}$$

$$\therefore x[n] = \exp(-2j\pi n/4)$$

$$+ \frac{j}{2}\exp(-j\pi n/4) + 1 - \frac{j}{2}\exp(j\pi n/4) + \exp(2j\pi n/4)$$

The Fourier Series coefficients are therefore:

$$a_{-2} = 1; \quad a_{-1} = \frac{j}{2}; \quad a_0 = 1; \quad a_1 = \frac{-j}{2}; \quad a_2 = 1$$

$x[n]$ has eight samples per period, so its Fourier Series must repeat every eight harmonics. Our analysis gives only five finite terms, therefore the other three must be zero. The real and imaginary parts are shown in part (b) of the figure. We have drawn more than one complete repetition of the spectrum, to emphasize its periodic nature.

Note that the real parts of a_k form an even function of k, and represent the cosine and DC terms in $x[n]$. The imaginary parts of a_k form an odd function of k, and represent the sine component in $x[n]$. The three known frequency components in $x[n]$ are clearly shown by these spectral diagrams.

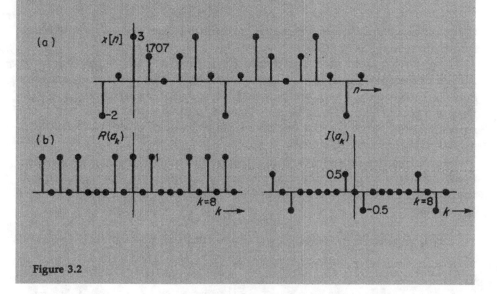

Figure 3.2

The worked example shows clearly how individual sine and cosine components contribute to the spectrum of a periodic digital signal. Let us extend the argument a little further by considering what happens in the case of an *odd* or *even* signal. If $x[n]$ is odd, such that $x[n] = -x[-n]$, it is antisymmetrical about $n = 0$. Since sine functions are also odd, $x[n)$ contains only sines — not cosines. Its spectral coefficients a_k must therefore be purely imaginary. Conversely, an even signal for which $x[n] = x[-n]$ is symmetrical about $n = 0$, and contains only cosines. The coefficients a_k are all real.

You may like to check these conclusions by loading odd and even signals into Computer Program no.7. For example, a suitable odd signal has sample values (remember that the first value corresponds to $n = 0$):

$$0, 2, 1, 3, -3, -1, -2$$

The program should give coefficients with zero (or vanishingly small) real parts. A suitable even signal is:

$$2, -1, 3, 1, 1, 3, -1$$

In this case all the imaginary parts should be vanishingly small.

So far we have concentrated on the real and imaginary parts of spectral coefficients, corresponding to the cosine and sine components of a signal. A widely-used alternative is to express each component in terms of amplitude and phase. Thus if the kth coefficient has real part $\Re(a_k)$ and imaginary part $\Im(a_k)$, its magnitude equals the root of the sum of the squares:

$$|a_k| = \{\Re(a_k)^2 + \Im(a_k)^2\}^{1/2} \tag{3.4}$$

and its phase angle is:

$$\phi_k = \arctan\left\{ \frac{\Im(a_k)}{\Re(a_k)} \right\} \tag{3.5}$$

Quite often we are more interested in magnitudes than phases, so the phase information may be omitted.

Computer Program no.8 in Appendix A1 calculates the magnitudes and phases of the coefficients a_k for a signal with 64 samples per period, and produces a screen plot. There are two main reasons for including the program. First, we would like you to become familiar with the magnitude and phase representation. Secondly, the use of 64 sample values gives more scope for investigating interesting signals.

We will first use Program no.8 to illustrate the magnitude and phase spectrum of a signal with just a few sine and cosine components. This complements the analysis given in Worked Example 3.1. The screen plot in Figure 3.3 shows the signal:

$$x[n] = \sin\left(\frac{2\pi n}{64}\right) + \cos\left(\frac{2\pi n}{16}\right) + 0.6 \cos\left(\frac{2\pi n}{8}\right) + 0.5 \sin\left(\frac{2\pi n}{4}\right)$$

$$\text{for } 0 \le n \le 63 \tag{3.6}$$

Its four components have 64, 16, 8 and 4 samples per period respectively. They correspond to the fundamental, 4th harmonic, 8th harmonic, and 16th harmonic. The composite signal $x[n]$ repeats once every 64 sample values.

The magnitudes of spectral coefficients a_0 to a_{63} inclusive are plotted in part (b) of the figure. As expected, only a_1, a_4, a_8, and a_{16} — and their 'mirror-images' — are nonzero. The mirror-image pattern means that coefficients a_0 to a_{32} are sufficient to define the spectrum completely.

The phases plotted in Figure 3.3(c) show which of the components are sines, and which are cosines. For example, a_1 and a_{63} have phases of $\pm \pi/2$, representing a sine; whereas a_4 and a_{60} both have zero phase, and denote a cosine. Of course, in the more general case there would be a mixture of sine and cosine components at each harmonic frequency, giving phases with intermediate values between 0 and $\pm \pi/2$.

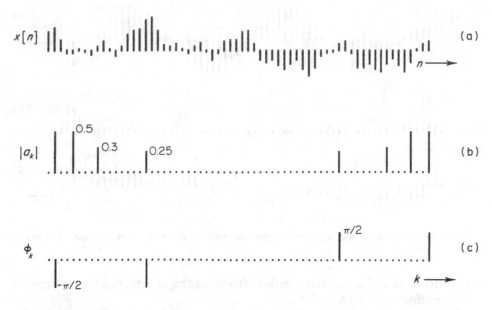

Figure 3.3 Computed spectral coefficients of a periodic signal with several sine and cosine components *(abscissa: 64 samples)*.

Such phase spectra may be a little hard to interpret, because the computer arctan function always returns a value between $\pm \pi/2$ (remember that $\tan\theta = \tan(\theta + \pi)$ for any value of θ). Furthermore, care must be taken in estimating the phase when the real or imaginary part of a component is very small, or zero. Otherwise spurious phase values may be obtained. Program no.8 includes a number of additional statements to prevent this happening.

The four frequency components of the signal in Figure 3.3 each complete an integral number of cycles, or periods, between $n=0$ and $n=64$. Since $x[n]$ is assumed to be repeated end-on-end, the natural periodicity of each component is preserved. There are no sudden discontinuities. The resulting spectrum is well-behaved, each component occupying a definite harmonic frequency.

The situation is however rather different when $x[n]$ contains sinusoids which do not display an exact number of periods between $n=0$ and $n=64$. Program no.8 can easily be modified to demonstrate the effect. Figure 3.4(a) shows a cosine with $2\frac{1}{2}$ periods. Repeated end-on-end, this signal displays sudden discontinuities. Furthermore, unlike an eternal cosine, it is not an even function of n.

Since the cosine goes through $2\frac{1}{2}$ periods within each period of $x[n]$, we may expect its spectral energy to be concentrated close to $k=2$ or $k=3$. This is confirmed by part (b) of the figure. However the spectrum is complicated: there is a lot of 'spreading' of energy due to the discontinuities, and the phase relationships between the various components are hard to visualize. It is clear

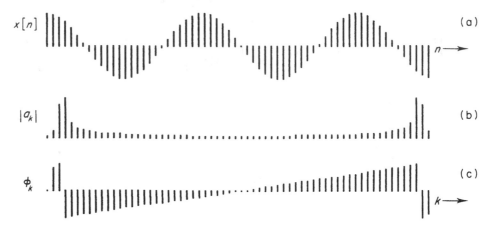

Figure 3.4 The spectrum of a periodic signal with sudden discontinuities *(abscissa: 64 samples)*.

that discontinuities can have major effects on the spectrum of an otherwise 'straightforward' signal.

We next use Program no.8 to investigate two other types of signal — the unit impulse, and the delayed unit impulse. We have already seen in the previous chapter that these signals are crucial in time-domain analysis and convolution; and their spectra are also of great interest.

Figure 3.5 shows the screen plot obtained when a unit impulse $\delta[n]$ forms the input signal for Program no.8. We must of course remember that all signals in this section are considered periodic, and our program assumes 64 samples per period. We are therefore finding the spectrum of an impulse train, not of an individual impulse. Nevertheless, the results give us some valuable insights.

We see that all the spectral coefficients are equal in this case. The spectral energy is evenly distributed in the frequency domain. This suggests why 'impulse testing' of an LTI processor is so effective: it simultaneously delivers an equal amount of all frequencies to the processor's input. The impulse

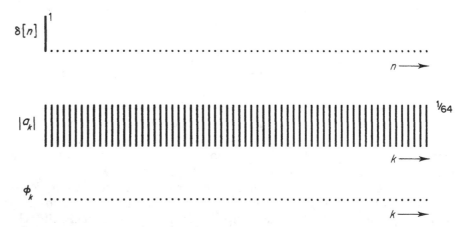

Figure 3.5 Computed spectrum of an impulse train *(abscissa: 64 samples)*.

response of the processor is therefore completely characteristic of its performance.

The figure also shows that the phase spectrum is zero. All the components are cosines. This must be so, because our periodic impulse train is an even function of n.

These results can also be readily explained in terms of the analysis equation of the discrete Fourier Series, equation (3.1):

$$a_k = \frac{1}{N} \sum_{n=0}^{N-1} x[n] \exp\left(-j2\pi kn/N\right)$$

For a train of unit impulses we have:

$$a_k = \frac{1}{N} \sum_{n=0}^{N-1} \delta[n] \exp\left(-j2\pi n/N\right) \tag{3.7}$$

Using the sifting property of the unit impulse (first mentioned in Section 2.2), we may write directly:

$$a_k = \frac{1}{N} \exp\left(-j2\pi kn/N\right)\Big|_{n=0} = \frac{1}{N} \exp(0) = \frac{1}{N} \tag{3.8}$$

This confirms that all the spectral coefficients are equal, and real.

Let us now look at the effects of time shift on a signal spectrum. The signal in Figure 3.6 is identical to that in Figure 3.5, apart from a delay of one sampling interval. The impulse now occurs at $n = 1$. We see that the spectral magnitudes are unaltered, but the phases have changed. In fact we now have a phase lag proportional to frequency (bearing in mind that the computer's arctan function always interprets a phase angle as being in the range $\pm\pi/2$). This is known as a *linear-phase characteristic*. You may like to explore the effects of placing the impulse at a different value of n. You will find that the *slope* of the phase characteristic is proportional to the time-shift introduced.

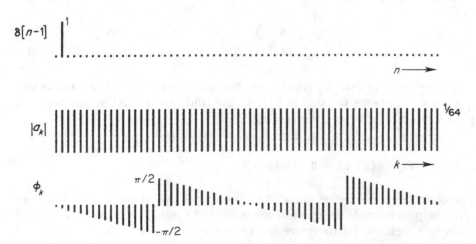

Figure 3.6 The spectrum of a delayed impulse train *(abscissa: 64 samples).*

To end this section, we consider one further aspect of digital signal spectra, and of the relationship between the time and frequency domains.

Parseval's theorem is well known in continuous-time Fourier analysis, and is covered in Appendix A2. It allows us to equate the total power or energy of a signal in the time and frequency domains. In the case of a real, periodic, digital signal, the theorem takes the form:

$$\frac{1}{N} \sum_{n=0}^{N-1} \{x[n]\}^2 = \sum_{k=0}^{N-1} |a_k|^2 \tag{3.9}$$

The left-hand side represents the average energy per sample value, measured over one period in the time domain. The right-hand side is a measure of the signal's total spectral energy, taken over one repetition, or period, in the frequency domain. Note that the energy of a spectral component is given by its squared magnitude, and is independent of phase.

Rather than prove equation (3.9), we will demonstrate it for two of the signals we have met in this section. The signal in Figure 3.2(a) gives:

$$\frac{1}{N} \sum_{n=0}^{N-1} \{x[n]\}^2 = \frac{1}{8}(9 + 2.914 + 0 + 2.914 + 9 + 0.086 + 4 + 0.086)$$

$$= \frac{28}{8} = 3.5$$

and also:

$$\sum_{k=0}^{N-1} |a_k|^2 = (1 + 0.25 + 1 + 0 + 0 + 0 + 1 + 0.25) = 3.5 \tag{3.10}$$

The impulse train of Figure 3.5 is much easier! Here we have:

$$\frac{1}{N} \sum_{n=0}^{N-1} \{x[n]\}^2 = \frac{1}{N} (1) = \frac{1}{N}$$

and:

$$\sum_{k=0}^{N-1} |a_k|^2 = N\left(\frac{1}{N}\right)^2 = \frac{1}{N} \tag{3.11}$$

It is worth noting that the relative phases, or time-shifts, of the various frequency components present in the signals could be altered in any way we please, without affecting the results.

3.2.2 PROPERTIES OF THE SERIES

The discrete Fourier Series possesses a number of useful properties. They can often help us estimate the spectral coefficients of a signal, or synthesize a signal from its spectrum. Furthermore the properties, considered as mathematical operations, shed additional light on the relationships between the time and frequency domains. We summarize the most important ones in this section. They are: linearity; time shifting; differentiation; integration; convolution; and modulation.

In the following discussion we use a double-headed arrow to denote the relationship between a signal and its spectrum. Thus $x[n] \leftrightarrow a_k$ signifies that the periodic digital signal $x[n]$ has spectral coefficients a_k. $x[n]$ is said to *transform* into a_k; a_k *inverse transforms* into $x[n]$. We should regard the time-domain and frequency-domain descriptions of the signal as entirely equivalent. Which we choose to work with on a particular occasion is essentially a matter of convenience.

The *linearity* property is straightforward, and may be stated as follows.

If

$$x_1[n] \leftrightarrow a_k \quad \text{and} \quad x_2[n] \leftrightarrow b_k$$

Then

$$Ax_1[n] + Bx_2[n] \leftrightarrow Aa_k + Bb_k \tag{3.12}$$

where A and B are constants. Thus the spectrum of two (or more) weighted, superposed, signals equals the weighted sum of their individual spectra. Remember, however, that the summation of spectra must take account of phase as well as magnitude.

The *time-shifting* property is as follows.

If

$$x[n] \leftrightarrow a_k$$

Then

$$x[n - n_0] \leftrightarrow a_k \exp(-j2\pi k n_0 / N) \tag{3.13}$$

This defines the effect on the spectrum of shifting $x[n]$ by n_0 sampling intervals. The exponential changes the phases of the coefficients, but not their magnitudes. For example, if we put $n_0 = N$ the signal is shifted by one complete period, and:

$$\exp(-j2\pi k n_0 / N) = \exp(-j2\pi k) = 1 \tag{3.14}$$

for all integer values of k. Therefore, as expected, the spectrum in unchanged.

When discussing time-shifts in relation to the discrete Fourier Series, we should be clear that the shifts involved are periodic, or *circular* — that is to say, a shift by n_0 sampling intervals is indistinguishable from a shift by $(n_0 + mN)$ intervals, where N is the period and m is an integer. An alternative way of expressing this idea is to say that all shifts are evaluated *modulo-N*. The modulo function may be thought of as producing the remainder, following an integer division. This has the effect of making an ascending sequence of numbers periodic, or cyclic. For example the following integer sequence, evaluated modulo-N for $N = 2$, 4, and 8, gives the results:

$$n \quad = \quad 0 \;\; 1 \;\; 2 \;\; 3 \;\; 4 \;\; 5 \;\; 6 \;\; 7 \;\; 8 \;\; 9 \;\; 10 \;\; \ldots$$
$$n \text{ modulo-2} = (n)_2 = \quad 0 \;\; 1 \;\; 0 \;\; 1 \;\; 0 \;\; 1 \;\; 0 \;\; 1 \;\; 0 \;\; 1 \;\; 0 \;\; \ldots$$
$$n \text{ modulo-4} = (n)_4 = \quad 0 \;\; 1 \;\; 2 \;\; 3 \;\; 0 \;\; 1 \;\; 2 \;\; 3 \;\; 0 \;\; 1 \;\; 2 \;\; \ldots$$
$$n \text{ modulo-8} = (n)_8 = \quad 0 \;\; 1 \;\; 2 \;\; 3 \;\; 4 \;\; 5 \;\; 6 \;\; 7 \;\; 0 \;\; 1 \;\; 2 \;\; \ldots$$

Note that the modulo function is often signified using brackets and a subscript. Thus $(n)_8$ specifies that n is being evaluated modulo-8.

Returning to our main theme, you may recall that we have previously met an example of time-shifting. The impulse in Figure 3.6 was shifted forward one sampling interval, compared with Figure 3.5. For $n = 1$, and $N = 64$, equation (3.13) gives:

$$x[n-1] \leftrightarrow a_k \exp(-j2\pi k/64) \tag{3.15}$$

The exponential shows that the time shift introduces a phase term proportional to frequency, with a shift of $-2\pi/64$ radian between adjacent harmonics. You may like to check this against Figure 3.6.

The *differentiation* property of the discrete Fourier Series may be expressed as:

If

$$x[n] \leftrightarrow a_k$$

Then

$$x[n] - x[n-1] \leftrightarrow a_k \{1 - \exp(-j2\pi k/N)\} \tag{3.16}$$

Note that we are interpreting 'differentiation' as forming the first-order difference of $x[n]$. This gives a simple estimate of the slope of the signal. The above result then follows directly from the linearity and time-shifting properties.

Provided the coefficient a_0 of a periodic digital signal is zero, the following *integration* property holds good:

If

$$x[n] \leftrightarrow a_k$$

Then

$$\sum_{k=-\infty}^{n} x[k] \leftrightarrow a_k \{1 - \exp(-j2\pi k/N)\}^{-1} \tag{3.17}$$

We are defining 'integration' as forming the running sum of $x[n]$. The result is only itself periodic if $x[n]$ has zero average, or DC, value — that is, $a_0 = 0$. Otherwise the summation grows or reduces without limit, and invalidates expression (3.17). The integration property is essentially the opposite, or *inverse*, of the differentiation property defined by expression (3.16).

Let us now consider the convolution property. If the digital signals $x_1[n]$ and $x_2[n]$ have the same period, and:

If

$$x_1[n] \leftrightarrow a_k \quad \text{and} \quad x_2[n] \leftrightarrow b_k$$

Then

$$\sum_{m=0}^{N-1} x_1[m] \, x_2[n-m] \leftrightarrow N a_k b_k \tag{3.18}$$

The left-hand side of the expression denotes a convolution over one period.

This ensures convergence of the summation. The operation is called *circular convolution,* or *periodic convolution,* and is often given the symbol Ⓧ. We may visualize circular convolution as the placing of the N samples of $x_1[n]$ around the circumference of a cylinder, and the N samples of $x_2[n]$ in *reverse order* around another, concentric, cylinder. One cylinder is rotated, and coincident samples of $x_1[n]$ and $x_2[n]$ are multiplied and summed. Expression (3.18) shows that such *time-domain convolution is equivalent to frequency-domain multiplication.*

We have previously noted that the time and frequency-domain descriptions of a signal are essentially equivalent. They are like the two sides of a coin. It should therefore cause little suprise that time-domain multiplication is equivalent to frequency-domain convolution. This is summarized by the *modulation* property of the discrete Fourier Series. It may be stated as follows.

If

$$x_1[n] \leftrightarrow a_k \quad \text{and} \quad x_2[n] \leftrightarrow b_k$$

Then

$$x_1[n] \, x_2[n] \leftrightarrow \sum_{m=0}^{N-1} a_m b_{k-m} \tag{3.19}$$

Note the essential symmetry between expressions (3.18) and (3.19), which underlines the reciprocal nature, or *duality,* of the time and frequency domains.

You will find several problems based on the properties of the discrete Fourier Series at the end of this chapter. Furthermore, for convenience, we summarize the main properties in Table 3.1 (to be found at the end of this chapter).

3.3 THE FOURIER TRANSFORM OF APERIODIC DIGITAL SEQUENCES

3.3.1 DERIVATION AND PROPERTIES

Most practical digital signals are *aperiodic* — that is, they are not strictly repetitive. Good examples are fluctuations in the daily price of gold (Figure 1.3), and the midday temperature recorded at a certain place (Figure 2.11). Even when a signal is more or less periodic — such as a heartbeat signal (EKG) — it generally displays fluctuations of amplitude or timing. From the communications engineering viewpoint, a certain amount of randomness or uncertainty is essential if a signal is to convey useful information. It is therefore important to appreciate how Fourier analysis applies to aperiodic sequences. The relevant technique is the Fourier Transform.

We should straightaway add that the discrete Fourier Series, as applied to periodic digital signals in the previous section, is by no means redundant! It is very closely related to the transform, and to computer analysis using the DFT. We shall return to this important matter in Chapter 7.

There are several ways of developing the Fourier Transform for a digital sequence. A common approach is via the continuous-time Fourier Transform, as used in analog signal and system analysis and described in Appendix A2. However, since our book is concerned with DSP, we prefer a digital approach.

We will therefore start with the discrete Fourier Series equations, and modify them to cope with aperiodic signals.

The analysis equation of the discrete Fourier Series, already defined by equation (3.1) is:

$$a_k = \frac{1}{N} \sum_{n=0}^{N-1} x[n] \exp(-j2\pi kn/N) \tag{3.20}$$

This applies, of course, to a strictly periodic signal with period N. Each spectral coefficient a_k is found by multiplying the signal $x[n]$ by an exponential of the relevant frequency, and summing over one period. Although the equation specifies the period between $n=0$ and $(N-1)$, any other complete period will do equally well.

Now suppose that we 'stretch' adjacent repetitions of the signal apart, filling the gaps between them with zeros. This is illustrated by Figure 3.7. Part (a) shows an arbitrary periodic signal with five samples per period ($N=5$). In part (b) we have separated adjacent repetitions, creating a signal for which $N=12$. By continuing this process we could, in principle, make $N \to \infty$. The signal would have just five finite sample values, centered at $n=0$, the neighboring repetitions having moved away towards $\pm\infty$. We would be left with an aperiodic signal.

What happens to the spectral coefficients a_k if we stretch the signal in this way? First, we note that they must become smaller, because of the $(1/N)$ multiplier in equation (3.20). Secondly, it is clear that they must come closer together in frequency, because N also appears in the denominator of the exponential. In the limit as $N \to \infty$, the various harmonics therefore bunch together extemely closely and have vanishingly-small amplitudes. We must think in terms of a continuous, rather than discrete, distribution of spectral energy.

Although each spectral coefficient becomes vanishingly small as $N \to \infty$, the product Na_k remains finite. Let us write it as X. We will also write $(2\pi k/N)$ as

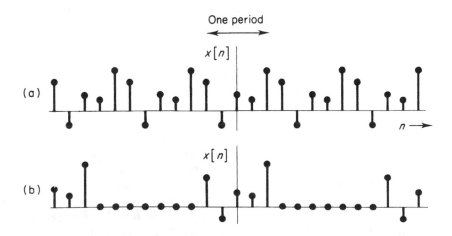

Figure 3.7 'Stretching' a periodic signal.

Ω, and think of it as a continuous frequency variable. Equation (3.20) becomes:

$$X = Na_k - \sum_{n=0}^{N-1} x[n] \exp(-j\Omega n) \qquad (3.21)$$

The limits of summation should also be changed, to take account of the fact that $x[n]$ is now aperiodic. In general, $x[n]$ will exist for both positive and negative values of n, so for generality we sum between $n = \pm\infty$. A further minor change is to write X as $X(\Omega)$, making clear that it is a function of the frequency Ω. Thus:

$$X(\Omega) = \sum_{n=-\infty}^{\infty} x[n] \exp(-j\Omega n) \qquad (3.22)$$

This important equation defines the Fourier Transform $X(\Omega)$ of the aperiodic signal $x[n]$.

Using similar arguments and substitutions, we can develop the *inverse transform* from the synthesis equation of the discrete Fourier Series. The inverse transform tells us how to derive the signal $x[n]$ from its spectrum $X(\Omega)$. The synthesis equation, equation (3.2), is:

$$x[n] = \sum_{k=0}^{N-1} a_k \exp(j2\pi kn/N) \qquad (3.23)$$

In this case it is helpful to substitute Ω_0 for $2\pi/N$. Ω_0 is the first harmonic, or fundamental, frequency. Thus $\Omega = k\Omega_0$ and:

$$x[n] = \sum_{k=0}^{N-1} \left\{ \frac{X(k\Omega_0)}{N} \right\} \exp(jk\Omega_0 n) \qquad (3.24)$$

Furthermore, since $1/N = \Omega_0/2\pi$ we have:

$$x[n] = \frac{1}{2\pi} \sum_{k=0}^{N-1} X(k\Omega_0) \exp(jk\Omega_0 n) \, \Omega_0 \qquad (3.25)$$

Now as $N \to \infty$ and $\Omega_0 \to 0$, the summation becomes an integration. Since the spectrum of a digital signal is always periodic, we integrate over one spectral period — equivalent to an interval of 2π in Ω. Furthermore we may write Ω_0, which becomes vanishingly small, as $d\Omega$. We finally obtain:

$$x[n] = \frac{1}{2\pi} \int_{2\pi} X(\Omega) \exp(j\Omega n) \, d\Omega \qquad (3.26)$$

Equations (3.22) and (3.26) are key results, and constitute a discrete-time *Fourier Transform pair*. The first is an *analysis equation,* showing how an aperiodic digital signal can be expressed in terms of imaginary exponentials (or sines and cosines). The second shows how $x[n]$ can be *synthesized,* or regenerated, from its spectrum $X(\Omega)$. If you are familiar with the Fourier Transform applied to analog signals, you will notice many similarities. However, there is one major difference. The spectrum of a digital signal is always repetitive, unlike that of an analog signal. This is an inevitable consequence of sampling, and reflects the ambiguity of digital signals.

You may find these Fourier Transform equations a little hard to visualize. So we straightaway evaluate the transforms of two simple aperiodic signals, and discuss their main features.

Example 3.2   Find the Fourier Transforms of the aperiodic digital signals shown in Figure 3.8. Sketch their magnitudes over the range $-2\pi < \Omega < 2\pi$, and comment on their form.

Solution

(a)

$$X(\Omega) = \sum_{n=-\infty}^{\infty} x[n] \exp(-j\Omega n)$$

$$= \sum_{n=-\infty}^{\infty} 0.2\{\delta[n-2] + \delta[n-1] + \delta[n] + \delta[n+1] + \delta[n+2]\} \exp(-j\Omega n)$$

Using the sifting property of the unit impulse, we may write directly:

$$X(\Omega) = 0.2\{\exp(-j2\Omega) + \exp(-j\Omega) + 1 + \exp(j\Omega) + \exp(j2\Omega)\}$$
$$= 0.2 (1 + 2 \cos \Omega + 2 \cos 2\Omega)$$

The spectrum is real since $x[n]$ is an even function. It is sketched in the lower part of Figure 3.8(a).

We see that the signal is relatively rich in low frequencies below $\Omega = 2\pi/5$, although there are 'sidelobes' above this which contain signficiant energy. The spectrum is periodic in Ω, repeating every 2π. If we think of $x[n]$ as representing an 'underlying' analog signal, then only frequency components between $\Omega = \pm\pi$ have been adequately sampled according to the Sampling Theorem. The repetitions of the spectrum are simply a consequence of sampling.

$X(\Omega)$ shows how the spectral components of $x[n]$ are continuously distributed in the frequency domain. We should therefore think of it as a *frequency density* function.

(b)

$$X(\Omega) = \sum_{n=-\infty}^{\infty} x[n] \exp(-j\Omega n)$$

$$= 0.5 + 0.25 \exp(-j\Omega) + 0.125 \exp(-j2\Omega) + \cdots$$

$$= 0.5 \sum_{n=0}^{\infty} \{0.5 \exp(-j\Omega)\}^n = \frac{0.5}{1 - 0.5 \exp(-j\Omega)}$$

The magnitude of the spectrum is given by:

$$|X(\Omega)| = \frac{0.5}{\{(1-0.5\cos\Omega)^2 + (0.5\sin\Omega)^2\}^{1/2}}$$

$$= \frac{0.5}{(1-\cos\Omega + 0.25\cos^2\Omega + 0.25\sin^2\Omega)^{1/2}}$$

$$= \frac{0.5}{(1.25 - \cos\Omega)^{1/2}}$$

Note that if $\Omega=0$, $|X(\Omega)| = 1$ and is a maximum; if $\Omega=\pi$, $|X(\Omega)| = \frac{1}{3}$ and is a minimum. The function is sketched in the lower part of the figure. Like the spectrum in part (a), $X(\omega)$ is richest in low frequencies close to $\Omega=0$; but there are no 'spot' frequencies at which signal energy is completely absent. Remember that $X(\Omega)$ is a complex function of Ω in this case, so a complete spectral representation would require a plot of phase as well as magnitude.

We have chosen two signals of predominantly low-frequency content. It would, of course, be quite possible to select signals rich in high frequencies (which, in the context of digital signals, means frequencies close to $\Omega=\pi$), or rich in intermediate frequencies.

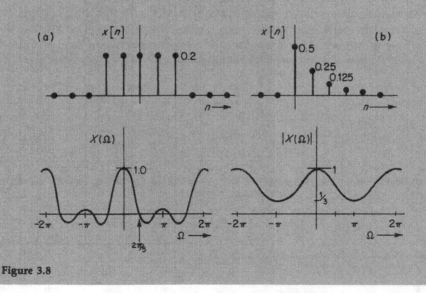

Figure 3.8

Towards the end of Section 3.2.1, we found the spectral coefficients of a unit impulse, and of a delayed unit impulse (see Figures 3.5 and 3.6). However, at that stage we were dealing with periodic signals, and each impulse represented one member of a repetitive train. For completeness, and because of their great importance in DSP, we now look at the corresponding results for isolated impulses.

The Fourier Transform of an *isolated* unit impulse at $n = 0$ is given by:

$$X(\Omega) = \sum_{n=-\infty}^{\infty} x[n] \exp(-j\Omega n) = \sum_{n=-\infty}^{\infty} \delta[n] \exp(-j\Omega n) \qquad (3.27)$$

Again using the sifting property of the unit impulse, we obtain directly:

$$X(\Omega) = \exp(-j\Omega n) \Big|_{n=0} = 1 \qquad (3.28)$$

Thus $\delta[n]$ contains an equal amount of all frequencies. It could be synthesized from an infinite set of cosines, all of vanishingly small, but equal, amplitudes. The spectrum is said to be *white,* just as white light contains an equal mixture of all colors of the rainbow. The unit impulse and its spectrum are shown in Figure 3.9(a). Since the spectrum is real, it is completely represented by a single diagram.

The delayed unit impulse in part (b) of the figure has the spectrum:

$$X(\Omega) = \sum_{n=-\infty}^{\infty} \delta[n-1] \exp(-j\Omega n) = \exp(-j\Omega n) \Big|_{n=1}$$
$$= \exp(-j\Omega) \qquad (3.29)$$

Its magnitude is still unity, but there is a phase shift proportional to frequency. A complete representation requires two diagrams. You may like to compare this spectrum with the one shown in Figure 3.6.

We conclude that isolated impulses possess very wide spectral distributions. By contrast, an eternal sinusoid or exponential has all its energy concentrated at a single frequency. This illustrates the antithesis between limitation in the time and frequency domains: a time-limited signal has a narrow spectrum, and vice versa.

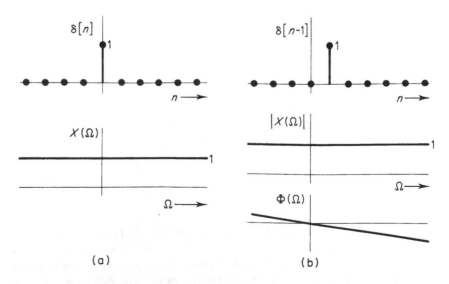

(a)   (b)

Figure 3.9 Spectra of isolated impulses.

The foregoing examples show that it is quite easy to find Fourier Transforms — at least for certain simple signals. We could certainly extend our 'library' of results to other types of signal. However, the normal approach is to use a look-up table, as supplied in many books and references on Fourier analysis. We include a few important transforms and inverse transforms — known as *transform pairs* — as part of Table 3.2 at the end of this chapter.

In Section 3.2.2 we discussed some important properties of the discrete Fourier Series, and summarized them in Table 3.1. Not surprisingly, the Fourier Transform possesses an equivalent set of properties, also included in Table 3.2. If you compare the two tables, you will see many parallels. From the point of view of our work in this chapter, the most important properties of the transform are *linearity*, *time-shifting* and *convolution*. These may be stated as follows:

Linearity

If
$$x_1[n] \leftrightarrow X_1(\Omega) \quad \text{and} \quad x_2[n] \leftrightarrow X_2(\Omega)$$

Then
$$ax_1[n] + bx_2[n] \leftrightarrow aX_1(\Omega) + bX_2(\Omega) \tag{3.30}$$

Time-shifting

If
$$x[n] \leftrightarrow X(\Omega)$$

Then
$$x[n - n_0] \leftrightarrow X(\Omega) \exp(-j\Omega n_0) \tag{3.31}$$

Convolution

If
$$x_1[n] \leftrightarrow X_1(\Omega) \quad \text{and} \quad x_2[n] \leftrightarrow X_2(\Omega)$$

Then
$$x_1[n] * x_2[n] \leftrightarrow X_1(\Omega)X_2(\Omega) \tag{3.32}$$

In particular, we note that a time-shift is equivalent to multiplying by an imaginary exponential in the frequency domain; and that time-domain convolution is equivalent to frequency-domain multiplication. We shall find these properties of great value in the next section.

3.3.2 FREQUENCY RESPONSES OF LTI PROCESSORS

We have so far used the Fourier Transform to investigate the spectra of aperiodic digital signals. We now turn to another extremely useful application — its ability to describe the frequency-domain performance of LTI processors.

The key relationships defining an LTI system in the time and frequency domains are summarized by Figure 3.10. In the time domain, the input signal $x[n]$ is *convolved* with the impulse response $h[n]$ to produce the output signal $y[n]$. The convolution property of the transform tells us that the equivalent frequency-domain process must be a *multiplication*. The output signal spectrum $Y(\Omega)$ is simply the product of the input spectrum $X(\Omega)$ and a function $H(\Omega)$ representing the system. $H(\Omega)$ is known as the system's *frequency response*.

When multiplying $X(\Omega)$ by $H(\Omega)$ we must take proper account of phase as well as magnitude. In general we may write $X(\Omega)$ in the *polar* form:

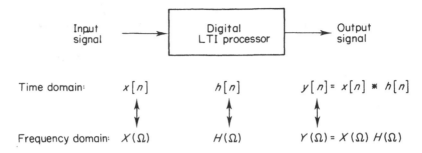

Figure 3.10 Time- and frequency-domain descriptions of signal flow through an LTI processor.

$$X(\Omega) = |X(\Omega)| \exp (j\Phi_X(\Omega)) \tag{3.33}$$

where $|X(\Omega)|$ is its magnitude and $\Phi_X(\Omega)$ is its phase. Similarly $H(\Omega)$ may be expressed as:

$$H(\Omega) + |H(\Omega)| \exp (j\Phi_H(\Omega)) \tag{3.34}$$

where $|H(\Omega)|$ denotes the magnitude of the response, often referred to as the *gain*, and $\Phi_H(\Omega)$ is the phase shift imposed by the system. The product of $X(\Omega)$ and $H(\Omega)$ is:

$$X(\Omega)H(\Omega) = |X(\Omega)|\ |H(\Omega)| \exp (j\{\Phi_X(\Omega) + \Phi_H(\Omega)\}) \tag{3.35}$$

Hence we *multiply* the magnitudes and *add* the phases. This follows the normal rules of complex arithmetic.

The impulse response $h[n]$ and frequency response $H(\Omega)$ are a Fourier Transform pair. The relationship is quite easy to demonstrate. If we deliver a unit impulse to the processor input, then $x[n] = \delta[n]$. Now the spectrum of a unit impulse is unity (see equation (3.28)), so $X(\Omega)=1$. Therefore:

$$Y(\Omega) = X(\Omega)H(\Omega) = H(\Omega) \tag{3.36}$$

Since $y[n]$ clearly equals $h[n]$ in this case it follows that $h[n]$ transforms into $H(\Omega)$.

Just as the impulse response of an LTI system completely characterizes it in the time domain, so the frequency response completely characterizes it in the frequency domain. The two descriptions are equivalent. As we have seen, an impulse contains an equal amount of all frequencies. Used as an input signal, it simultaneously probes the system's response to all possible input frequencies.

We can now explore the frequency responses of some of the practical processors described in earlier chapters. For example, you may remember that in Section 2.4.1 we used a 5-point moving-average filter to smooth a temperature record (see Figure 2.11). The filter's impulse response, or weighting function, was:

$$h[n] = 0.2\{\delta[n-2] + \delta[n-1] + \delta[n] + \delta[n+1] + \delta[n+2]\} \tag{3.37}$$

The corresponding frequency response is:

$$H(\Omega) = \sum_{n=-\infty}^{\infty} h[n] \exp (-j\Omega n) \tag{3.38}$$

Using the sifting property of the unit impulse function we obtain directly:

$$H(\Omega) = 0.2\{\exp(-j2\Omega) + \exp(-j\Omega) + 1 + \exp(j\Omega) + \exp(j2\Omega)\}$$
$$= 0.2(1 + 2\cos\Omega + 2\cos 2\Omega) \tag{3.39}$$

We sketched the spectrum of a *signal* having just this form in Worked Example 3.2. Therefore the frequency response $H(\Omega)$ of our 5-point moving-average filter must be identical to the function $X(\Omega)$ already shown in Figure 3.8(a). It displays simple low-pass properties because, over the frequency range $\Omega = 0$ to π, it transmits low frequencies most strongly. This accounts for the smoothing action of the filter.

Note also that this particular filter is *zero-phase*. $H(\Omega)$ is real, denoting no phase shift at any frequency. This occurs because $h(n)$ is an even function of n, and the filter is noncausal. If we were to shift $h[n]$ forward so that it began at $n = 0$, the filter would become causal, imposing a phase shift proportional to frequency. Such a *linear-phase characteristic*, equivalent to a pure time delay, is often considered the ideal characteristic for a causal processor.

Another good example is the simple recursive filter used in Chapter 2 to smooth the output from a thermocouple. Its impulse response (see Figure 2.16(b)) is:

$$h[n] = 0.5\,\delta[n] + 0.25\,\delta[n-1] + 0.125\,\delta[n-2] + \cdots \tag{3.40}$$

Once again, we have already found the corresponding spectral function in Worked Example 3.2. Thus:

$$H(\Omega) = \frac{0.5}{1 - 0.5\exp(-j\Omega)} \tag{3.41}$$

Its magnitude, $|H(\Omega)|$, is the same as $|X(\Omega)|$ shown in Figure 3.8(b). It, too, displays a simple low-pass characteristic. A complete representation of $H(\Omega)$ would also involve its phase characteristic $\Phi_H(\Omega)$ — which is not linear in this case.

An alternative way of finding the frequency response of a digital processor is via its difference equation. We have previously noted that LTI processors are characterized by difference equations of the general form (see equation (2.25)):

$$\sum_{k=0}^{N} a_k\, y[n-k] = \sum_{k=0}^{M} b_k\, x[n-k] \tag{3.42}$$

We must be careful to remember that the terms a_k here represent recursive *multiplier coefficients*. We have also used the symbol a_k in this chapter to denote the spectral coefficients of a discrete Fourier Series. Furthermore, we are now using N to denote the *order* of the system — not the period of a periodic signal. Such are the difficulties of finding enough symbols for a book on DSP!

We may write the Fourier Transforms of both sides of equation (3.42) as follows:

$$\sum_{k=0}^{N} a_k \exp(-jk\Omega)\, Y(\Omega) = \sum_{k=0}^{M} b_k \exp(-jk\Omega)\, X(\Omega) \tag{3.43}$$

This result stems directly from the linearity and time-shifting properties of the transform. Now $Y(\Omega) = X(\Omega)\, H(\Omega)$, so that:

$$H(\Omega) = \frac{Y(\Omega)}{X(\Omega)} = \frac{\sum\limits_{k=0}^{M} b_k \exp(-jk\Omega)}{\sum\limits_{k=0}^{N} a_k \exp(-jk\Omega)} \tag{3.44}$$

Equation (3.44) is quite general, and allows us to find $H(\Omega)$ for any nonrecursive or recursive LTI processor.

Example 3.3   Figure 3.11(a) shows a high-pass filter. Find its frequency response $H(\Omega)$, and sketch its magnitude and phase over the range $0 < \Omega < \pi$. What is the gain of the filter at $= \pi$? Confirm its value by convolving the impulse response of the filter with a sinusoid at the appropriate frequency. Finally, comment on the contribution of the input term $-x[n-1]$ to the filter's frequency response.

Solution   By inspection, we may write down the following recurrence formula relating input and output signals:

$$y[n] = -0.8y[n-1] + x[n] - x[n-1]$$
$$\therefore y[n] + 0.8y[n-1] = x[n] - x[n-1]$$

Referring back to equation (3.42) we see that the only nonzero filter coefficients are:

$$a_0 = 1, \quad a_1 + 0.8, \quad b_0 = 1, \quad b_1 = -1$$

Therefore equation (3.44) becomes:

$$H(\Omega) = \frac{\{1\exp(-j0)\} + \{-1\exp(-1\Omega)\}}{\{1\exp(-j0)\} + \{0.8\exp(-j\Omega)\}} = \frac{1 - \exp(-j\Omega)}{1 + 0.8\exp(-j\Omega)}$$

$$\therefore H(\Omega) = \frac{1 - \cos\Omega + j\sin\Omega}{1 + 0.8\cos\Omega - 0.8j\sin\Omega}$$

Hence:

$$|H(\Omega)| = \frac{\{(1 - \cos\Omega)^2 + \sin^2\Omega\}^{1/2}}{\{(1 + 0.8\cos\Omega)^2 + 0.64\sin^2\Omega\}^{1/2}} = \left\{ \frac{1 - 2\cos\Omega}{1.64 - 1.6\cos\Omega} \right\}^{1/2}$$

and the phase function is:

$$\Phi_H(\Omega) = \arctan\left(\frac{\sin\Omega}{1 - \cos\Omega}\right) - \arctan\left(\frac{-0.8\sin\Omega}{1 + 0.8\cos\Omega}\right)$$

These are plotted over the range $0 < \Omega < \pi$ in part (b) of Figure 3.11. We should remember that both functions are periodic in Ω, but that their behaviuour over the range shown defines the filter's action on any adequately sampled signal. Note that the filter is high-pass, with a peak gain at $\Omega = \pi$ equal to 10.

The impulse response is easily found from the recurrence formula, by delivering a unit impulse $\delta[n]$ at the input and evaluating the output term by term. Thus:

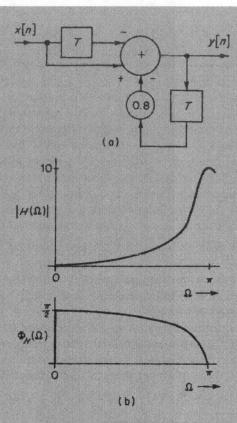

Figure 3.11   (b)

$$h[0] = 1 \qquad\qquad\qquad h[1] = -0.8(1) - 1 = -1.8$$
$$h[2] = -0.8(-1.8) = 0.8(1.8) \qquad h[3] = -0.8^2(1.8)$$
$$h[4] = 0.8^3(1.8) \qquad\qquad h[5] = -0.8^4(1.8)$$

and so on. A discrete sinusoid input signal at the frequency $\Omega = \pi$ has two samples per cycle. Therefore at this frequency successive input samples are equal but opposite. For convenience, let us assume them to be ± 1. The graphical interpretation of discrete-time convolution involves laying the impulse response $h[n]$ out backwards beneath the input signals, cross-multiplying and summing all finite products. In this case, it is clear that this is equivalent to summing all impulse response terms, but with alternate ones inverted. When the impulse response is moved along by one sampling interval, and the process repeated, we get an equal but opposite result. Hence successive output samples from the filter must be $\pm G$, where:

$$G = 1 + 1.8 + 0.8(1.8) + 0.8^2(1.8) + 0.3^3(1.8) + \ldots$$
$$\therefore G = 1 + 1.8(1 + 0.8 + 0.8^2 + 0.8^3 + \ldots)$$
$$= 1 + \frac{1.8}{(1 - 0.8)} = 10$$

But we have assumed input samples to be ± 1. Therefore G is the gain of the filter for a steady-state discrete sinusoid of frequency $\Omega = \pi$. This derivation neatly confirms that the peak gain of the filter is 10, and ties up its frequency and time domain performances. Note, however, that the gain refers to steady-state inputs and does not describe the transients which occur when an input signal is first applied, or removed.

The inclusion of the input term $-x[n-1]$ provides a true null in the frequency response at $\Omega = 0$. This may be desirable if an input signal contains a DC level which must be suppressed. There are at least two possible explanations. Referring to the frequency response $H(\Omega)$, we see that the nonrecursive input term produces a numerator equal to $\{1- \exp(-j\Omega)\}$, which equals zero at $\Omega = 0$. In the time-domain, we recall that forming the first-order difference $\{x[n] - x[n-1]\}$ is equivalent to differentiation. This gives zero response at $\Omega = 0$, because a DC level has zero slope.

In practice it is tedious to estimate and draw $H(\Omega)$ 'by hand' — especially when the processor is a complicated one. It is much better to let a digital computer do the work for us.

As an example, Computer Program no. 9 in Appendix A1 calculates and plots the frequency response of the bandpass filter first mentioned in Section 1.1.2 (see Figure 1.5). This filter displays its maximum response to input frequencies having about ten samples per period. The difference equation is:

$$y[n] = 1.5\, y[n-1] - 0.85\, y[n-2] + x[n] \tag{3.45}$$

Using equation (3.44) we readily obtain:

$$H(\Omega) = \frac{1}{1 - 1.5\, \exp(-j\Omega) + 0.85\, \exp(-j2\Omega)}$$

$$= \frac{1}{(1 - 1.5\, \cos\Omega + 0.85\, \cos2\Omega) + j(1.5\, \sin\Omega - 0.85\, \sin2\Omega)} \tag{3.46}$$

The gain and phase functions are therefore:

$$|H(\Omega)| = \frac{1}{\{(1 - 1.5\, \cos\Omega + 0.85\, \cos2\Omega)^2 + (1.5\, \sin\Omega - 0.85\, \sin2\Omega)^2\}^{1/2}} \tag{3.47}$$

and:

$$\Phi_H(\Omega) = \arctan\left(\frac{1.5\, \sin\Omega - 0.85\, \sin2\Omega}{1 - 1.5\, \cos\Omega + 0.85\, \cos2\Omega}\right) \tag{3.48}$$

The computer plots are shown in Figure 3.12, for 320 equally-spaced values of Ω in the range 0 to π. (We are, of course, representing continuous functions by sets of frequency-domain 'samples'). The form of $|H(\Omega)|$ confirms that this filter has a simple bandpass characteristic, with its maximum gain at about $\Omega = 0.2\pi$. Now $\Omega = \pi$ corresponds to a sinusoid with two samples per period; so $\Omega = 0.2\pi$ corresponds to ten samples per period, as expected.

The phase characteristic is harder to explain and visualize. This is particularly true because the computer arctan function always returns an angle between

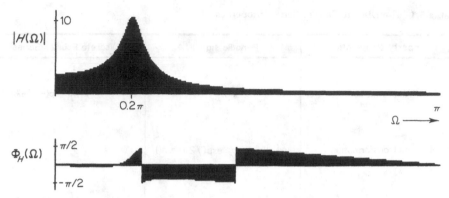

Figure 3.12 Computed magnitude and phase response of a simple bandpass filter *(abscissa: 320 samples)*.

$\pm \pi/2$. When the phase passes through $\pm \pi/2$ there is a sudden jump of π (it happens twice in the figure). Ignoring this effect, we see that the phase shift imposed by the filter changes most rapidly where the gain is also showing rapid variation. This is generally true of an LTI processor.

It is straightforward to modify Program no.9 to cope with other digital processors — such as the bandstop filter used to suppress unwanted supply-frequency 'hum' in an EKG signal (Figure 1.4). The filter's difference equation is:

$$y[n] = 1.8523y[n-1] - 0.94833y[n-2]$$
$$+ x[n] - 1.9021\,x[n-1] + x[n-2] \qquad (3.49)$$

Using equation (3.44) we obtain:

$$H(\Omega) = \left\{ \frac{1 - 1.9021\exp(-j\Omega) + \exp(-2j\Omega)}{1 - 1.8523\exp(-j\Omega) + 0.94833\exp(-2j\Omega)} \right\} \qquad (3.50)$$

Figure 3.13 shows the computed magnitude of this function. Over most of the range, the filter transmits input frequency components with unity gain. But there is a deep notch at $\Omega = 0.1\pi$ which must, of course, coincide with the mains-supply frequency. $\Omega = 0.1\pi$ corresponds to 20 samples per period. Therefore if the supply frequency is 60 Hz, we must use a sampling frequency of 1200 Hz; if the supply frequency is 50 Hz, we must sample at 1000 Hz.

You should now be confident about *analyzing* the frequency-domain performance of LTI processors. However, we are not quite ready to tackle the problems of *synthesis*, or design. We will get closer to this in the next chapter on the *z*-transform, and begin the task in earnest in Chapter 5.

Figure 3.13 The magnitude response of a narrowband 'notch' filter *(abscissa: 320 samples)*.

Table 3.1 The Discrete Fourier Series: properties

Property or operation	Periodic signal	Discrete Fourier Series
Transformation	$x[n]$	$a_k = \frac{1}{N} \sum_{n=0}^{N-1} x[n]\ \exp(-j2\pi kn/N)$
Inverse transformation	$x[n] = \sum_{k=0}^{N-1} a_k\ \exp(j2\pi kn/N)$	a_k
Linearity	$Ax_1[n] + Bx_2[n]$	$Aa_k + Bb_k$
Time-shifting	$x[n - n_0]$	$a_k\ \exp(-j2\pi kn_0/N)$
Time-differentiation	$x[n] - x[n-1]$	$a_k\{1 - \exp(-j2\pi k/N)\}$
Time-integration	$\sum_{k=-\infty}^{n} x[k], a_0 = 0$	$a_k\{1 - \exp(-j2\pi k/N)\}^{-1}$
Convolution	$\sum_{m=0}^{N-1} x_1[m]\, x_2[n-m]$	$Na_k b_k$
Modulation	$x_1[n]\, x_2[n]$	$\sum_{m=0}^{N-1} a_m\, b_{k-m}$
Real time-function	$x[n]$	$a_k = a_{-k}^{*}$ $\mathrm{Re}(a_k) = \mathrm{Re}(a_{-k})$ $\mathrm{Im}(a_k) = -\mathrm{Im}(a_{-k})$

Table 3.2 The Fourier Transform of aperiodic digital signals: properties and pairs

Property or operation	Aperiodic signal	Fourier Transform
Transformation	$x[n]$	$X(\Omega) = \sum\limits_{n=-\infty}^{\infty} x[n]\,\exp(-j\Omega n)$
Inverse transformation	$x[n] = \dfrac{1}{2\pi}\int_{2\pi} X(\Omega)\exp(j\Omega n)\,d\Omega$	$X(\Omega)$
Linearity	$ax_1[n] + bx_2[n]$	$aX_1(\Omega) + bX_2(\Omega)$
Time-shifting	$x[n - n_0]$	$X(\Omega)\exp(-j\Omega n_0)$
Time-differentiation	$x[n] - x[n-1]$	$X(\Omega)\{1 - \exp(-j\Omega)\}$
Convolution	$x_1[n] * x_2[n]$	$X_1(\Omega)\,X_2(\Omega)$
Modulation	$x_1[n]\,x_2[n]$	$\dfrac{1}{2\pi}\int_{2\pi} X_1(\lambda)\,X_2(\Omega-\lambda)\,d\lambda$

Waveform	Aperiodic signal $x[n]$	Spectrum $X(\Omega)$
Unit impulse	$\delta[n]$	1
Shifted unit impulse	$\delta[n - n_0]$	$\exp(-j\Omega n_0)$
Unit step	$u[n]$	$\{1 - \exp(-j\Omega)\}^{-1}$ $+ \sum\limits_{k=-\infty}^{\infty} \pi\,\delta(\Omega - 2\pi k)$
Exponential	$a^n u[n],\ |a| < 1$	$\{1 - a\exp(-j\Omega)\}^{-1}$
Rectangular pulse	$x[n] = 1,\ |n| \leqslant m$ $x[n] = 0,\ |n| > m$	$\dfrac{\sin\{(m + \frac{1}{2})\Omega\}}{\sin\left(\frac{\Omega}{2}\right)}$

PROBLEMS

SECTION 3.2.1

Q3.1 Find the spectral coefficients a_k for the following periodic digital signals:

(a) $x[n] = 5 + \sin\left(\dfrac{n\pi}{2}\right) + \cos\left(\dfrac{n\pi}{4}\right)$

(b) $x[n] = \cos\left(\dfrac{n\pi}{2} - \dfrac{\pi}{4}\right)$

(c) $x[n] = 2n$ for $0 \le n \le 3$, then repeats.

Q3.2 Use Computer Program no.7 in Appendix A1 to estimate the real and imaginary parts of spectral coefficients a_0 to a_6, for a periodic signal $x[n]$ with the following sample values:

$$x[0] = 2 \qquad x[1] = -4 \qquad x[2] = 1$$
$$x[3] = -2 \qquad x[4] = 3 \qquad x[5] = -2$$
$$x[6] = 2$$

Why is a_0 equal to zero? Why are the magnitudes of a_3 and a_4 relatively large? If the signal has been sampled at 1 MHz, what frequency is represented by the coefficient a_2?

Q3.3 Load the sample values of an odd signal into Computer Program no.7 in Appendix A1, and check that its spectral coefficients are purely imaginary.

Q3.4 Repeat Q3.3 for an even signal, checking that the spectral coefficients are real.

Q3.5 Predict the magnitudes and phases of the various spectral coefficients of the periodic signal:

$$x[n] = 1 + \cos\left(\dfrac{\pi n}{32}\right) + \sin\left(\dfrac{\pi n}{4}\right), \quad 0 \le n \le 63$$

assuming that it repeats every 64 sample values. Check your predictions with the aid of Computer Program no.8 in Appendix A1.

Q3.6 Predict the phase spectrum of a train of unit impulses spaced 64 sampling intervals apart, with one member of the train occurring at $n = 5$. Check your prediction using Computer Program no.8 in Appendix A1.

Q3.7 Show that Parseval's theorem (equation (3.9) in the main text) holds good for the periodic signal defined in Q3.2.

SECTION 3.2.2

Q3.8 Sketch the periodic signal $x[n] = \sin n\pi/6$ in the range $0 \le n \le 18$. Use the differentiation property of the discrete Fourier Series to find the relative magnitude and phase of the first-order difference signal $\{x[n] - x[n-1]\}$. Confirm your results by sketching the latter signal.

Q3.9 Two periodic signals $x_1[n]$ and $x_2[n]$ repeat every seven sample values. One period of each signal is as follows (starting at $n = 0$):

$$x_1[n]: \quad 1, 0, 0.5, 0, 0, 0.5, 0$$
$$x_2[n]: \quad 0, 1, 0, 0, 0, 0, 1$$

Perform a periodic convolution of the two signals, defining the result $x_1[n] \circledast x_2[n]$ over the period $0 \le n \le 6$. Use Program no.7 in Appendix A1 to find the spectral coefficients of $x_1[n]$, $x_2[n]$, and $x_1[n] \circledast x_2[n]$. Check that they obey the convolution property of the discrete Fourier Series.

Q3.10 The product of the two signals defined in Q3.9 is zero for all values of n. Use the modulation property of the discrete-time Fourier Series to confirm that all the spectral coefficients of their product are also zero.

SECTION 3.3.1

Q3.11 Find an expression for the spectrum $X(\Omega)$ of each of the following aperiodic signals:

(a) $x[n] = \delta[n] + 2\delta[n-1] + \delta[n-2]$
(b) $x[n] = \delta[n+1] - \delta[n-1]$
(c) $x[n] = u[n+3] - u[n-4]$

SECTION 3.3.2

Q3.12   Find an expression for the frequency response $H(\Omega)$ of an LTI processor whose impulse response is defined as follows:

(a) $h[n] = 4\delta[n] + 2\delta[n-1] + \delta[n-2]$
(b) $h[n] = h[n-1] + \delta[n] - \delta[n-7]$
(c) $h[n] = 0.8\,h[n-1] + \delta[n]$

Q3.13   A simple high-pass filter has the recurrence formula:

$$y[n] = -0.9y[n-1] + 0.1x[n]$$

Find an expression for the frequency response $H(\Omega)$, and sketch its magnitude over the range $0 < \Omega < \pi$. What is the value of $|H(\Omega)|$ at (a) $\Omega = 0$, and (b) $\Omega = \pi$?

Q3.14   A bandpass filter is shown in Figure Q3.14. Find an expression for its frequency response magnitude $|H(\Omega)|$, and sketch the function in the range $0 < \Omega < \pi$.

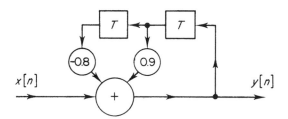

Figure Q3.14

Q3.15   Figure 3.13 in the main text shows the notch characteristic of a filter used to suppress mains-frequency interference in an EKG signal (see also equation (3.50)). Find the magnitude of its frequency response at (a) $\Omega = 0$, and (b) $\Omega = 0.1\pi$. (Note that $\Omega = 0.1\pi$ corresponds to the centre of the notch.)

Q3.16   Write a computer program similar to Program no.9 in Appendix A1, to calculate and plot the magnitude response of a filter with the following difference equation:

(a) $y[n] = -0.9\, y[n-1] + 0.1\, x[n]$ (see also Q3.13)

(b) $y[n] = 0.1 \sum_{k=0}^{9} x[n-k]$

(note: this is a moving-average low-pass filter.)

CHAPTER 4

Frequency-Domain Analysis: the z-Transform

4.1 INTRODUCTION

The z-transform offers a valuable set of techniques for the frequency analysis of digital signals and processors. It is also very useful in design. The z-transform and the Fourier Transform are closely related, and should be regarded as complementary to one another. However, the z-transform is inherently concerned with sampled data and systems, whereas digital Fourier techniques were developed from ideas which originated in the analog domain.

There are three main reasons for covering the z-transform in an introductory book on DSP:

It offers an extremely compact and convenient notation for describing digital signals and systems.

It is widely used by DSP designers, and in the DSP literature.

We shall see that the so-called *pole-zero* description of a processor is a great help in visualizing its stability and frequency response characteristics.

This chapter gives an introductory account of the z-transform, concentrating on aspects which will be most useful to us later in the book.

4.2 DEFINITION AND PROPERTIES OF THE TRANSFORM

The z-transform of a digital signal $x[n]$ is defined as:

$$X(z) = \sum_{n=0}^{\infty} x[n] z^{-n} \qquad (4.1)$$

Since the summation is taken between $n = 0$ and $n = \infty$, $X(z)$ is not concerned with the history of $x[n]$ prior to $n = 0$. The transform is said to be *unilateral*. This is rarely a disadvantage in practical DSP. In most cases we can consider a signal to start at (or after) $n = 0$, taken as our reference instant. Furthermore the impulse response $h[n]$ of any causal processor is zero for $n < 0$. So whether we are describing signals or LTI processors with the z-transform, the unilateral version is normally adequate.

The alternative *bilateral* version of the transform, with summation limits of $n = \pm\infty$, is occasionally needed. Its main disadvantage is its more awkward convergence conditions. In other words the mathematical conditions for the transform to exist and converge are more stringent, and they must be considered carefully. For this reason we concentrate on the unilateral transform in this book.

The z-transform defined by equation (4.1) is quite easy to visualize. $X(z)$ is essentially a *power series* in z^{-1}, with coefficients equal to successive values of the time-domain signal $x[n]$. Therefore if we express $X(z)$ as a power series, we can immediately regenerate the signal. This may not be the most economical way of performing the inverse transform — but it is always possible in principle.

Example 4.1

(a) Find the z-transform of the exponentially-decaying signal shown in Figure 4.1(a), expressing it as compactly as possible.

(b) Find, and sketch, the signal corresponding to the z-transform:

$$X(z) = \frac{1}{(z + 1.2)}$$

Solution

(a) The z-transform of the signal is given by:

$$X(z) = \sum_{n=0}^{\infty} x[n]\, z^{-n}$$

$$= 1 + 0.8z^{-1} + 0.64z^{-2} + 0.512z^{-3} + \cdots$$

$$= 1 + (0.8z^{-1}) + (0.8z^{-1})^2 + (0.8z^{-1})^3 + \cdots$$

$$= \frac{1}{1 - 0.8z^{-1}} = \frac{z}{z - 0.8}$$

(b) Recasting $X(z)$ as a power series in z^{-1}, we obtain:

$$X(z) = \frac{1}{(z + 1.2)} = \frac{z^{-1}}{(1 + 1.2z^{-1})} = z^{-1}(1 + 1.2z^{-1})^{-1}$$

$$= z^{-1}\{1 + (-1.2z^{-1}) + (-1.2z^{-1})^2 + (-1.2z^{-1})^3 + \cdots\}$$

$$= z^{-1} - 1.2z^{-2} + 1.44z^{-3} - 1.728z^{-4} + \cdots$$

Successive values of $x[n]$, starting at $n = 0$, are therefore:

$$0, 1, -1.2, 1.44, -1.728, \cdots$$

$x[n]$ is shown in part (b) of the figure.

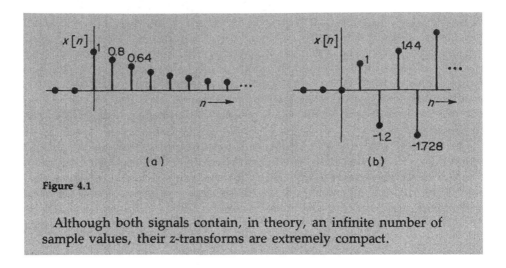

Figure 4.1

Although both signals contain, in theory, an infinite number of sample values, their z-transforms are extremely compact.

If we substitute $\exp(j\Omega)$ for z in equation (4.1), we obtain:

$$X(\Omega) = \sum_{n=0}^{\infty} x[n] \exp(-j\Omega n) \qquad (4.2)$$

Apart from a change in the lower limit of summation, this is identical to the Fourier Transform given by equation (3.22). It is therefore clear that the z-transform and the Fourier Transform are closely related.

An alternative, and probably simpler, way of thinking of z is as a *time-shift operator*. Multiplication by z is equivalent to a time *advance* by one sampling interval. Division by z is equivalent to a time *delay* by the same amount. For example, a unit impulse at $n = 0$ has the z-transform:

$$X(z) = \sum_{n=0}^{\infty} \delta[n]z^{-n} = z^{-n}\Big|_{n=0} = 1 \qquad (4.3)$$

A unit impulse delayed by n_0 sampling intervals has the transform:

$$X(z) = \sum_{n=0}^{\infty} \delta[n-n_0]z^{-n} = z^{-n}\Big|_{n=n_0} = z^{-n_0} \qquad (4.4)$$

Notice how the transform converts a time-shift into a simple algebraic manipulation in the frequency domain.

The z-transform gives us all the advantages of a frequency-domain approach to signal and system analysis. As far as DSP is concerned, one of the most important is its *convolution property*, which states that time-domain convolution is equivalent to frequency-domain multiplication. This property is common to the Fourier Transform and the z-transform. We can demonstrate it by a simple example. Let us take an arbirary, finite-length, signal $x[n]$ and deliver it to the input of an LTI processor with impulse response $h[n]$. Suppose that $x[n]$ and $h[n]$ have sample values:

$$
\begin{array}{lccccccccc}
n = 0 & 1 & 2 & 3 & 4 & 5 & 6 & 7 & 8 & \cdots \\
x[n] = 1 & -2 & 3 & -1 & -1 & 0 & 0 & 0 & 0 & \cdots \\
h[n] = 2 & 1 & -1 & 0 & 0 & 0 & 0 & 0 & 0 & \cdots
\end{array}
$$

In the time-domain, we may find the output signal $y[n]$ by *convolving* $x[n]$ with $h[n]$. You may remember that this is done by shifting a reversed version of $h[n]$ beneath $x[n]$, and summing all finite cross-products at each step. The following output sample sequence (starting at $n=0$) is readily obtained:

$$
y[n] = 2 \quad -3 \quad 3 \quad 3 \quad -6 \quad 0 \quad 1 \quad 0 \quad 0 \quad \cdots
$$

In the frequency-domain, we can describe the signal and the LTI processor by their z-transforms. The transforms of $x[n]$ and $h[n]$ are, by inspection:

$$
X(z) = 1 - 2z^{-1} + 3z^{-2} - z^{-3} - z^{-4}
$$

and

$$
H(z) = 2 + z^{-1} - z^{-2} \tag{4.5}
$$

where $H(z)$ is called the *transfer function* of the processor, and is the z-transform equivalent of its frequency response $H(\Omega)$. It is easy to show that the product $X(z) H(z)$ is:

$$
X(z)H(z) = 2 - 3z^{-1} + 3z^{-2} + 3z^{-3} - 6z^{-4} + z^{-6} \tag{4.6}
$$

The corresponding signal has sample values equal to the coefficients of the power series, namely:

$$
2 \quad -3 \quad 3 \quad 3 \quad -6 \quad 0 \quad 1 \quad 0 \quad 0 \quad \cdots
$$

This neatly confirms the result obtained by time-domain convolution. In general, we conclude that the inverse transform of $X(z)H(z)$ yields the output signal $y[n]$.

So far, we have recovered a signal by expanding its transform as a power series. We should now consider the question of inverse transformation more carefully. Formally, the inverse transform of a function $X(z)$ is defined as:

$$
x[n] = \frac{1}{2\pi j} \oint X(z)z^{n-1} \, dz \tag{4.7}
$$

where the circular symbol on the integral sign denotes a closed contour in the complex plane. Such *contour integration* is rather difficult, and beyond the scope of this book. Fortunately, several simpler approaches are available. As we have already noted, one is to express $X(z)$ as a power series. Another is to look up the function we need in a table of z-transform pairs. Such tables are common in the DSP literature, and we include a short one at the end of this chapter (Table 4.1). If the function is not listed, it may be possible to express it as the sum of two or more simpler functions which *do* appear in the table, using the algebraic method of *partial fractions*. We will demonstrate this for you by a worked example:

Example 4.2   A signal has the z-transform:

$$X(z) = \frac{1}{z(z-1)(2z-1)}$$

Use the method of partial fractions, together with Table 4.1 at the end of the chapter, to recover the signal $x[n]$. Check the result by expanding $X(z)$ as a power series in z^{-1}.

Solution   The method of partial fractions, covered in most algebra textbooks, involves expressing $X(z)$ in the form:

$$X(z) = \frac{1}{z(z-1)(2z-1)} \equiv \frac{A}{z} + \frac{B}{(z-1)} + \frac{C}{(2z-1)}$$

where A, B, and C are constants. It is always possible to do this, providing $X(z)$ is a *proper rational function*, with the numerator of lower degree than the denominator. Hence:

$$X(z) = \frac{A(z-1)(2z-1) + Bz(2z-1) + Cz(z-1)}{z(z-1)(2z-1)}$$

$$= \frac{A(2z^2-3z+1) + B(2z^2-z) + C(z^2-z)}{z(z-1)(2z-1)}$$

$$= \frac{z^2(2A+2B+C) + z(-3A-B-C) + A}{z(z-1)(2z-1)}$$

The numerator of this expression must be identically equal to that of the original function. Hence:

$$2A + 2B + C = 0$$
$$3A + B + C = 0$$
$$A = 1$$

These equations yield:

$$A = 1, \quad B = 1, \quad C = -4$$

Hence:

$$X(z) = \frac{1}{z} + \frac{1}{(z-1)} - \frac{4}{(2z-1)}$$

If you refer to Table 4.1, you will see that these partial fractions are not listed as they stand. However, the following functions do appear:

$$1, \frac{z}{(z-1)} \text{ and } \frac{z}{(z-\alpha)}.$$

Recalling that multiplication by z^{-1} is equivalent to a delay of one sampling interval, let us write:

$$X(z) = z^{-1} \left\{ 1 + \frac{z}{(z-1)} - \frac{2z}{(z-0.5)} \right\}$$

The three terms in brackets produce the inverse transform:

$$\delta[n] + u[n] - 2(0.5^n u[n])$$

So the required signal $x[n]$ is given by:

$$x[n] = \delta[n-] + u[n-1] - 2(0.5^{n-1} u[n-1])$$

Superposing the three components, we find that the first few values of $x[n]$ are:

$$
\begin{array}{cccccccc}
n = & 0 & 1 & 2 & 3 & 4 & 5 & \ldots \\
x[n] = & 0 & 0 & 0 & 0.5 & 0.75 & 0.875 & \ldots
\end{array}
$$

We can check the result by expressing $X(z)$ as a power series in z^{-1}. This may be done by a standard long division of the numerator by the denominator. We will evaluate just the first few terms. The denominator is:

$$z(z-1)(2z-1) = 2z^3 - 3z^2 + z$$

We therefore perform the long division:

$$
\begin{array}{r}
0.5z^{-3} + 0.75z^{-4} + 0.875z^{-5} + \cdots \\
\hline
2z^3 - 3z^2 + z \, \big/ \, 1 \qquad\qquad\qquad\qquad\qquad
\end{array}
$$

$$
\begin{array}{l}
1 \qquad -1.5z^{-1} + 0.5z^{-2} \\
\hline
\quad 1.5z^{-1} - 0.5z^{-2} \\
\\
\quad 1.5z^{-1} - 2.25z^{-2} + 0.75z^{-3} \\
\hline
\quad\quad\quad 1.75z^{-2} - 0.75z^{-3}
\end{array}
$$

$$\text{etc.}$$

The required power series is:

$$X(z) = 0.5z^{-3} + 0.75z^{-4} + 0.875z^{-5} \ldots$$

Its coefficients are equal to the sample values of $x[n]$, the first of them occurring at $n=3$. This checks with our previous result.

Our main reason for including partial fraction and long-division expansions is to demonstrate the various ways of finding an inverse z-transform. However, there is another, much more convenient, method which is particularly relevant to our own approach to DSP. We may derive $x[n]$ from $X(z)$ using a recursive algorithm; and if the algorithm is a complicated one, we can get a computer to do the work for us.

A good way of explaining the method is to assume that the z-transform represents an LTI *system* rather than a digital *signal*. So we will denote it by $H(z)$. The corresponding time function must correspond to the system's impulse response $h[n]$. Now in general:

$$Y(z) = X(z) H(z) \quad \text{or} \quad H(z) = \frac{Y(z)}{X(z)} \tag{4.8}$$

where $Y(z)$ and $X(z)$ are the z-transforms of the output and input signals respectively. These frequency-domain relationships are precisely equivalent to those for the Fourier Transform, discussed in Section 3.3.2.

As a simple example let us take the z-transform already used in worked Example 4.2. Thus:

$$H(z) = \frac{1}{z(z-1)(2z-1)} = \frac{Y(z)}{X(z)} \tag{4.9}$$

giving:

$$Y(z)\{z(z-1)(2z-1)\} = X(z)$$

or:

$$Y(z)\{2z^3 - 3z^2 + z\} = X(z)$$

$$\therefore 2z^3 Y(z) - 3z^2 Y(z) + zY(z) = X(z) \tag{4.10}$$

Recalling that multiplication by z is equivalent to a time advance by one sampling interval, multiplication by z^2 is equivalent to an advance by two sampling intervals — and so on — we may write down the equivalent difference equation by inspection:

$$2y[n+3] - 3y[n+2] + y[n+1] = x[n] \tag{4.11}$$

Since this is a recurrence formula which applies for all values of n, we may subtract 3 (or any other integer) from all terms in square brackets. Hence:

$$2y[n] - 3y[n-1] + y[n-2] = x(n-3]$$

or:

$$y[n] = 1.5y[n-1] - 0.5y[n-2] + 0.5x[n-3] \tag{4.12}$$

Equation (4.12) is the time-domain equivalent of the function $H(z)$ whose inverse transform is required.

To find the corresponding time function $h[n]$, we deliver a unit impulse $\delta[n]$ as input signal, and evaluate $h[n]$ term-by-term. Thus:

$$h[n] = 1.5h[n-1] - 0.5h[n-2] + 0.5\delta[n-3] \tag{4.13}$$

The term $0.5\delta[n-3]$ only contributes at $n=3$, when it has the value 0.5. The first nonzero value of $h[n]$ is therefore $h[3] = 0.5$. Subsequent values are generated recursively:

$$h[4] = 1.5h[3] - 0.5h[2] = 0.75$$
$$h[5] = 1.5h[4] - 0.5h[3] = 1.5(0.75) - 0.5(0.5) = 0.875$$

and so on. You may like to check that the results agree with those found in Worked Example 4.2.

Equation (4.13) hardly justifies the use of a computer. Let us therefore take a much more complicated z-transform and apply the same method. (The transform specified will also be useful to us later in the chapter.)

Suppose we need to find the inverse transform of the function:

$$\frac{z^2(z-1)\,(z^2+1)}{(z+0.8)\,(z^2+1.38593z+0.9604)\,(z^2-1.64545z+0.9025)} \tag{4.14}$$

We multiply out the numerator and denominator, and assume the function represents an LTI system. Thus:

$$H(z) = \frac{Y(z)}{X(z)}$$

$$= \frac{z^5 - z^4 + z^3 - z^2}{z^5+0.54048z^4-0.62519z^3-0.66354z^2+0.60317z+0.69341} \tag{4.15}$$

The corresponding difference equation is readily obtained:

$$y[n] = -0.54048y[n-1] + 0.62519y[n-2] + 0.66354y[n-3]$$
$$-0.60317y[n-4] - 0.69341y[n-5] + x[n]$$
$$-x[n-1] + x[n-2] - x[n-3] \tag{4.16}$$

Program no.10 in Appendix A1 estimates $y[n]$ when the input is a unit impulse. The result is shown in Figure 4.2. This is a relatively fast and convenient way of finding the inverse transform. Note, however, that it gives us a numerical sequence of sample values — rather than an analytical expression. Whether or not this is a disadvantage will depend upon the application.

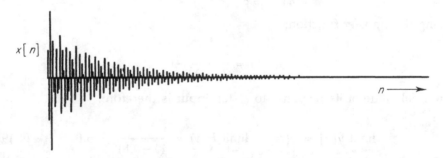

Figure 4.2 An inverse z-transform evaluated by computer *(abscissa: 320 samples)*.

The z-transform, like the Fourier Transform, has a number of useful properties. We have demonstrated its convolution property earlier in this section, and have described the role of z as a time-shift operator. The important properties — including *linearity, time-shifting,* and *convolution* — are listed in Table 4.2. If you compare them with the corresponding properties of the Fourier Transform in Table 3.2, you will notice obvious similarities. In particular, z is equivalent to exp $(j\Omega)$.

Another feature of the z-transform, which may be regarded as a property, is its *final-value theorem*. It may be stated as follows:

if

$$x[n] \leftrightarrow X(z)$$

then

$$\underset{n \to \infty}{\text{limit}}\, x[n] = \underset{z \to 1}{\text{limit}}\left(\frac{z-1}{z}\right) X(z) \qquad (4.17)$$

The theorem is also listed in Table 4.2. A useful application is in finding the final, steady-state, response of a system to a step input. We note from Table 4.1 that the unit step function $u[n]$ has the z-transform $z/(z-1)$. Hence if $u[n]$ is the input signal to a system with transfer function $H(z)$, the output signal has the transform:

$$Y(z) = \left(\frac{z}{z-1}\right) H(z)$$

Using the expression (4.17), the final value of $y[n]$ is therefore:

$$\underset{n \to \infty}{\text{limit}}\, y[n] = \underset{z \to 1}{\text{limit}}\left(\frac{z-1}{z}\right)\left(\frac{z}{z-1}\right) H(z) = \underset{z \to 1}{\text{limit}}\, H(z) \qquad (4.18)$$

The steady-state response is therefore found from $H(z)$ simply by putting $z = 1$. Since z is equivalent to $\exp(j\Omega)$, this is the same as finding the zero-frequency (DC) response by putting $\Omega = 0$ in a frequency response expression.

A simple example is the system previously explored in Worked Example 2.2, and Figure 2.7. The difference equation of the first-order low-pass filter was:

$$y[n] - 0.8\, y[n-1] = x[n]$$

giving the transfer function:

$$H(z) = \frac{z}{(z-0.8)}$$

The final value of its response to a step input is therefore:

$$\underset{n \to \infty}{\text{limit}}\, y[n] = y[\infty] = \underset{z \to 1}{\text{limit}}\, H(z) = \frac{1}{(1-0.8)} = 5.0 \qquad (4.19)$$

The result agrees with part (d) of Figure 2.7.

We hope that you now feel familiar with the basic notation and properties of the z-transform. Our next task is to introduce the important idea of z-plane poles and zeros, and illustrate their value in analysis and design.

4.3 z-PLANE POLES AND ZEROS

4.3.1 DESCRIBING SIGNALS AND SYSTEMS BY POLES AND ZEROS

A z-transform used to describe a real digital signal or an LTI system is always a *rational function* of the frequency variable z. In other words it can always be written as the ratio of numerator and denominator polynomials in z:

$$X(z) = \frac{N(z)}{D(z)} \qquad (4.20)$$

This is true whether $X(z)$ represents an input or output signal, or the transfer function of a processor. Apart from a gain factor K it follows that the transform may be completely specified by the roots of $N(z)$ and $D(z)$. Thus, in general, we can express $X(z)$ in the form:

$$X(z) = \frac{N(z)}{D(z)} = \frac{K(z-z_1)\,(z-z_2)\,(z-z_3)\,\cdots}{(z-p_1)\,(z-p_2)\,(z-p_3)\,\cdots} \qquad (4.21)$$

The constants $z_1, z_2, z_3 \ldots$ are called the *zeros* of $X(z)$, because they are the values of z for which $X(z)$ is zero. Conversely $p_1, p_2, p_3 \ldots$ are known as the *poles* of $X(z)$, giving values of z for which $X(z)$ tends to infinity. It is found that whenever the corresponding time function is real, then the poles and zeros are themselves either real, or occur in complex conjugate pairs.

A very useful representation of a z-transform is obtained by plotting its poles and zeros in the complex plane (Argand diagram). The plane is then referred to as the *z-plane*. We shall see a little later that it is quite easy to visualize the spectrum of a signal, or the frequency response of an LTI system, from such a diagram. It also gives a good indication of the degree of stability of a system.

As well as listing some useful z-transform pairs, Table 4.1 at the end of the chapter shows the corresponding z-plane poles and zeros. For example, the transform:

$$u[n] \;\leftrightarrow\; \frac{z}{(z-1)}$$

may be characterized by a zero at the origin of the z-plane, and a pole on the real axis at the point $z = (1, j0)$. Note that a zero is shown as an open circular symbol, and a pole as a cross, or asterisk. Also, for reasons which will become clear below, we normally draw a circle of unit radius centered at the z-plane origin. This is called the *unit circle*. As a further example, we see from the table that a sinusoidal signal, switched on at $n = 0$, has a zero at the origin and a complex conjugate pole-pair on the unit circle.

Example 4.3 Plot the z-plane poles and zeros of the following z-transforms:

(a)

$$X(z) = \frac{z^2(z-1.2)\,(z+1)}{(z-0.5 + j0.7)\,(z-0.5 - j0.7)\,(z-0.8)}$$

(b)

$$X(z) = (z^5 - 1)\,(z^2 + 1)$$

Solution
(a) The poles and zeros of this function are shown in Figure 4.3(a), and

should be self-explanatory. Note that the two coincident zeros at $z = 0$ are often referred to as a *second-order* zero.

(b) This function has zeros only. The five roots of $(z^5 - 1) = 0$ are equally spaced around the unit circle. The two roots of $(z^2 + 1) = 0$ are at $z = \pm j$.

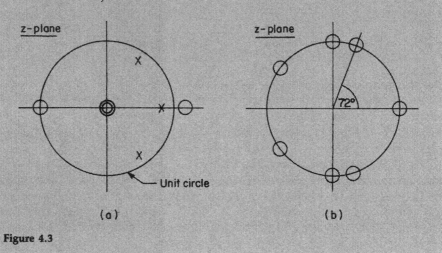

Figure 4.3

When a z-transform represents the transfer function of a system, the positions of its poles give important information about system stability. Suppose we have a digital processor with a single real pole at $z = \alpha$. Hence:

$$H(z) = \frac{Y(z)}{X(z)} = \frac{1}{(z - \alpha)} \qquad (4.22)$$

or:

$$zY(z) - \alpha Y(z) = X(z) \qquad (4.23)$$

The processor's difference equation is:

$$y[n+1] - \alpha y[n] = x[n]$$

or:

$$y[n] = \alpha y[n-1] + x[n-1] \qquad (4.24)$$

Its impulse response is given by:

$$h[n] = \alpha h[n-1] + \delta[n-1] \qquad (4.25)$$

Successive terms of $h[n]$, starting at $n = 0$, are:

$$0, 1, \alpha, \alpha^2, \alpha^3, \alpha^4 \ldots$$

$h[n]$ follows a real exponential envelope. If the magnitude of α is less than unity, the envelope decays towards zero as $n \to \infty$; but if the magnitude of α is greater than unity, it grows without limit. We conclude that the system is only stable if $|\alpha| < 1$. This means that the pole must lie *inside the unit circle* in the z-plane.

Let us take another example — a processor with a pair of poles on the imaginary axis at $z = \pm j\alpha$. The transfer function is:

$$H(z) = \frac{Y(z)}{X(z)} = \frac{1}{(z - j\alpha)\,(z + j\alpha)} = \frac{1}{(z^2 + \alpha^2)} \qquad (4.26)$$

Hence:

$$z^2 Y(z) + \alpha^2 Y(z) = X(z) \qquad (4.27)$$

The corresponding difference equation is:

$$y[n] = -\alpha^2 y[n-2] + x[n-2] \qquad (4.28)$$

giving an impulse response with successive terms:

$$0,\ 0,\ 1,\ 0,\ -\alpha^2,\ 0,\ \alpha^4,\ 0,\ -\alpha^6,\ 0 \ldots$$

Once again, this grows without limit if $|\alpha| > 1$. We conclude that, for a stable system, the poles must be inside the unit circle.

The argument also applies to complex poles lying away from the real and imaginary axes. If their radius (measured from the z-plane origin) is greater than unity, they denote an unstable system.

Although we have developed the argument for *systems*, similar ideas apply to *signals*. If a digital signal has one or more z-plane poles lying outside the unit circle, it will grow without limit as $n \to \infty$.

Since stability is intimately related to the radius of z-plane poles, it is often helpful to express their locations in polar coordinates. Suppose we have a processor with a complex conjugate pole-pair, as shown in Figure 4.4(a). The poles are at radius r, and make angle $\pm\theta$ with the positive real axis. Their locations are therefore $z = r\exp(j\theta)$ and $z = r\exp(-j\theta)$ and we may write:

$$H(z) = \frac{Y(z)}{X(z)} = \frac{1}{\{z - r\exp(j\theta)\}\,\{z - r\exp(-j\theta)\}}$$

$$= \frac{1}{(z^2 - 2rz\cos\theta + r^2)} \qquad (4.29)$$

The corresponding difference equation is:

$$y[n] = 2r\cos\theta\, y[n-1] - r^2 y[n-2] + x[n-2] \qquad (4.30)$$

The processor will only be stable if $r < 1$.

We met a z-transform with denominator factors similar to those of equation (4.29) at the end of the previous section. If you refer back to equation (4.14), you will see that the z-transform we considered had a real pole at $z = -0.8$, and two complex conjugate pole pairs with denominator factors:

$$(z^2 + 1.38593z + 0.9604)$$

and:

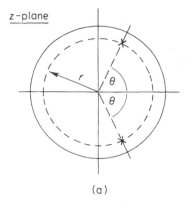 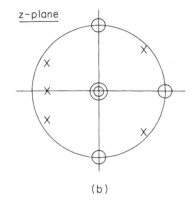

(a)   (b)

Figure 4.4 z-plane poles and zeros.

$$(z^2 - 1.64545z + 0.9025)$$

Comparing with equation (4.29) we see that, for the first pole pair:

$$r^2 = 0.9604 \quad \text{and} \quad 2r \cos\theta = -1.38593$$
$$\text{giving } r = 0.98 \quad \text{and} \quad \theta = 45°$$

For the second pole pair:

$$r^2 = 0.9025 \quad \text{and} \quad 2r \cos\theta = 1.64545$$
$$\text{giving } r = 0.95 \quad \text{and} \quad \theta = 150°$$

The complete pole-zero pattern of this z-transform is plotted in part (b) of Figure 4.4. Since all poles lie inside the unit circle, we know that the corresponding time function must decay to zero as $n \to \infty$. This is confirmed by the computed result already shown in Figure 4.2.

Before we end this introduction to z-plane poles and zeros, two further points should be mentioned. First, our comments about stability and the unit circle apply only to *poles*. There are no corresponding restrictions on zeros, which can be placed anywhere in the z-plane.

The second point concerns zeros (or poles) at the origin of the z-plane. These produce a pure time advance (or delay), but have no other effect on the characteristics of the processor or signal. Consider, for example, the processor defined by equations (4.29) and (4.30). It has two z-plane poles, but no zeros. We see from its difference equation that each output value $y[n]$ depends on the input $x[n-2]$, implying a time delay of two sampling intervals. The impulse response begins at $n=2$, not at $n=0$. Such a delay is unnecessary, and is normally considered undesirable. We can easily correct it by placing a second-order zero at the origin. The transfer function becomes:

$$H(z) = \frac{z^2}{(z^2 - 2rz \cos\theta + r^2)} \qquad (4.31)$$

giving the difference equation:

$$y[n] = 2r \cos\theta \, y[n-1] - r^2 y[n-2] + x[n] \qquad (4.32)$$

In general we achieve a minimum-delay system by ensuring that there is an equal number of poles and zeros. Of course, if a function has more zeros than poles, we must add poles at the origin. Otherwise the corresponding time function will begin before $n = 0$, denoting a noncausal system.

4.3.2 GEOMETRICAL EVALUATION OF THE FOURIER TRANSFORM IN THE z-PLANE

In general we may think of z as a frequency variable which can take real, imaginary, or complex values. Furthermore we have already noted that z is equivalent to $\exp(j\Omega)$ in Fourier notation. Therefore if we put $z = \exp(j\Omega)$, where Ω is real, we are effectively converting from a z-transform to the exponentials (or sines and cosines) of Fourier analysis.

To understand the relationship between the z-transform and Fourier transform more fully, we should consider where values of $z = \exp(j\Omega)$ lie in the z-plane. Since $\exp(j\Omega)$ always has unit magnitude, irrespective of the value of Ω, they must all lie *on the unit circle*. If we set $\Omega = 0$, then $\exp(j\Omega) = 1$, and we are at the point $z = (1, j0)$ on the real axis. As Ω increases, we move anti-clockwise around the unit circle. The angle made with the positive real axis equals the value of Ω. We reach the point $z = (-1, j0)$ on the negative real axis when $\Omega = \pi$. By the time $\Omega = 2\pi$ we are back again at our starting point. If Ω further increases from 2π to 4π, we make another complete journey around the unit circle. Values of Ω corresponding to several points on the circle are illustrated in Figure 4.5.

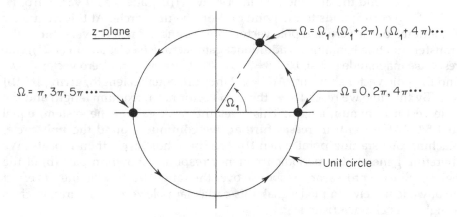

Figure 4.5 Sinusoidal frequencies corresponding to various points on the unit circle in the z-plane.

The figure helps explain why the z-transform gives such a compact description of digital signals and systems. We know from our work on Fourier analysis that their spectral characteristics always repeat indefinitely along the frequency axis at intervals of 2π. This is a consequence of sampling. The z-transform takes the effect into account automatically, since any 2π interval is equivalent to one complete revolution around the unit circle in the z-plane.

We can use this new perspective to infer the frequency response of an LTI processor. We will start with a simple example. Suppose we have a processor with a *z*-plane pole at $z = -0.8$ and a zero at $z = 0.8$, as shown in Figure 4.6(a). Hence:

$$H(z) = \frac{(z-0.8)}{(z+0.8)} \qquad (4.33)$$

Substituting $\exp(j\Omega)$ for *z* gives the frequency response of the system:

$$H(\Omega) = \frac{(\exp(j\Omega) - 0.8)}{(\exp(j\Omega) + 0.8)} \qquad (4.34)$$

We may now use the normal rules of complex arithmetic to interpret $H(\Omega)$ in terms of geometrical vectors drawn in the *z*-plane.

At a particular value of sinusoidal frequency (say $\Omega = \Omega_1$) the numerator of $H(\Omega)$ may be represented by a *zero vector* Z_1 drawn from the zero to the relevant point on the unit circle. The denominator may be represented by a *pole vector* P_1 drawn from the pole to the same point. The magnitude of the frequency response is now given by the magnitude (length) of the zero vector, divided by the magnitude (length) of the pole vector. The phase shift equals the difference between the phase angles of the two vectors (both measured with respect to the positive real axis). For the frequency illustrated, we see that $|H(\Omega)|$ must be about 0.6, with a phase shift $\Phi_H(\Omega)$ of about $110° - 35° = 75°$.

We can extend the argument to infer how $|H(\Omega)|$ alters as Ω varies from 0 to 2π. $\Omega = 0$ corresponds to the point $z = 1$ on the unit circle. At this frequency the lengths of the zero and pole vectors are 0.2 and 1.8 respectively. Since the transfer function contains no additional gain factor *k* (see equation (4.21)), the response magnitude is just $0.2/1.8 = 0.111$. As Ω increases, the zero vector grows and the pole vector shortens. At $\Omega = \pi/2$ they are equal in length, giving $|H(\Omega)| = 1$. By the time we reach $\Omega = \pi$, the zero vector has maximum length and the pole vector minimum length. This gives the peak gain of the system, equal to $1.8/0.2 = 9.0$. As Ω increases further, we continue around the unit circle, reaching our starting point when $\Omega = 2\pi$. The whole cycle then repeats. We therefore generate the periodic frequency response shown in part (b) of the figure. (Since a processor's response over the range $0 < \Omega < \pi$ defines its effect on any adequately sampled signal, this particular pole-zero configuration gives a high-pass characteristic.)

Geometric evaluation of the Fourier Transform in the *z*-plane can be extended to more complex systems (or signals), with a greater number of poles and zeros. We draw a vector from each pole and zero to a point on the unit circle representing the sinusoidal frequency of interest. Then:

The magnitude of the spectral function equals the *product* of all zero-vector lengths, divided by the *product* of all pole-vector lengths. (An additional gain factor may have to be taken into account. However, it only alters the scale of the function, not the shape.)

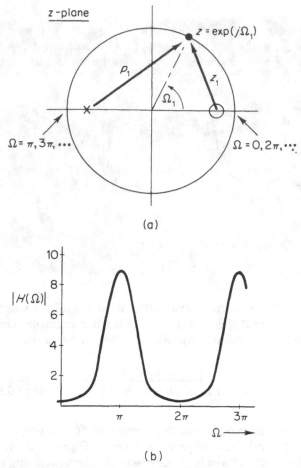

$$z\text{-plane}$$

(a)

Figure 4.6 Visualizing the frequency response of an LTI processor.

The phase equals the *sum* of all zero-vector phases, minus the *sum* of all pole-vector phases.

You can probably visualize the effect of poles or zeros close to the unit circle. As the frequency varies and we move around the unit circle, the spectral magnitude function peaks whenever we pass close to a pole. It goes through a minimum whenever we pass close to a zero. Note that zeros can occur actually on the unit circle, giving rise to true nulls at the corresponding frequencies. However, the poles of a stable system are always inside the unit circle, so its response must be finite at all frequencies.

Example 4.4 In Worked Example 4.3 we plotted the poles and zeros of two z-transforms (see Figure 4.3). Let us now assume these transforms represent transfer functions of LTI processors. Without doing any detailed analysis, make approximate sketches of the corresponding frequency-response magnitude characteristics.

Solution The sketches in Figure 4.7 should be self-explanatory. Note that zeros (or poles) at the origin of the z-plane do not affect the magnitude of a spectral function, since the corresponding vectors always have unit length.

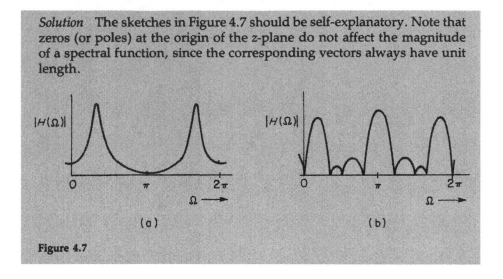

Figure 4.7

The technique of geometrical evaluation is so useful for visualization and design that we will end this section with a further example. Let us consider the rather complicated z-transform already met in Section 4.2:

$$X(z) = \frac{z^2(z-1)\,(z^2+1)}{(z+0.8)\,(z^2+1.38593z+0.9604)\,(z^2-1.64545z+0.9025)}$$

(4.35)

We previously showed how to find its inverse transform using a computer program (Figure 4.2), and drew its poles and zeros (Figure 4.4(b)). We can now pull together the various threads of our discussion by visualizing its magnitude spectrum, and relating it to the time-domain signal.

First, we note that there are zeros at $z=1$ and $z=\pm j$. Over the frequency range 0 to π, these produce true nulls at $\Omega=0$ and $\Omega=\pi/2$. (There is also a second-order zero at the origin, which has no effect on the spectral magnitude.) Secondly, there is a complex conjugate pole-pair close to the unit circle at $\Omega=0.25\pi$ (45°); another at $\Omega=0.833\pi$ (150°); and a single real pole at $z=-0.8$. The spectrum must therefore be relatively rich in sinusoidal components close

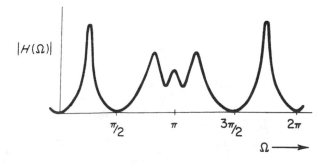

Figure 4.8

to $\Omega = 0.25\pi$, $\Omega = 0.833\pi$, and $\Omega = \pi$. We expect its overall shape to be as shown in Figure 4.8.

In the time domain, $\Omega = 0.25\pi$ corresponds to a sinusoid with eight samples per period; $\Omega = 0.833\pi$ has 2.4 samples per period; $\Omega = \pi$ has two samples per period. If you examine Figure 4.2 very carefully you may just about be able to detect these three major signal components!

4.3.3 FIRST- AND SECOND-ORDER LTI SYSTEMS

We have already met several examples of first and second-order systems. The purpose of this section is to summarize their performance from the pole-zero, frequency response, and impulse response points of view.

First and second-order systems are important in their own right as elementary processors and filters. They may also be considered as building blocks for more complicated systems. Suppose we write the transfer function of a high-order system in terms of its poles and zeros:

$$H(z) = \frac{Y(z)}{X(z)} = \frac{K(z-z_1)\,(z-z_2)\,(z-z_3)\,(z-z_4)\,\cdots}{(z-p_1)\,(z-p_2)\,(z-p_3)\,(z-p_4)\,\cdots} \qquad (4.36)$$

Then $H(z)$ may in principle be built up by *cascading* a number of first and second-order subsystems defined by:

First-order: $$H_1(z) = \frac{(z-z_1)}{(z-p_1)} \qquad (4.37)$$

Second-order: $$H_2(z) = \frac{(z-z_2)\,(z-z_3)}{(z-p_2)\,(z-p_3)} \qquad (4.38)$$

$H_1(z)$ has one real pole and one real zero. $H_2(z)$ has two poles (either both real, or a complex conjugate pair), and two zeros. Cascading has the effect of *multiplying* the individual transfer functions together to produce the overall function $H(z)$. In the time domain, the individual impulse responses are *convolved* to produce the overall impulse response $h[n]$.

The frequency-selective properties of first and second-order systems can be controlled by appropriate choice of pole-zero locations. Poles are particularly effective in this respect, because when placed close to the unit circle they produce sharp, well-defined, peaks in the frequency response. Indeed, frequency selectivity is quite often achieved using poles only. An equal number of zeros can then be placed at the z-plane origin to ensure that the impulse response begins at $n = 0$.

In this section we therefore focus on first and second-order systems with their zeros at the origin. In line with the notation used previously in this book, the location of a first-order real pole is written as $z = \alpha$; and of a complex-conjugate pole-pair as $z = r\exp(j\theta)$ and $z = r\exp(-j\theta)$. Hence equations (4.37) and (4.38) become:

$$H_1(z) = \frac{z}{(z-\alpha)} \qquad (4.39)$$

and

$$H_2(z) = \frac{z^2}{\{z - r\exp(j\theta)\}\{z - r\exp(-j\theta)\}}$$

$$= \frac{z^2}{z^2 - 2rz\cos\theta + r^2} \tag{4.40}$$

First-order systems need little explanation, since we have already met a number of them in this book (see, for example, Figures 2.5 and 3.11). The single real pole must be inside the unit circle to ensure stability. If it is on the *positive* real axis, we get an elementary (but often useful) recursive low-pass filter with a peak response at $\Omega = 0$. If the pole is on the *negative* real axis, we get the equivalent *high-pass* filter with a peak response at $\Omega = \pi$. The pole-zero configuration, frequency response (magnitude and phase), and impulse response of typical low-pass and high-pass systems are illustrated in Figure 4.9. (We have only drawn the frequency responses for $0 < \Omega < \pi$, because this range defines a system's effect on any adequately-sampled signal.)

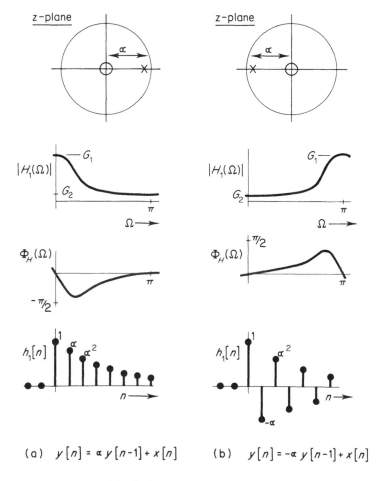

(a) $y[n] = \alpha\, y[n-1] + x[n]$ (b) $y[n] = -\alpha\, y[n-1] + x[n]$

Figure 4.9 Characteristics of first-order systems.

The frequency response of the low-pass system is given by:

$$H_1(\Omega) = H_1(z)\Big|_{z=\exp(j\Omega)} = \frac{\exp(j\Omega)}{\exp(j\Omega)-\alpha}, \quad 0<\alpha<1 \tag{4.41}$$

Its peak magnitude, or gain, occurs at $\Omega=0$ and is given by:

$$G_1 = \left|\frac{\exp(0)}{\exp(0)-\alpha}\right| = \frac{1}{(1-\alpha)} \tag{4.42}$$

The minimum gain occurs at $\Omega=\pi$ and equals:

$$G_2 = \left|\frac{\exp(j\pi)}{\exp(j\pi)-\alpha}\right| = \frac{1}{(1+\alpha)} \tag{4.43}$$

G_1 and G_2 are marked on the figure. The high-pass system has the same values — but they occur, of course, at different frequencies. Both types of system have impulse responses which follow decaying exponential envelopes. However, in the high-pass case successive sample values are inverted.

If we move the pole of a first-order system closer to the unit circle, there are three principal effects:

the peak gain increases
the bandwidth decreases
the impulse responses decays more slowly.

The last two of these underline the general antithesis between frequency limitation and time limitation in LTI systems: the more restricted the spectral function, the more spread out the corresponding time function — and vice versa.

Let us now turn to second-order systems specified by equation (4.40). There is a second-order zero at the origin, and a complex-conjugate pole pair at $z=r\exp(\pm j\theta)$, as shown in Figure 4.10. The frequency at which the peak gain occurs (the *center-frequency*) is determined by the parameter θ; and the selectivity, or *bandwidth*, of the system by the parameter r.

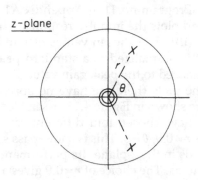

Figure 4.10 The z-plane pole-zero configuration of a second-order system.

We can readily derive expressions for the impulse and frequency responses from equation (4.40). Dividing numerator and denominator by z^2 gives:

$$H_2(z) = \frac{Y(z)}{X(z)} = \frac{1}{(1 - 2r\cos\theta\, z^{-1} + r^2 z^{-2})} \qquad (4.44)$$

and hence the difference equation:

$$y[n] = 2r\cos\theta\, y[n-1] - r^2\, y[n-2] + x[n] \qquad (4.45)$$

Substituting $\exp(j\Omega)$ for z in equation (4.44) gives the frequency response:

$$H_2(\Omega) = \frac{1}{1 - 2r\cos\theta\, \exp(-j\Omega) + r^2\, \exp(-j2\Omega)} \qquad (4.46)$$

The magnitude, or gain, function is:

$$|H_2(\Omega)| = \frac{1}{\{(1 - 2r\cos\theta\,\cos\Omega + r^2\,\cos2\Omega)^2 + (2r\cos\theta\,\sin\Omega - r^2\,\sin2\Omega)^2\}^{1/2}} \qquad (4.47)$$

Assuming the peak gain occurs when $\Omega = \theta$, its value is:

$$|H_2(\Omega)|_{max} = G_1$$

$$= \frac{1}{\{(1 - 2r\cos^2\theta + r^2\,\cos2\theta)^2 + (2r\cos\theta\,\sin\theta - r^2\,\sin2\theta)^2\}^{1/2}} \qquad (4.48)$$

For convenience let us write:

$$(1 - 2r\cos^2\theta + r^2\,\cos2\theta) = A$$

and

$$(2r\cos\theta\,\sin\theta - r^2\,\sin2\theta) = B$$

Then:

$$G_1 = (A^2 + B^2)^{-1/2} \qquad (4.49)$$

These expressions are rather complicated, so we will use a computer program to evaluate them. Program no.11 in Appendix A1 requests values for r and θ, then estimates and plots the impulse response $h[n]$ and the frequency response magnitude $|H(\Omega)|$. The chosen values of r and θ affect the scale of $h[n]$ substantially, so it is normalized to a standard peak value for plotting. Similarly $|H(\Omega)|$ is normalized to the peak gain value given by equation (4.49). The vertical scales of the plots therefore have no special significance.

Some typical results are shown in Figure 4.11, assembled into a single diagram to aid comparison. We have chosen four different combinations of r and θ. Part (a) of the figure has $r = 0.9$, $\theta = 0$. This is a low-pass system with a second-order pole on the real axis in the z-plane. Its performance is the same as two cascaded first-order systems. The choice of $r = 0.9$ gives a moderately selective frequency response.

In part (b), $r = 0.99$ and $\theta = 25°$. The poles are much closer to the unit circle, giving a very selective frequency-domain characteristic. In the time domain, the impulse response is prolonged. The frequency of oscillation corresponds to $\theta = 25°$, giving about fourteen samples per cycle.

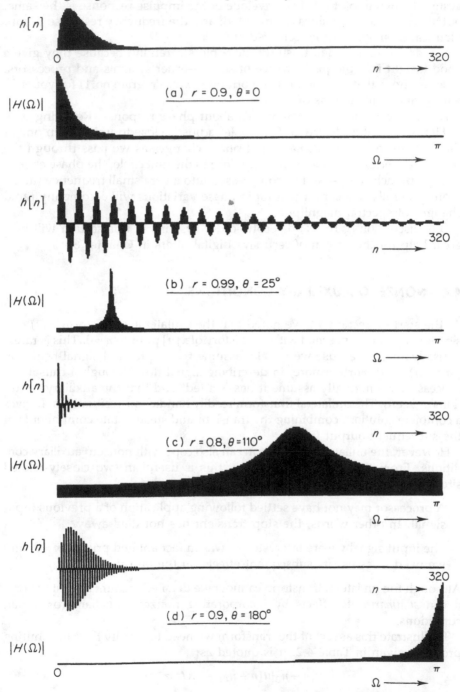

Figure 4.11 Computed impulse and frequency responses of several second-order systems (*abscissa: 320 samples*).

Part (c) shows the results for $r = 0.8$, $\theta = 110°$. This system is much less selective in the frequency domain, so its impulse response is short. Finally, part (d) has $r = 0.9$ and $\theta = 180°$, producing a high-pass counterpart of the low-pass system shown in part (a). The envelope of the impulse response is the same, but alternate sample values are inverted; and the frequency response hump, identical in shape, is now centered at $\Omega = \pi$.

You are recommended to study these plots carefully, because they give a good insight into the performance of second-order systems and processors. Also, we hope that you have the opportunity to try Program no.11 for yourself, with other combinations of r and θ.

We should finally say a few words about phase response. Referring back to Figure 4.10, it is clear that if the pole radius is close to unity, then one of the pole vectors must display rapid phase changes as we pass through the center frequency. For poles extremely close to the unit circle, the phase change must approach 180° — and be compressed into a very small frequency range. More generally, we expect that rapid phase variations will accompany rapid changes of spectral magnitude.

Our discussion of first and second-order systems in this section will be a great help for the design of recursive digital filters in Chapter 6.

4.4 NONZERO AUXILIARY CONDITIONS

At the start of Section 4.2 we noted that the unilateral z-transform $X(z)$ of a signal $x[n]$ is not concerned with the history of $x[n]$ prior to $n = 0$. This is rarely a disadvantage, because we usually assume that a practical signal begins at (or after) $n = 0$. Furthermore, in describing signal flow through a causal LTI processor, we normally assume it has 'settled' and that the auxiliary conditions are zero. The unilateral transform readily handles such situations. It gives a complete solution, combining the transient and steady-state components of the subsequent output signal.

However, the unilateral z-transform can also cope with nonzero auxiliary conditions. From the DSP point of view, this is useful in two closely-related situations:

A processor may not have settled following application of a previous input signal. In other words, the stop transient has not died away.

The input signal we are interested in was in fact applied prior to $n = 0$, and we need to assess its subsequent effects on the output.

Although the unilateral transform cannot take detailed account of past history, it can *summarize* its effects by incorporating nonzero auxiliary (or initial) conditions.

To illustrate this aspect of the transform we need to modify the time-shifting property given in Table 4.2. It is quoted as:

$$x[n - n_0]u[n - n_0] \;\leftrightarrow\; X(z)z^{-n_0}$$

Thus a signal shifted by n_0 sampling intervals has its z-transform multiplied by z^{-n_0}. Note, however, that the shifted signal is multiplied by the shifted step function $u[n - n_0]$, to ensure that it is zero prior to $n = n_0$. If we are to use the z-transform to solve difference equations with nonzero initial conditions, we need to remove this restriction. Let us therefore consider a signal $x_1[n]$ equal to a shifted version of a general signal $x[n]$, which is not necessarily zero prior to $n = 0$:

$$x_1[n] = x[n-1]$$

Therefore:

$$X_1(z) = \sum_{n=0}^{\infty} x[n-1]z^{-n} = x[-1] + \sum_{n=1}^{\infty} x[n-1]z^{-n}$$

$$\therefore X_1(z) = x[-1] + z^{-1}\left\{ \sum_{n=0}^{\infty} x[n]z^{-n} \right\} = x[-1] + z^{-1} X(z) \qquad (4.50)$$

By similar arguments it may be shown that:

If

$$x_2[n] = x[n-2]$$

Then

$$X_2(z) = x[-2] + x[-1]z^{-1} + z^{-2}X(z) \qquad (4.51)$$

and so on.

These results can be illustrated by a simple example. Consider a first-order recursive low-pass filter with the difference equation:

$$y[n] - \alpha y[n-1] = x[n] \qquad (4.52)$$

We wish to apply an input signal at $n = 0$, allowing for the possibility that the filter has not settled following some previous excitation. Using equation (4.50), and taking z-transforms of both sides, we have:

$$Y(z) - \alpha\{y[-1] + z^{-1}Y(z)\} = X(z)$$

$$\therefore Y(z)\{1 - \alpha z^{-1}\} = X(z) + \alpha y[-1]$$

$$\therefore Y(z) = \frac{X(z) + \alpha y[-1]}{(1 - \alpha z^{-1})} \qquad (4.53)$$

This general result holds for any form of input signal and any value of $y[-1]$. Note that if the initial condition is zero, then:

$$Y(z) = \frac{X(z)}{(1 - \alpha z^{-1})} \quad \text{or} \quad \frac{Y(z)}{X(z)} = \frac{1}{(1 - \alpha z^{-1})} = \frac{z}{(z - \alpha)} \qquad (4.54)$$

The ratio of output to input transforms equals the transfer function $H(z)$ in this case. However when $y[-1]$ is nonzero, the ratio of $Y(z)$ to $X(z)$ is *not* equal to $H(z)$.

Let us now take the particular case $y[-1] = -1/\alpha$, and deliver a unit impulse at $n = 0$. Hence $x[n] = \delta[n]$ and $X[z] = 1$. Equation (4.53) becomes:

$$Y(z) = \frac{1 + \alpha(-1/\alpha)}{(1 - \alpha z^{-1})} = \frac{0}{(1 - \alpha z^{-1})} = 0 \qquad (4.55)$$

The output signal for $n > 0$ is therefore zero, as shown by Figure 4.12. The result may surprise you at first. What has happened is that the effects of the initial condition and the input impulse have exactly cancelled each other out. The initial condition, on its own, would have generated a decaying exponential transient with successive values (starting at $n = 0$):

$$\left(\frac{-1}{\alpha}\right)\alpha, \ \left(\frac{-1}{\alpha}\right)\alpha^2, \ \left(\frac{-1}{\alpha}\right)\alpha^3 \ \ldots \ \text{or} \ -1, \ -\alpha, \ -\alpha^2 \ \ldots$$

The impulse input, on its own, would have generated the filter's impulse response:

$$1, \ \alpha, \ \alpha^2 \ \ldots$$

Since the filter is an LTI system the two effects are superposed, giving zero output for $n > 0$. Of course this is a special case. More generally, the initial condition would *modify* the response to the input signal, without cancelling it completely.

Our discussion demonstrates that an LTI system's response to a unit impulse is only its 'true' impulse response $h[n]$ if the initial conditions are zero.

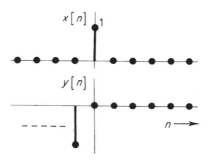

Figure 4.12 The effect of nonzero initial conditions on the impulse response of a first-order filter.

Likewise, the ratio of output to input z-transforms only equals the transfer function $H(z)$ for zero initial conditions. These are important conclusions.

You may like to check that the output sample sequence in the figure can also be derived term-by-term using equation (4.52), assuming that $y[-1] = -1/\alpha$ and $x[0] = 1$. In this case a direct time-domain approach is certainly simpler than solution by the z-transform. Our aim in showing you the latter is to prove that the unilateral transform is by no means powerless when it comes to nonzero initial conditions!

A first-order processor's past history can be summarized in terms of a single initial condition. In the case of a second-order system, we need the two initial condition terms $y[-1]$ and $y[-2]$, and so on. We therefore end this section with a rather more complicated example involving a difference equation of second order.

Example 4.5 Part (a) of Figure 4.13 shows a second-order system. Use the z-transform to find its response to the input signal $x[n] = \delta[n]$ when (a) the initial conditions are zero, and (b) the initial conditions are $y[-1] = -1.25$ and $y[-2] = -0.52\,083$.

Solution The difference equation of the system is, by inspection:

$$y[n] + 0.2y[n-1] - 0.48y[n-2] = x[n]$$

Using equations (4.50) and (4.51) we take the z-transform of both sides of the equation, allowing for nonzero initial conditions:

$$Y(z) + 0.2\{y[-1] + z^{-1}Y(z)\} - 0.48\{y[-2] + y[-1]z^{-1}$$
$$+ z^{-2}Y(z)\} = X(z)$$

This gives:

$$Y(z) = \frac{X(z) - 0.2y[-1] + 0.48y[-2] + 0.48y[-1]z^{-1}}{(1 + 0.2z^{-1} - 0.48z^{-2})}$$

(a) *Zero initial conditions.* Since $x[n] = \delta[n]$, $X(z) = 1$. All other numerator terms are zero, so that:

$$Y(z) = \frac{1}{(1 + 0.2z^{-1} - 0.48z^{-2})} = \frac{z^2}{(z^2 + 0.2z - 0.48)}$$

The denominator may be factorized to give:

$$Y(z) = \frac{z^2}{(z + 0.8)(z - 0.6)}$$

This z-transform does not appear in Table 4.1, so we use the partial fraction expansion

$$\frac{z^2}{(z + 0.8)(z - 0.6)} = \frac{Az}{(z + 0.8)} + \frac{Bz}{(z - 0.6)}$$

which yields $A = 0.5714$ and $B = 0.4286$. Using the table, we may now write the output signal directly:

$$y[n] = 0.5714(-0.8)^n u[n] + 0.4286(0.6)^n u[n]$$

This is clearly the impulse response of the system, and is illustrated in part (b) of the figure.

(b) *Nonzero initial conditions.* Inserting the given initial conditions into the formula for $Y(z)$, we obtain:

$$Y(z) = \frac{1 - 0.2(-1.25) + 0.48(-0.52083) + 0.48(-1.25)z^{-1}}{(1 + 0.2z^{-1} - 0.48z^{-2})}$$

$$= \frac{1 + 0.25 - 0.25 - 0.6z^{-1}}{(1 + 0.8z^{-1})(1 - 0.6z^{-1})} = \frac{1}{(1 + 0.8z^{-1})} = \frac{z}{(z + 0.8)}$$

Therefore:

$$y[n] = (-0.8)^n u[n]$$

This output signal is shown in part (c) of the figure, together with the two initial-condition values which precede $n = 0$.

We see that, with zero initial conditions, we have obtained the normal impulse response of the system $h[n]$. However, the nonzero initial conditions chosen for part (b) of the problem make the numerator of $Y(z)$ equal to $(1 - 0.6z^{-1})$, which cancels a similar term in the denominator. This has the effect of suppressing one of the components in the system's natural response. Although this is again rather a special case, it emphasizes that the response of an LTI system to a unit impulse is only a complete and valid description of the system when the initial conditions are zero. Otherwise, the response may be quite different in form. Finally, we should note that the z-transform has produced the complete solution to the problem 'all in one go'. Transient and steady-state components, plus the 'after-effects' of nonzero initial conditions, are all superimposed — and may be hard to disentangle.

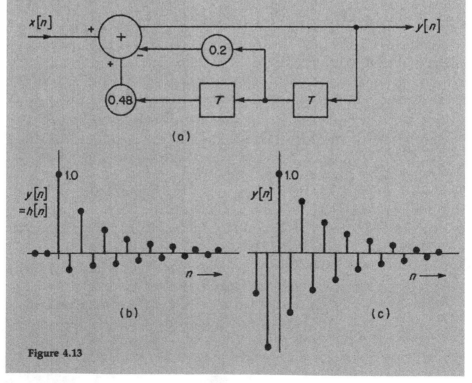

(a)

(b)   (c)

Figure 4.13

Table 4.1 The unilateral z-Transform: pairs

Waveform	Signal $x[n]$	Spectrum $X(z)$	z-plane poles and zeros
Unit impulse	$\delta[n]$	1	Unit circle
Unit step	$u[n]$	$\dfrac{z}{(z-1)}$	
Unit ramp	$r[n]$	$\dfrac{z}{(z-1)^2}$	2nd order
Exponential	$a^n\, u[n]$	$\dfrac{z}{(z-a)}$	
	$(1-a^n)\, u[n]$	$\dfrac{z(1-a)}{(z-a)(z-1)}$	
Cosine	$\cos n\Omega_0\, u[n]$	$\dfrac{z(z-\cos\Omega_0)}{(z^2-2z\cos\Omega_0+1)}$	
Sine	$\sin n\Omega_0\, u[n]$	$\dfrac{z\sin\Omega_0}{(z^2-2z\cos\Omega_0+1)}$	
Damped sine	$a^n \sin n\Omega_0 u[n]$	$\dfrac{az\sin\Omega_0}{(z^2-2az\cos\Omega_0+a^2)}$	

Table 4.2 The unilateral z-Transform: properties

Property or operation	Signal	z-transform
Transformation	$x[n]$	$\sum\limits_{n=0}^{\infty} x[n]z^{-n}$
Inverse transformation	$\frac{1}{2\pi j}\oint X(z)z^{n-1}dz$	$X(z)$
Linearity	$a_1 x_1[n] + a_2 x_2[n]$	$a_1 X_1(z) + a_2 X_2(z)$
Time-shifting	$x[n-n_0]\,u[n-n_0]$	$X(z)z^{-n_0}$
Time-differentiation	$x[n]-x[n-1]$	$X(z)(1-z^{-1})$
Time-integration	$\sum\limits_{k=0}^{n} x[k]$	$X(z)\left(\frac{z}{z-1}\right)$
Convolution	$x_1[n] * x_2[n]$	$X_1(z)\,X_2(z)$
Final-value theorem	$\underset{n\to\infty}{\text{Limit}}\,x[n]$	$\underset{z\to 1}{\text{Limit}}\,\left(\frac{z-1}{z}\right)X(z)$

PROBLEMS

SECTION 4.2

Q4.1 Find the z-transforms of the signals shown in Figure Q4.1, expressing them as compactly as possible.

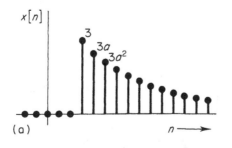

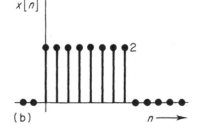

Figure Q4.1

Q4.2 Expand the following z-transforms as power series in z^{-1}, and write down their first five sample values (starting at $n = 0$):

$$\text{(a)} \quad X(z) = \frac{1}{(z - 0.5)}$$

$$\text{(b)} \quad X(z) = \frac{z}{(z + 1.1)}$$

$$\text{(c)} \quad X(z) = \frac{(z + 1)}{(z - 1)}$$

Q4.3 Using only the answers to Problem Q4.1, and the properties of the z-transform, find the z-transform of the signal shown in Figure Q4.3.

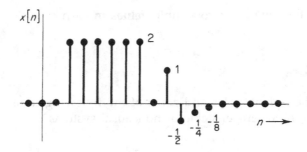

Figure Q4.3

Q4.4 Two digital signals are as follows:

$$x_1[n] = \delta[n] - \delta[n-2] + \delta[n-3]$$
$$x_2[n] = 2\delta[n-1] + \delta[n-2] - \delta[n-3]$$

Write down their respective z-transforms $X_1(z)$ and $X_2(z)$. Convolve the two signals to form a third signal $x_3[n]$, and show that its z-transform equals $X_1(z)X_2(z)$.

Q4.5 A signal $x[n]$ begins at $n = 0$ and has six finite sample values:

$$1, 2, 3, 1, -1, 1$$

It forms the input to an LTI processor whose impulse response $h[n]$ begins at $n = 0$ and has three finite sample values:

$$1, 1, 1$$

Convolve $x[n]$ with $h[n]$ to find the output signal $y[n]$. Check that the z-transform of $y[n]$ equals the product of the transforms of $x[n]$ and $h[n]$.

Q4.6 Using the table of z-transforms at the end of the chapter, and partial fraction expansions if necessary, find the signals corresponding to the following z-transforms:

$$\text{(a)} \quad X(z) = \frac{0.5z}{z^2 - z + 0.5}$$

$$\text{(b)} \quad X(z) = \frac{(z - 0.5)}{z(z - 0.8)(z - 1)}$$

Q4.7 Write computer programs to estimate and plot the inverse z-transforms of the following functions:

(a) The function in Problem Q4.6(b) above.

$$\text{(b)} \quad X(z) = \frac{(z + 1)(z^2 + 1.5z + 0.9)}{(z + 0.7)(z^2 - 1.6z + 0.95)}$$

Print out the first five nonzero sample values in each case.

SECTION 4.3.1

Q4.8 Find the z-plane poles and zeros of the following transfer functions. Which, if any, represent unstable or noncausal systems?

$$\text{(a)} \quad H(z) = \frac{z^2 - z - 2}{z^2 - 1.3z + 0.4}$$

$$\text{(b)} \quad H(z) = \frac{z^2 - z + 1}{z^2 + 1}$$

$$\text{(c)} \quad H(z) = \frac{z^3 - z^2 + z - 1}{z^2 - 0.25}$$

$$\text{(d)} \quad H(z) = \frac{z^9 - 1}{(z - 1)z^8}$$

Q4.9 Find and sketch the z-plane poles and zeros of signals having the following z-transforms:

$$\text{(a)} \quad X(z) = \frac{z^5 - 2}{z^{10} - 0.8}$$

$$\text{(b)} \quad X(z) = \frac{z^2 + 1.5z + 0.9}{z^2 - 1.5z + 1.1}$$

Would either signal grow without limit as $n \to \infty$ and, if so, why?

Q4.10 Using Table 4.1 at the end of the chapter, find the signals corresponding to the pole-zero configurations shown in Figure Q4.10.

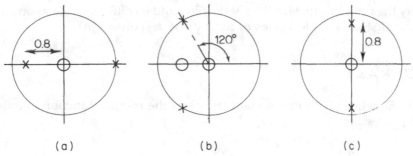

(a)   (b)   (c)

Figure Q4.10

SECTION 4.3.2

Q4.11 Using the technique of geometrical evaluation of the Fourier Transform in the z-plane, make rough sketches of the spectral magnitude characteristics of signals with the pole-zero configurations shown in Figure Q4.11.

  If the pole-zero configurations refer to LTI filters, rather than signals, what type of filter does each represent?

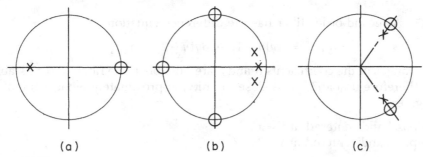

(a)   (b)   (c)

Figure Q4.11

Q4.12 Find and sketch the poles and zeros of the transfer function:

$$H(z) = \frac{z^3 - z^2 + 0.8z - 0.8}{z^3 + 0.8z^2}$$

Visualize and sketch the magnitude of the system's frequency response over the range $0 < \Omega < \pi$.

Q4.13 Using the smallest possible number of z-plane poles and zeros, design a digital filter with the following performance:

(a) complete rejection at $\Omega = 0$,
(b) complete rejection at $\Omega = \pi/3$,
(c) a narrow passband at $\Omega = 2\pi/3$ as a result of poles placed at radius $r = 0.9$ in the z-plane, and
(d) no unnecessary delay in the output signal.

Specify the transfer function $H(z)$ of the filter and its difference equation. Also find the first six sample values of its impulse response $h[n]$.

SECTION 4.3.3

Q4.14 Equation (4.39) in the main text gives the transfer function of a first-order processor:

$$H_1(z) = \frac{z}{(z-\alpha)}$$

In a particular application a low-pass filter is required, and α is set to 0.8. To improve discrimination against high frequencies, the z-plane zero is placed at $z = -1$ rather than at the origin.

(a) Specify the filter's difference equation.
(b) Estimate the first five sample values of its impulse response.
(c) Visualize and sketch its frequency response over the range $0 < \Omega < \pi$, in both magnitude and phase.

Q4.15 A second-order filter has the difference equation:

$$y[n] = \alpha y[n-1] - \beta y[n-2] + x[n]$$

What ranges of the coefficients α and β are possible if the filter is to be stable (and therefore usable)? If β is close to unity, approximately what value of α will give:

(a) a passband centered at $\Omega = \pi/3$
(b) a passband centered at $\Omega = 2\pi/3$?

Q4.16 Use Computer Program no.11 in Appendix A1 to investigate the impulse and frequency responses of various second-order LTI systems.

SECTION 4.4

Q4.17 A second-order bandpass filter has the difference equation:

$$y[n] = y[n-1] - 0.8\, y[n-2] + x[n]$$

When an input signal is applied at $n = 0$, the initial (auxiliary) conditions are $y[-1] = 1$ and $y[-2] = 1$. What contribution do these initial conditions make to the subsequent output signal? How is the contribution related to the filter's impulse response? Why is it often desirable to ensure zero initial conditions?

Q4.18 A digital processor is shown in Figure Q4.18. Find the z-transform of its output signal when a unit impulse $\delta[n]$ is delivered to its input side, given the following initial conditions:

(a) $y[-1] = y[-2] = 0$
(b) $y[-1] = -2,\ y[-2] = 2$

Which of your results represents the true transfer function of the processor?

What initial conditions would cause the output signal to be zero for $n > 0$, if a unit impulse is delivered to the input at $n = 0$? How can you account for this effect?

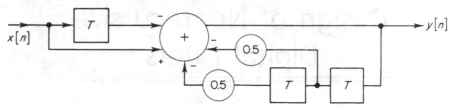

Figure Q4.18

CHAPTER 5
Design of Nonrecursive Digital Filters

5.1 INTRODUCTION

Earlier chapters have covered sufficient DSP theory to allow us to design various types of digital filter. In this and the next chapter we look at nonrecursive and recursive techniques respectively. Chapter 8 will discuss an alternative frequency-domain approach based upon the Fast Fourier Transform.

Before considering particular types of nonrecursive filter, we should make a few general points. First, almost any DSP algorithm or processor can reasonably be described as a 'filter'. However, the term is most commonly used for systems which transmit (or reject) well-defined frequency ranges. Typical, idealized, magnitude characteristics of four digital filter categories are shown in Figure 5.1. Remember that performance over the range $0 < \Omega < \pi$ defines the filtering action on any adequately-sampled signal. For example a filter with a response peak at $\Omega = \pi$ (and therefore also at $\Omega = 3\pi, 5\pi, \ldots$) is described as 'high-pass'.

Although we emphasize spectral magnitude functions in this book, there are other ways of specifying a digital filter. For example, there is a class of processors known as *all-pass*. They are designed to have a flat magnitude characteristic, but a phase characteristic which varies with frequency in some desired manner. Alternatively, we may be interested in a particular form of impulse response or step response; or we may wish to optimize the detection of a signal in the presence of random 'noise'. A wide variety of design techniques have been devised for tackling such problems. It is therefore important to realize that our own account of digital filtering will be selective, and introductory.

The general form of difference equation for a causal LTI processor was first given in equation (2.25):

$$\sum_{k=0}^{N} a_k y[n-k] = \sum_{k=0}^{M} b_k x[n-k] \tag{5.1}$$

In a nonrecursive filter the output depends only on present and previous inputs, so the difference equation may be written in the form:

$$y[n] = \sum_{k=0}^{M} b_k x[n-k] \tag{5.2}$$

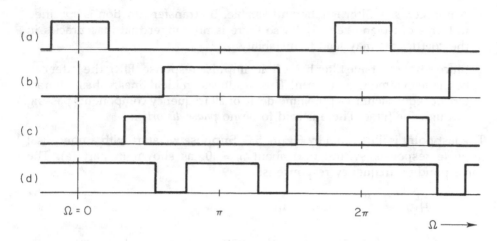

Figure 5.1 Idealized digital filter frequency responses: (a) low-pass, (b) high-pass, (c) bandpass, and (d) bandstop.

Such a filter implements the convolution sum directly, and the coefficients b_k are simply equal to successive terms in its impulse response, as shown by Figure 5.2. Since the number of coefficients must be finite, a practical nonrecursive filter is often referred to as *FIR (finite impulse response)*.

The transfer function and frequency response corresponding to equation 5.2 are respectively:

$$H(z) = \sum_{k=0}^{M} b_k z^{-k} \tag{5.3}$$

and:

$$H(\Omega) = \sum_{k=0}^{M} b_k \exp(-jk\Omega) \tag{5.4}$$

The art of designing a nonrecursive filter is to achieve an acceptable performance using as few coefficients b_k as possible. Practical filters typically need between (say) 10 and 150 coefficients. This makes them slower in operation than most recursive designs. However, there are two major compensating advantages:

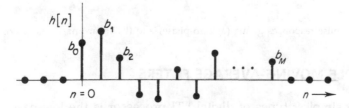

Figure 5.2 Impulse response coefficients of a nonrecursive filter.

A nonrecursive filter is inherently stable. Its transfer function is specified in terms of z-plane zeros only, so there is no danger that inaccuracies in the coefficients may lead to instability.

Since a nonrecursive filter has a finite impulse response (FIR), the latter can be made symmetrical in form. This produces an ideal linear-phase characteristic, equivalent to a pure time-delay of all frequency components passing through the filter. There is said to be no *phase-distortion*.

The last point is illustrated by Figure 5.3. Suppose we start with a noncausal impulse response, symmetrical about $n = 0$, as shown in part (a). The corresponding frequency response is:

$$H(\Omega) = \sum_{k=-M}^{M} b_k \exp(-jk\Omega)$$

$$= b_0 + 2b_1 \cos\Omega + 2b_2 \cos2\Omega + \dots 2b_M \cos M\Omega$$

$$= b_0 + 2 \sum_{k=1}^{M} b_k \cos k\Omega \tag{5.5}$$

$H(\Omega)$ is a real function of Ω, implying a zero-phase filter (no phase shift at any frequency). To make the filter causal we shift $h[n]$ by M sampling intervals as shown in part (b) of the figure. The effect is to delay the output by the same amount, converting the zero-phase characteristic into a pure linear-phase one. The magnitude of the frequency response is unaffected.

Although nonrecursive filters do not *have* to display linear-phase characteristics, the majority of practical designs take advantage of the possibility — which is not available in recursive filters based upon z-plane poles.

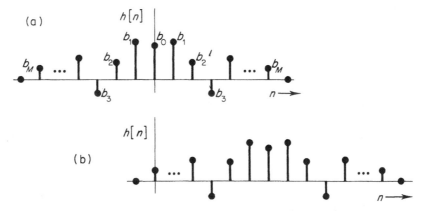

Figure 5.3 Impulse responses giving (a) zero-phase, and (b) linear-phase characteristics.

5.2 SIMPLE MOVING-AVERAGE FILTERS

One of the simplest types of digital LTI processor is the low-pass moving-average filter. We have already met such filters several times in this book. In

Chapter 1 we demonstrated the use of a 200-day moving-average for smoothing financial or stock market data (see Figure 1.3). Chapter 2 covered various aspects of their time-domain performance, including steady-state and transient responses (see Figures 2.11, 2.12, and 2.13). Our aim in this section is to examine such filters from the frequency response and z-transform viewpoints, and to show that equivalent high-pass and band-pass designs can readily be derived from a low-pass *prototype*.

Let us start with the impulse response shown in Figure 5.4. Its $(2M + 1)$ coefficients are all equal, and it is symmetrical about $n = 0$. In line with our comments in the previous section, it must represent a zero-phase system. The frequency response is given by:

$$H(\Omega) = \frac{1}{(2M + 1)} \{1 + 2\cos\Omega + 2\cos2\Omega + \ldots 2\cos M\Omega\} \qquad (5.6)$$

If we shift $h[n]$ forward in time to begin at $n = 0$, we obtain a causal version of the filter with a pure linear-phase characteristic. The magnitude characteristic is unchanged. The filter gives a low-pass filtering or smoothing action, the degree of smoothing increasing with the parameter M.

Computer Program no. 12 in Appendix A1 uses equation (5.6) to estimate and plot $|H(\Omega)|$ over the range $0 < \Omega < \pi$, for any desired value of M. Two typical plots are reproduced at the top of Figure 5.5. Part (a) represents a 5-term filter ($M = 2$), and part (b) a 21-term filter ($M = 10$). In both cases the response magnitude has its peak value of 1.0 at $\Omega = 0$. The width of the main lobe is inversely proportional to the number of impulse response terms, and can be chosen to suit a particular application.

Clearly, such magnitude responses are far removed from the ideal low-pass characteristic of Figure 5.1(a). The main lobes are by no means rectangular, and there are substantial unwanted *sidelobes* (if M is large, the form of $H(\Omega)$ tends to a sinc function, with a first sidelobe level about 22 per cent of the main lobe). In spite of these drawbacks, simple moving-average filters are often used in undemanding applications.

The z-plane poles and zeros of the two filters are shown in parts (c) and (d) of the figure. The 5-term filter has four zeros spaced around the unit circle; the 21-term filter has twenty zeros (there are also poles at the origin, but these do not contribute to the filtering action). In both cases there is a 'missing' zero at $z = 1$, which accounts for the passband centered at $\Omega = 0$. Since the zeros lie actually on the unit circle, they give rise to a series of true nulls in the corresponding frequency response.

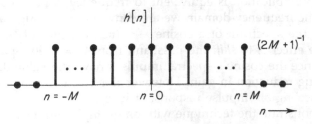

Figure 5.4 Impulse response of a noncausal, low-pass, moving-average filter.

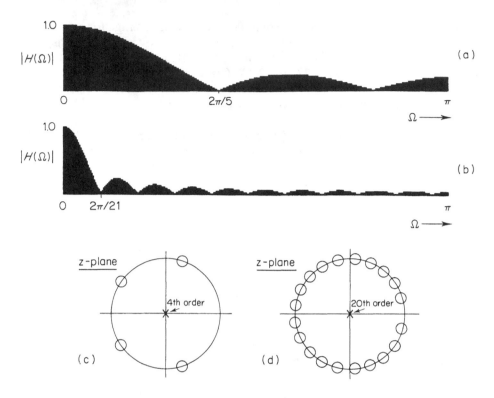

Figure 5.5 Frequency response magnitude characteristics of low-pass moving-average filters: (a) 5-term, and (b) 21-term. Parts (c) and (d) show their respective z-plane pole-zero configurations *(abscissa: 320 samples).*

We are treating moving-average filters as nonrecursive processors in this section. However they may also be implemented recursively, giving speed advantages which increase with the value of M. We have already mentioned this possibility in Section 1.1.2, and will return to it again at the end of Section 6.4.

Let us now consider how to derive a simple high-pass or band-pass design from our low-pass prototype. The basic idea may be summarized as follows. We multiply, or *modulate*, the original impulse response by the signal $\cos n\Omega_0$, where Ω_0 is the desired center-frequency of the new filter. The modulation property of the Fourier Transform, discussed in Section 3.3.1, tells us that such time-domain multiplication is equivalent to frequency-domain convolution. Therefore in the frequency-domain we are convolving the original low-pass function with the spectrum of a cosine – which consists of 'impulses' at $\Omega = \pm\Omega_0$. The effect is to *shift* the passband from $\Omega = 0$ (low-pass) to $\Omega = \pm\Omega_0$. (Also, since the cosine's spectral impulses have the value 0.5, there is a corresponding reduction in gain. However this can easily be made up, if required, by scaling all impulse response values.)

We can demonstrate the technique with an example. Suppose we wish to

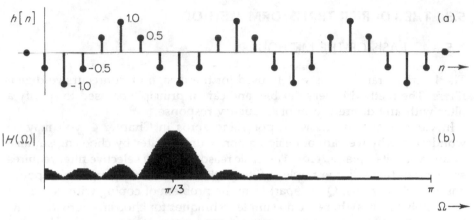

Figure 5.6 Deriving a simple bandpass filter from a low-pass prototype: (a) impulse response, and (b) frequency response magnitude function (*abscissa: 320 samples*).

design a causal band-pass filter with its peak response at $\Omega = \pi/3$, and a main-lobe width similar to the low-pass characteristic of Figure 5.5(b). We start with a noncausal impulse response of the type already shown in Figure 5.4, setting $M = 10$. We now multiply all its terms by the appropriate cosine, giving the impulse response:

$$h[n] = \frac{1}{(2M + 1)} \cos\left(\frac{n\pi}{3}\right), \quad -10 \leq n \leq 10 \tag{5.7}$$

$$= 0, \text{ elsewhere.}$$

$h[n]$ is next shifted forward to begin at $n = 0$, as shown in Figure 5.6(a). A scaling factor of 2 may also be applied, if required.

Program no. 12 in Appendix A1 can easily be modified to accept such a cosine-modulated impulse response, and used to plot $|H(\Omega)|$ for the new filter (we suggest this as a problem for you at the end of the chapter). The result is shown in part (b) of the figure.

The technique can be used to give a range of filters with different center frequencies and bandwidths. A high-pass filter is produced if the original impulse is multiplied by $\cos n\pi$, which has the effect of inverting (changing the sign of) successive terms. We have seen the effect before — for example, in the low-pass and high-pass impulse responses of Figure 4.11.

Although these nonrecursive filters can be useful for undemanding DSP applications, they have two major disadvantages. We have already noted that their frequency response magnitude characteristics are far from ideal. Secondly, they may be said to tackle the design problem 'the wrong way round'. Rather than start with a simple form of impulse response, and ask what frequency response it produces, we must know how to calculate the impulse response which best approximates a *specified* frequency response. We consider this problem in the following sections.

5.3 THE FOURIER TRANSFORM METHOD

5.3.1 BASIS OF THE METHOD

The Fourier Transform is widely used for the design of nonrecursive digital filters. The method is very flexible, and can in principle be used to specify a filter with any desired form of frequency response.

In view of our emphasis on poles and zeros in Chapter 4, you may be wondering why we cannot design a nonrecursive filter by choosing suitable locations for its z-plane zeros. The basic reason is that a selective filter requires so many of them. For example, a filter with 75 terms in its impulse response has 74 z-plane zeros. Quite apart from the problem of coping with very high order polynomials, there are no simple techniques for choosing zero locations to meet a desired frequency response characteristic. We must use a different approach.

The basis of the Fourier Transform method is easily explained. The transform equations, derived and discussed in Section 3.3.1, take the form:

$$X(\Omega) = \sum_{n=-\infty}^{\infty} x[n] \exp(-j\Omega n) \tag{5.8}$$

and

$$x[n] = \frac{1}{2\pi} \int_{2\pi} X(\Omega) \exp(j\Omega n) \, d\Omega \tag{5.9}$$

The second equation, the *inverse* transform, is relevant here. Rewriting it to describe an LTI processor, rather than a signal, we have:

$$h[n] = \frac{1}{2\pi} \int_{2\pi} H(\Omega) \exp(j\Omega n) \, d\Omega \tag{5.10}$$

If we start with a desired frequency response $H(\Omega)$, equation (5.10) shows how to derive the corresponding impulse response $h[n]$. The sample values of $h[n]$ give the required multiplier coefficients b_k for our nonrecursive filter.

The approach is conceptually straightforward, but there are two main practical difficulties. First, the integral in equation (5.10) is not always easy to solve − especially if $H(\Omega)$ has a complicated form. We will therefore concentrate on the simple, idealized, filter magnitude characteristics already illustrated in Figure 5.1, and will assume linear phase responses. Also, we will use a computer to help estimate the values of $h[n]$.

The second difficulty concerns the number of terms in $h[n]$. Our choice of $H(\Omega)$ may result in an impulse response with a large number of terms, giving a very uneconomic filter. We must clearly have some way of limiting the number of coefficients, and settling for a compromise between time-domain and frequency-domain performance.

Let us start by considering the ideal low-pass filter characteristic of Figure 5.7. This is the desired $H(\Omega)$. We have defined it over the range $\Omega = -\pi$ to

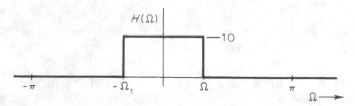

Figure 5.7 An ideal low-pass filter characteristic.

π, rather than 0 to 2π, to simplify the integral in equation (5.10). The filter is required to transmit frequency components between $\Omega = \pm\Omega_1$ with unity gain, and to cut off other components completely. We start by assuming a zero-phase system for which $H(\Omega)$ is real. Equation (5.10) specifies integration over any convenient 2π interval. Thus:

$$h[n] = \frac{1}{2\pi} \int_{-\pi}^{\pi} H(\Omega) \exp(j\Omega n)\, d\Omega$$

$$= \frac{1}{2\pi} \int_{-\Omega_1}^{\Omega_1} 1.\ \exp(j\Omega n)\, d\Omega = \frac{1}{2\pi} \left[\frac{\exp(j\Omega n)}{jn} \right]_{-\Omega_1}^{\Omega_1}$$

$$= \frac{1}{2\pi jn} \{\exp(j\Omega_1 n) - \exp(-j\Omega_1 n)\}$$

$$\therefore\ h[n] = \frac{1}{n\pi} \sin(n\Omega_1) = \frac{\Omega_1}{\pi} \operatorname{sinc}(n\Omega_1) \tag{5.11}$$

The impulse response is therefore of $\sin x/x$, or sinc, form. Such functions often arise in linear signal and systems theory. Generally speaking, a 'rectangular pulse' in either the time or frequency domain transforms into a sinc function in the other domain. We have now specified a rectangular form of frequency response, and we find that the impulse response terms follow a sinc function envelope.

Example 5.1 Find and sketch the impulse responses of ideal, zero-phase, low-pass digital filters with cut-off frequencies of (a) $\Omega_1 = \pi/5$, and (b) $\Omega_1 = \pi/2$.

Solution Using equation (5.11) we have, in the two cases:

(a) $h[n] = \dfrac{1}{n\pi} \sin\left(\dfrac{n\pi}{5}\right)$

and

(b) $h[n] = \dfrac{1}{n\pi} \sin\left(\dfrac{n\pi}{2}\right)$

The coefficient $h[0]$ is a little awkward to find, because the numerator and denominator are both zero. Resorting to l'Hospital's rule we have, in case (a):

$$h[0] = \frac{\dfrac{d}{dn}\left\{\sin\left(\dfrac{n\pi}{5}\right)\right\}}{\dfrac{d}{dn}\left(n\pi\right)}\Bigg|_{n=0} = \frac{\dfrac{\pi}{5}\cos\left(\dfrac{n\pi}{5}\right)}{\pi}\Bigg|_{n=0} = 0.2$$

Similarly in case (b) we find that h[0] = 0.5. Other values are readily calculated, and we list $h[0]$ and the eight values to either side of it in the table below (given to 6 decimal places):

	(a)	(b)
$h[0]$	0.2	0.5
$h[1] = h[-1]$	0.1 870 98	0
$h[2] = h[-2]$	0.1 513 65	−0.1 061 03
$h[3] = h[-3]$	0.1 009 10	0
$h[4] = h[-4]$	0.0 467 74	0.0 636 62
$h[5] = h[-5]$	0	0
$h[6] = h[-6]$	−0.0 311 83	−0.0 454 73
$h[7] = h[-7]$	−0.0 432 47	0
$h[8] = h[-8]$	−0.0 378 41	0.0 353 68

The impulse responses are drawn over the range $-16 \le n \le 16$ in Figure 5.8.

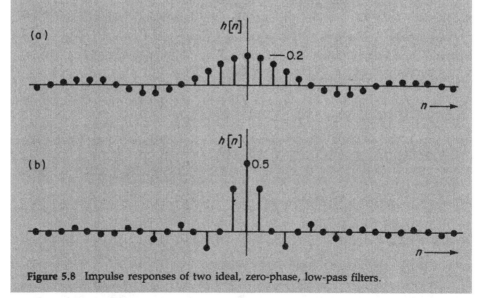

Figure 5.8 Impulse responses of two ideal, zero-phase, low-pass filters.

Although the impulse responses of Figure 5.8 decay to either side of $n=0$, they theoretically continue for ever in both directions. This reflects the general antithesis between band limitation and time limitation: since we have chosen a frequency response $H(\Omega)$ with an infinitely sharp cut-off, the time-domain response continues for ever. To realize such a filter we must clearly limit, or *truncate*, the impulse response in some way. The obvious approach is to ignore the small sample values in its 'tails'. We can then shift $h[n]$ to begin at $n=0$, giving a causal, linear-phase, filter.

There is a compromise to be made here. The more samples of $h[n]$ we include, the closer we get to the desired form of $H(\Omega)$; but the less economic the filter becomes, and the greater its time delay. In practice we must settle for an *approximation* to the ideal frequency response.

In the previous section we showed how a bandpass or high-pass filter could be derived from a low-pass prototype, by modulating $h[n]$ with a cosine signal at the relevant frequency. The same technique can be used again. We multiply the unshifted version of $h[n]$ by $\cos n\Omega_0$, where Ω_0 is the required center frequency of the new filter. Combining this idea with equation (5.11), the impulse response coefficients of an ideal filter with centre frequency Ω_0 and bandwidth $2\Omega_1$ are given by:

$$h[n] = \frac{1}{n\pi} \sin (n\Omega_1) \cos(n\Omega_0) \qquad (5.12)$$

Substituting the values of $h[n]$ for the coefficients b_k in equation (5.5), we also have:

$$H(\Omega) = \frac{\Omega_1}{\pi} + 2 \sum_{k=1}^{\infty} h[k] \cos (k\,\Omega) \qquad (5.13)$$

In practice we will truncate the impulse response to $(2M+1)$ terms — that is, $h[0]$ and M terms to either side of it — and shift it to begin at $n=0$. The frequency response magnitude characteristic is then:

$$|H(\Omega)| = \frac{\Omega_1}{\pi} + 2 \sum_{k=1}^{M} h[k] \cos(k\,\Omega) \qquad (5.14)$$

Program no. 13 in Appendix A1 implements equations (5.12) and (5.14) for any values of Ω_0, Ω_1, and M. It prints out the coefficients $h[0]$ to $h[M]$ inclusive, then plots $|H(\Omega)|$ on the screen for $0<\Omega<\pi$. It may be used to design a wide variety of low-pass, high-pass, and bandpass filters.

Note that Ω_0 and Ω_1 are entered as angles in degrees rather than radians. Thus $\Omega_0 = 0°$ corresponds to a low-pass design, $\Omega_0 = 180°$ to a high-pass design, and so on. Also, the bandwidth of a low-pass or high-pass filter must be entered as *twice* its value in the range $0<\Omega<\pi$. We hope you will be able to try the program for yourself.

Figure 5.9 shows screen plots of $|H(\Omega)|$ for three bandpass designs, all with center-frequency $\Omega_0 = 60°$ and bandwidth $\Omega_1 = 30°$. The plots are for different values of M, representing impulse responses truncated to 21 terms, 51 terms, and 151 terms respectively. All are approximations to the ideal, rectangular, characteristic. However the 'goodness of fit' clearly improves with increasing M. You may like to compare this figure with the simple 21-term filter of Figure 5.6, which has the same center-frequency and a comparable bandwidth.

You perhaps feel that our discussion of filter performance has been rather abstract, so we will now demonstrate the action of a bandpass design in the

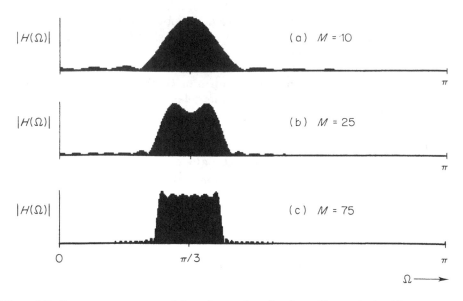

Figure 5.9 Frequency responses of three linear-phase bandpass filters, obtained by truncating the 'ideal' impulse response *(abscissa: 320 samples)*.

time domain. Before we begin, we should outline the type of practical situation in which bandpass filters are valuable. One of the best known is electronic communications, where relatively low-frequency signals or data are often shifted to a higher frequency band for transmission. In this way the transmission frequency can be chosen to suit the transmission medium (for example, cable, optical fiber, or radio). Also, we can often transmit a number of signals simultaneously by using slightly different bands for each (a technique known as *frequency-division multiplexing*).

At the receiving end, a signal must be separated from its neighbors with a suitable bandpass filter. The filter can also be useful for reducing interference or 'noise' picked up during transmission. An important advantage of a linear-phase design is that phase relationships are preserved, so the signal does not suffer phase distortion. We will see what this means a little later.

We can demonstrate these ideas by considering the situation shown in Figure 5.10. We have a signal with two discrete frequency components Ω_A and Ω_B. For the purposes of transmission it has been combined with other signals at frequencies Ω_C and Ω_D, and we need to recover the wanted signal by bandpass filtering. For convenience we have labeled the frequencies of the various components in degrees.

The filter magnitude characteristic shown in the figure is suitable. Components Ω_A and Ω_B fall within its passband, whereas Ω_C and Ω_D coincide with response nulls. The characteristic is in fact our old friend the 21-term linear-phase design already illustrated in Figure 5.9(a).

Computer Program no. 14 in Appendix A1 generates relevant signals, implements the bandpass filter by a direct convolution, and plots the signals on the screen. The filter's impulse response values, derived from Program no. 13, are entered as data.

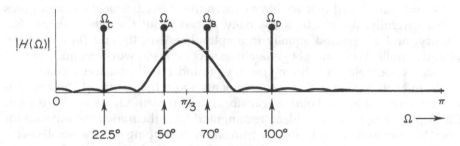

Figure 5.10

The screen plot is reproduced in Figure 5.11. Part (a) shows the 'wanted' signal, consisting of two cosine components at frequencies Ω_A and Ω_B (containing 36/5 and 36/7 samples per period respectively). We have switched the signal on and off to observe the effects of filter transients. The switching instants have been chosen to make the signal symmetrical in form.

In part (b) of the figure components Ω_C and Ω_D have been added, contaminating the wanted signal. (They contain 16 and 3.6 samples per period respectively.) The bandpass filter must accept this composite signal as its input $x[n]$, and reject components Ω_C and Ω_D.

The filtered output $y[n]$ is shown in part (c) of the figure. The unwanted components have been suppressed. The wanted signal displays start-up and stop transients, which last as long as the filter's impulse response — in this case, 21 sampling intervals. However the central portion of $y[n]$, labeled 'C' in the figure, is identical in shape to the original signal in part (a). Not only have the amplitudes of the Ω_A and Ω_B components been treated equally, but their relative phases have been preserved. (If you look carefully, you will also notice that the output signal has been delayed by half the duration of the filter's impulse response.)

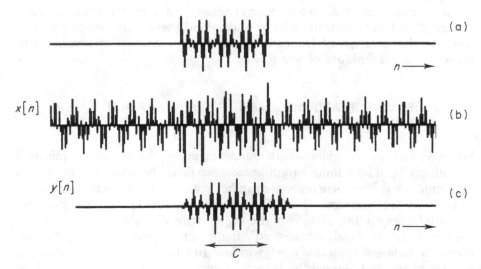

Figure 5.11 Time-domain performance of a linear-phase bandpass filter *(abscissa: 320 samples).*

It is only fair to add that we have chosen the above frequency components rather carefully. Ω_A and Ω_B are equally placed about the filter's center frequency, and are treated equally in amplitude. Also, Ω_C and Ω_D have been placed at nulls. We could not generally expect things to work out quite as well as this! Nevertheless we hope you have found the illustration valuable.

We end this section with some comments about the nature of the approximation involved in the Fourier Transform design method. As we have seen, the impulse response of an ideal 'rectangular' filter theoretically continues for ever. We can make it realizable by truncation, 'chopping off' the small coefficients in the 'tails'. The more coefficients we include, the better the frequency-domain approximation.

The Fourier Transform design method gives the best approximation in a *least-squares* sense. We may explain this statement as follows. Suppose we have specified a *desired* frequency response function $H_D(\Omega)$, but after impulse response truncation we get an *actual* response $H_A(\Omega)$. Let us define the overall error between desired and actual responses as:

$$e = \int_{2\pi} |H_D(\Omega) - H_A(\Omega)|^2 \, d\Omega \qquad (5.15)$$

That is, we take account of the *squared-magnitude* of the difference between $H_D(\Omega)$ and $H_A(\Omega)$ over one complete period in the frequency domain. All frequencies are treated equally. By using the *squared* magnitude we accentuate large errors compared with small ones, agreeing with the commonsense idea of 'goodness of fit'.

For a given number of impulse response coefficients, the Fourier Transform method minimizes the value of e. Those of you who are familiar with continuous-time Fourier analysis will recognize least-squares approximation as one of its central features. We discuss it more fully in Appendix A2.

Minimization of e is a valuable design criterion, but it is by no means the only possible one. For example, a designer may be more interested in controlling the sidelobe levels of a filter, or in achieving a sharp transition between passband and stopband. In order to understand these possibilities we must consider the question of truncation from a more general viewpoint, and introduce the techniques of windowing.

5.3.2 TRUNCATION AND WINDOWING: RECTANGULAR AND TRIANGULAR WINDOWS

When we truncate an infinite-length impulse response, the process is equivalent to multiplying it by a finite-length *window function*. The window determines how much of the impulse response can be 'seen', and is sometimes referred to as an *observation window*. This is illustrated by Figure 5.12. Part (a) represents an infinite-length impulse response $h_d[n]$ which is the inverse Fourier Transform of the ideal, or 'desired', frequency response $H_D(\Omega)$. Part (b) shows a *rectangular* (or *uniform*) window function $w[n]$. When $h_d[n]$ is multiplied by $w[n]$, we get the truncated impulse response in part (c). This 'actual' impulse response is labeled $h_a[n]$. Clearly, by altering the length of

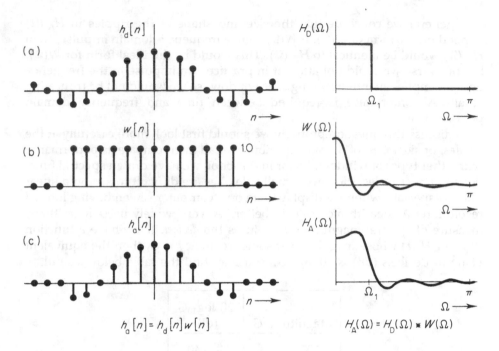

Figure 5.12 Windowing in the time and frequency domains.

$w[n]$ we can control the number of terms in $h_a[n]$. It is very important to realize that we have derived $h_a[n]$ by *time-domain multiplication*.

You may have noticed that we are treating the impulse responses and the window function as symmetrical about $n=0$. This is convenient since the corresponding spectral functions are real (zero-phase). Later we can easily shift the truncated impulse response to begin at $n=0$, defining a causal, linear-phase, filter.

The modulation property of the Fourier Transform tells us that time-domain multiplication is equivalent to *frequency-domain convolution* (see Table 3.2). Therefore the above windowing process must produce an actual frequency response $H_A(\Omega)$ which is the convolution of $H_D(\Omega)$ with the spectrum of the window $W(\Omega)$. Hence $H_A(\Omega) = H_D(\Omega) * W(\Omega)$. We see this on the right-hand side of the figure. The spectrum of a rectangular window tends to the sinc, or $\sin x / x$, form. Its convolution with $H_D(\Omega)$ gives an approximation to the desired frequency response containing a number of fluctuations or *ripples*. These distort the shape of the passband, and produce unwanted sidelobes.

If we increase the length of the rectangular window, its spectrum becomes narrower. The ripples in $H_A(\Omega)$ bunch more closely around the nominal cut-off frequency Ω_1. Also, the transition from passband to stopband becomes sharper. (You may like to refer back to Figure 5.9, which shows such effects clearly.) However the window spectrum is still of sinc form, so lengthening the window does not reduce the ripple *magnitudes*. J.W. Gibbs showed in about 1900 that the maximum ripple in the region of a sudden transition is around 9 per cent, regardless of the length of the window. The effect is known as the *Gibbs phenomenon*.

In general we conclude that the size and shape of the ripples in H_A (Ω) depend on the form of $W(\Omega)$. If $W(\Omega)$ were a frequency-domain impulse, then H_A (Ω) would be identical to H_D (Ω). This would be the ideal form for $W(\Omega)$, but of course we could not attain it in practice. An impulse in the frequency domain implies an infinitely long time window, so we would not be truncating at all! A compromise is required between time and frequency domain performance.

To discuss this question properly we should first look more carefully at the ripples, or sidelobes, of the sinc function. We can then compare its performance with other types of window. So far in this book we have drawn spectral functions to linear scales. However small sidelobes are difficult to see on a linear plot — especially when we display them on a computer screen having limited resolution. A logarithmic scale is better. A very widely-used logarithmic measure of spectral magnitude (or gain) is the *decibel*. If we have a function $W(\Omega)$ or $H(\Omega)$ whose magnitude at some frequency is G, then the equivalent value in decibels (dB) is $20 \log_{10}G$. You may find the table below useful.

Magnitude G	$20 \log_{10}G$ (dB)
100	40
10	20
1	0
0.1	-20
0.01	-40
0.001	-60

It is often convenient to normalize the function we are plotting to unity, giving a logarithmic plot with a maximum of 0 dB.

Program no. 15 in Appendix A1 produces a decibel plot of the spectrum of a rectangular window with $(2M+1)$ terms. The spectrum is given by:

$$W(\Omega) = \sum_{n=-\infty}^{\infty} w[n] \exp(-j\Omega n) = \sum_{n=-M}^{M} 1 \exp(-j\Omega n)$$

$$= \exp(-j\Omega M) + \ldots 1 + \ldots \exp(j\Omega M)$$

$$= 1 + 2\{\cos(\Omega) + \cos(2\Omega) + \ldots \cos(M\Omega)\}$$

$$(5.16)$$

The program estimates $|W(\Omega)|$ for 320 values of Ω in the range $0 < \Omega < \pi$, normalizes to unity, and converts to decibels (the natural logarithm function is used; this is multiplied by 0.4343 to give the logarithm to base 10). The range of the plot is 0 to -50 dB. You may choose the value of M for yourself.

Two typical computer plots, with axes added, are reproduced in Figure 5.13. In part (a) $M=10$, representing a 21-term window; in part (b) $M=25$, giving a 51-term window. The sidelobe structure is clearly shown by the dB scale. We see that the first sidelobe is about '13.5 dB down' on the main lobe, and

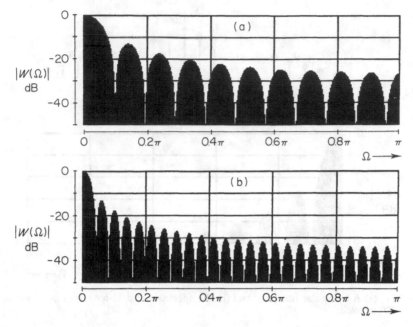

Figure 5.13 Spectra of rectangular windows with (a) 21 terms, and (b) 51 terms (*abscissa: 320 samples*).

that there are many sidelobes greater than about -30 dB. A window with this sidelobe performance is unsuitable for most digital filtering applications.

In spite of the above problems, we must be careful not to dismiss the rectangular window completely! In the previous section we saw that rectangular truncation gives the best approximation to the desired frequency response in a least-squares sense. That is, it gives the least squared, integrated, error between $H_D(\Omega)$ and $H_A(\Omega)$ for a given number of impulse response terms. If we choose another type of window, we will have to forfeit this advantage. However, we should be able to trade it against improved ripple and sidelobe performance.

The basic reason for the rectangular window's poor sidelobe levels stems from the ever-present antithesis between time-limitation and band-limitation. Since the window suddenly 'chops off' in the time domain, it tends to spread out in the frequency domain. Intuitively we may expect that a more 'gentle' window, tapered towards its edges, will give better results.

The *triangular* function at the top of Figure 5.14 offers a simple form of tapering. Let us again consider it to have $(2M+1)$ terms. The outside ones in the tails have value unity, and the central value is $(M+1)$. Its spectrum is given by:

$$(M+1) + 2\{M \cos(\Omega) + (M-1) \cos(2\Omega) + \ldots \cos(M\Omega)\} \qquad (5.17)$$

The peak value occurs when $\Omega = 0$ and equals:

$$(M+1) + 2\{M + (M-1) + \ldots 1\} = (M+1)^2 \qquad (5.18)$$

If we divide all terms by $(M+1)^2$ we obtain a triangular, or *Bartlett*, window with $(2M+1)$ terms and a spectrum with a peak value of unity. Hence:

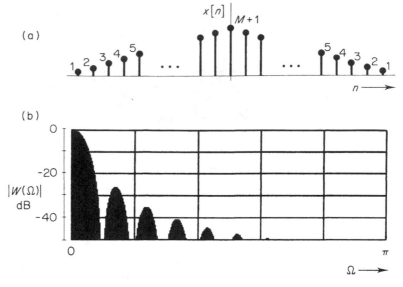

Figure 5.14 (a) A triangular function, and (b) the spectrum of a 41-term triangular window (*abscissa: 320 samples*).

$$w[n] = \frac{(M+1) - |n|}{(M+1)^2}, \quad -M \le n \le M \tag{5.19}$$

$$= 0 \text{ elsewhere.}$$

The window's spectrum is:

$$W(\Omega) = \frac{1}{(M+1)} + \frac{2}{(M+1)^2} \{M \cos(\Omega) + (M-1) \cos(2\Omega) + \ldots \cos(M\Omega)\} \tag{5.20}$$

An alternative, and very helpful, way of looking at a triangular window is as the self-convolution of a rectangular window. If a rectangular window with $(M+1)$ terms is convolved with itself, the result is a triangular window with $(2M+1)$ terms. Now time-domain convolution is equivalent to frequency-domain multiplication. Therefore the spectrum of a triangular window with $(2M+1)$ terms equals the *square* of that of a rectangular window with $(M+1)$ terms. Plotted on *logarithmic* scales, the triangular window has sidelobe levels *half* as great as those of the rectangular window. For example, if you refer back to Figure 5.13, you will see that a rectangular window has a first sidelobe about 13.5 dB down on the main lobe. We may therefore expect a triangular window to give a first sidelobe level of about -27 dB.

It is quite straightforward to modify Computer Program no. 15 to accommodate equation (5.20) (we suggest this as a problem for you at the end of the chapter). The resulting plot is shown in Figure 5.14 for the case $M = 20$ (a 41-term triangular window). As argued above, the first sidelobe level is indeed about -27 dB; and the sidelobe pattern is the same as for a rectangular window with $M = 10$ — except that all decibel values are reduced by a factor of 2.

A disadvantage of the triangular window is that, for a given window length, its main spectral lobe is twice as wide as that of the rectangular window. Convolved with a desired frequency response $H_D(\Omega)$, this causes a broadening of the transition region between passband and stopband. We shall have more to say about the trade-off between main-lobe width and sidelobe levels at the start of the next section.

In summary, the triangular window offers considerably better sidelobe levels than the rectangular window, at the expense of increased main lobe width. However, a first sidelobe level of about -27 dB is still too large for many digital filtering applications. The search for better truncation windows continues!

5.3.3   THE VON HANN AND HAMMING WINDOWS

In the previous section we noted that the spectrum of a truncation window should ideally be a frequency-domain impulse. In other words it should have a very narrow main lobe, and no sidelobes. In practice a compromise is necessary. Windows with narrow main lobes tend to have large sidelobes, and vice versa. We have already seen this in the case of the rectangular and triangular functions. For a given window length, the triangular function offers better sidelobe levels, but has a broader main lobe.

When a window is used in nonrecursive filter design, its spectrum $W(\Omega)$ is convolved with the desired frequency response $H_D(\Omega)$ to produce the actual frequency response $H_A(\Omega)$. This has already been illustrated by Figure 5.12. In general we may summarize the effects of the convolution as follows:

The width of the transition region from passband to stopband in $H_A(\Omega)$ depends on the width of the main lobe in $W(\Omega)$. A sharp transition (normally considered desirable) requires a narrow main lobe.

The ripples in $H_A(\Omega)$ depend on the sidelobe levels of $W(\Omega)$. The smaller the sidelobes, the better the ripple performance.

Since all practical windows involve a compromise between main lobe width and sidelobe levels, there must be a trade-off between a sharp passband—stopband transition and low ripple levels in the actual filter. Incidentally, a rectangular window gives the sharpest passband—stopband transition for a given window length. Therefore it should be used when ripple performance is not a major consideration.

Many different windows have been devised over the years. Of these the *von Hann* and especially the *Hamming* designs have been widely accepted by DSP designers. Both have a main spectral lobe similar to that of the triangular window, but offer considerably smaller sidelobe levels.

The von Hann function is named after its originator. Perhaps unfortunately, it is also widely referred to as the *Hanning* window. You will find both names used in the DSP literature. A window of this type with $(2M + 1)$ terms is defined by:

$$w[n] = 0.5 + 0.5 \cos\left(\frac{n\pi}{M+1}\right), \qquad -M \le n \le M$$

$$= 0 \text{ elsewhere.} \tag{5.21}$$

It consists of one period of a sampled cosine, plus a DC level which makes all the sample values positive. A 21-term von Hann window ($M = 10$) is shown in Figure 5.15(a). Note that the cosine shape gives a smoother tapering action than the triangular window. We shall investigate its spectrum a little later.

By altering the relative proportions of the DC and cosine components in the von Hann window, R.W. Hamming found that he could further improve sidelobe levels. The Hamming window is defined as:

$$w]n] = 0.54 + 0.46 \cos\left(\frac{n\pi}{M}\right), \qquad -M \le n \le M$$

$$= 0 \text{ elsewhere.} \tag{5.22}$$

(Actually the optimum proportions are slightly dependent on the value of M. Values of 0.54 and 0.46 are usually quoted, being very close to optimum for M greater than about 10.) A 21-term Hamming window is illustrated in part (b) of the figure.

Note that equations (5.21) and (5.22) may both be written as:

$$w[n] = A + B \cos\left(\frac{n\pi}{C}\right), \qquad -M \le n \le M$$

$$= 0 \text{ elsewhere.} \tag{5.23}$$

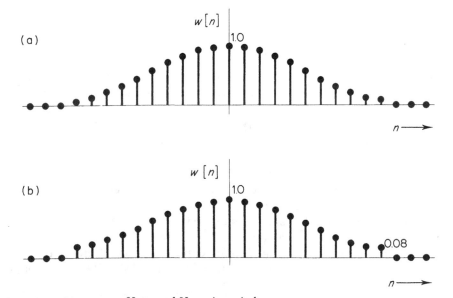

Figure 5.15   21-term von Hann and Hamming windows.

For the von Hann window, $A = 0.5$, $B = 0.5$, and $C = M+1$; for the Hamming window, $A = 0.54$, $B = 0.46$, and $C = M$. The corresponding spectral function is:

$$W(\Omega) = w[0] + 2 \sum_{k=1}^{M} w[k] \cos(k\,\Omega) \qquad (5.24)$$

Computer Program no. 16 in Appendix A1 first estimates the values of $w[n]$ according to equation (5.23), for a von Hann or Hamming window of any length. It then uses equation (5.24) to find 320 values of $W(\Omega)$ in the range $0 < \Omega < \pi$, and produces a decibel plot on the screen.

Figure 5.16 shows the plots for $M = 25$ (51-term windows). We have also included a plot for a 51-term triangular window at the top of the figure, to aid comparison. Its first sidelobe level of about -27 dB is similar to Figure 5.14. The von Hann window gives a first sidelobe level of about -32 dB, with subsequent sidelobes below -40 dB. The Hamming window is best of all, with all its sidelobes below -40 dB. If you look very carefully you will see that the main lobe of the von Hann and Hamming windows is slightly greater than

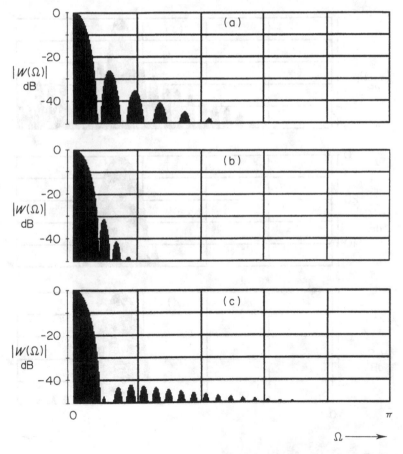

Figure 5.16 Spectra of 51-term windows: (a) triangular, (b) von Hann, and (c) Hamming (*abscissa: 320 samples*).

that of the triangular window. However, this is a small price to pay for the much improved sidelobe performance.

The Hamming window has the best performance of the three, and is widely used in nonrecursive filter design.

Our next computer program brings together most of the ideas covered in this chapter so far. Program no. 17 in Appendix A1 may be used to design a wide range of nonrecursive digital filters, based on rectangular, von Hann, or Hamming truncation windows. It is essentially an amalgamation of previous programs 13 and 16.

The program requests filter center-frequency, bandwidth, window length and window type as input data. (Note that the bandwidth of a low-pass or high-pass design must be entered as *twice* its value in the range $0 < \Omega < \pi$, as in Program no. 13.) It produces a decibel plot of the filter's frequency response; and it prints out the impulse response values, scaled to give unity peak gain (0 dB). The latter feature compensates for the scaling effect of the chosen window. The impulse response is considered to be symmetrical about $n = 0$, and

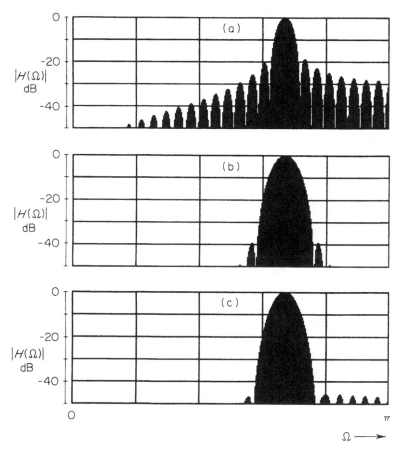

Figure 5.17 Frequency responses of three nonrecursive bandpass filters designed by the window method *(abscissa: 320 samples).*

values $h[0]$ (the central value) to $h[M]$ (one of the 'tails') are given. To make the impulse response causal, its 'other half' must be added, and you must then shift it so that it begins at $n = 0$. If you are in any doubt about this procedure, you may like to refer back to Figure 5.3.

Figure 5.17 shows some typical results, emphasizing the differences caused by the various windows. The specification is for a bandpass filter with center-frequency $\Omega = 2\pi/3$ (120°) and bandwidth $\pi/18$ (10°), having a 51-term impulse response ($M = 25$). Part (a) shows the rectangular-window version of the filter. Its largest sidelobe level is about -20 dB (roughly 10 per cent of the main lobe on a linear scale), and is an example of the Gibbs phenomenon. The -20 dB figure is the result of convolving the window spectrum (having a first sidelobe level of -13.5 dB) with the rectangular spectrum of the 'desired' filter.

Part (b) of the figure illustrates the von Hann version of the filter. Sidelobes are reduced at the expense of a broader main lobe. Part (c) shows the Hamming version. All sidelobes are better than about -46 dB (0.5 per cent of the main lobe on a linear scale).

A disadvantage of the von Hann and Hamming filters is that their main lobes are a lot wider than the 10° specified. To some extent we could narrow them by requesting a smaller bandwidth in the first place. However a limit is soon reached, because a filter's main lobe cannot be narrower than that of its truncation window. Further bandwidth reduction can only be achieved by accepting a longer window (that is, a greater value of M).

For good measure, we end the section with another example of a Hamming filter designed with Program no. 17. Figure 5.18 shows the characteristic of a 101-term low-pass filter with a bandwidth of 0.4π (72°). It has a rapid transition from passband to stopband, with no sidelobes above -50 dB. Comparing with Figure 5.5, our ability to design effective nonrecursive filters seems to have progressed quite well!

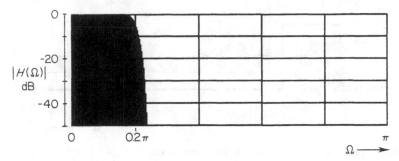

Figure 5.18 Frequency response of a 101-term low-pass filter based upon the Hamming window *(abscissa: 320 samples)*.

5.3.4 THE KAISER WINDOW

The windows we have met so far have fixed shapes. Each window gives a particular tradeoff between the width of the main spectral lobe, and sidelobe levels.

The major contribution of J.F. Kaiser to window design was to suggest a function in which the tradeoff can be adjusted by the DSP designer.

The Kaiser window is defined as:

$$w[n] = \frac{I_0\left(\alpha\sqrt{1 - \left(\frac{n}{M}\right)^2}\right)}{I_0(\alpha)}, \qquad -M \le n \le M$$

$$= 0 \text{ elsewhere.} \tag{5.25}$$

I_0 is the *modified Bessel function* of the first kind and of zero order. Note that when $n = 0$, the numerator and denominator of the above expression both equal $I_0(\alpha)$. Hence the central window value $w[0]$ is always unity. The value of α controls the degree of taper towards the edges of the window. If $\alpha = 0$ there is no taper and we get a rectangular window; if $\alpha = 5.44$ the window is similar to the Hamming function. Other values of α offer the designer a whole range of tradeoffs between main lobe width and sidelobe performance.

We can explain Kaiser's design approach in terms of Figure 5.19. It shows an ideal filter characteristic $H_D(\Omega)$, together with an acceptable *ripple level* $\pm\delta$ in both passband and stopband (expressed as a fraction), and an acceptable *transition width* Δ. The aim is to produce an actual filter characteristic $H_A(\Omega)$ which avoids the shaded regions of the figure, using as few coefficients as possible.

Note that in this case we have shown the stopband ripple as going both positive and negative. Of course, if we plotted $|H_D(\Omega)|$ rather than $H_D(\Omega)$, the ripples would appear positive.

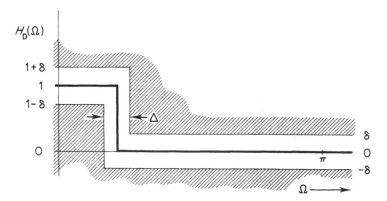

Figure 5.19 Specifying the design of a Kaiser-window filter.

The design of the Kaiser window is based on the following findings:

The parameter α depends on the allowable ripple value δ. This is because α controls the taper of the window, and hence its sidelobe levels.

For a given value of ripple (and hence α), the transition width Δ is related to the window length. Hence if Δ is specified, we can find the parameter M.

Knowing both α and M, the coefficients $w[n]$ are found using equation (5.25).

Do not worry if all this seems rather complicated! A computer can evaluate the various parameters and functions, using formulae devised by Kaiser. These are well summarized in a book by Hamming (see bibliography).

The ripple level is first expressed as an attenuation in decibels:

$$A = -20 \log_{10} \delta \tag{5.26}$$

The parameter α is now found using the empirical formulae:

$$
\begin{aligned}
\alpha &= 0.1102(A - 8.7) && \text{if } A \geq 50 \\
\alpha &= 0.5842(A - 21)^{0.4} + 0.07886(A - 21) && \text{if } 21 < A < 50 \\
\alpha &= 0 && \text{if } A \leq 21
\end{aligned}
\tag{5.27}
$$

Note that α is zero for attenuation values below 21 dB. This corresponds to a rectangular window, which produces ripples of around -21 dB according to the Gibbs phenomenon.

Given the value of A and the transition width Δ (expressed as a fraction of 2π), M is found using:

$$M \geq \frac{A - 7.95}{28.72\Delta} \tag{5.28}$$

This generally gives a fractional result, so M is rounded up to the nearest integer. If the window length $(2M+1)$ is unacceptably long, we can revise the original specification — by increasing the allowable ripple and/or the transition width.

We now use α and M in equation (5.25) to find the window coefficients $w[n]$. The function $I_0(x)$ may be expanded as a power series:

$$I_0(x) = 1 + \sum_{n=1}^{\infty} \left[\left(\frac{x}{2} \right)^n \frac{1}{n!} \right]^2$$

$$= 1 + \frac{x^2}{4} + \frac{x^2}{4} \frac{x^2}{4} \frac{1}{2^2} + \frac{x^4}{16} \frac{x^2}{4} \frac{1}{3^3} + \dots \tag{5.29}$$

The series converges satisfactorily for all values of x relevant here, if more than about ten terms are included. We shall use a program loop to sum twenty terms.

Program no. 18 in Appendix A1 incorporates equations (5.25) thru (5.29), and designs a Kaiser window filter with any desired center-frequency, bandwidth, ripple, and transition width. The program is very similar to Program no. 17. Note that the impulse response values, which equal the coefficients of the nonrecursive filter, are scaled before print-out to give a filter with a maximum gain of 0 dB.

It is difficult to decide how to present the frequency response plot in this case. A linear plot would be useful for checking performance against Figure 5.19. However many practical filters require very small ripple levels, which are almost invisible on a linear plot. So we have settled for a decibel plot of response magnitude, allowing easy comparison with the filters described earlier in the chapter.

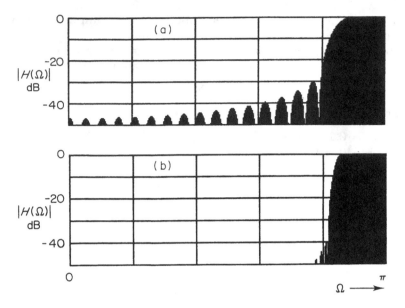

Figure 5.20 Frequency responses of high-pass filters with 60° bandwidth, based upon the Kaiser window; (a) $\delta = 0.0316$ (-30 dB), $\Delta = 15°$, $\alpha = 2.1176$ and $M = 19$; (b) $\delta = 0.01$ (-40 dB), $\Delta = 7.5°$, $\alpha = 3.3954$ and $M = 54$ *(abscissa: 320 samples)*.

Figure 5.20 shows two examples — this time of high-pass filters. Both have a bandwidth of 60° (or 30° over the range 0 to π), but the transition widths and ripple levels are different. The results help emphasize the flexibility of the Kaiser window. We hope you will have the opportunity to try the program for yourself.

5.4 EQUIRIPPLE FILTERS

Our work in the previous sections has shown that practical filter design is essentially an approximation problem. Starting with a desired form of frequency response, we try to approximate it to an acceptable degree of accuracy. The Fourier design method uses various window functions — rectangular, triangular, von Hann, Hamming, Kaiser — to give different compromises between the width of the transition band, and the size of ripples and sidelobes.

If you refer back to the actual frequency responses obtained with different forms of window — for example, Figures 5.17 and 5.20 — you will see that the largest ripples and sidelobes generally occur near the transition from passband to stopband. As we move away from the transition region, the error between desired and actual responses becomes smaller. This raises the interesting possibility that, if the error can be distributed more equally over the range $0 \leq \Omega \leq \pi$, we may be able to achieve a better overall compromise between ripple levels, transition bandwidth, and filter order. Of course, such a compromise will not produce the best approximation in the least-squares sense — this is only achieved by the Fourier method with a rectangular window. But it should offer other features which are valuable to the DSP designer.

Equiripple filters exploit the above possibility. The aim is to find an approximation giving acceptable levels of ripple throughout the passband and stopband — rather than just meeting the specification at one frequency, and greatly exceeding it elsewhere.

Nonrecursive filters of this type are designed in quite a different way from the filters we have met previously. Essentially, the frequency response function $H(\Omega)$ is examined for local maxima and minima (extrema), corresponding to passband and stopband ripples. Using an iterative algorithm, the filter is then adjusted to meet the equiripple specification. Unfortunately, the mathematical techniques are complicated, and beyond the scope of this book. Our aim here is simply to summarize the approach, so that if you meet nonrecursive equiripple filters later you will at least appreciate their main features. Further theoretical background, plus a useful list of research references, is given in the book by Oppenheim and Schafer (see Bibliography).

Figure 5.21 illustrates the specification for a low-pass equiripple filter. In the passband ($0 \leq \Omega \leq \Omega_p$), the acceptable level of ripple is $\pm\delta_1$; in the stopband ($\Omega_s < \Omega < \pi$), the acceptable ripple is $\pm\delta_2$. The width of the transition band is ($\Omega_s - \Omega_p$). The figure also shows a typical filter response which just meets the specification. (Note that the number of ripples is directly related to the order of the filter, and hence to the length of its impulse response $h[n]$.) The ripple peaks and troughs occur at $\Omega_1, \Omega_2, \Omega_3, \ldots$. In most respects this tolerance scheme is like that of a Kaiser-window filter (see Figure 5.19). However it has the added flexibility of allowing different ripple levels in passband and stopband.

As with other nonrecursive techniques, equiripple filter design normally starts by assuming an even impulse response, symmetric about $n=0$. This gives zero-phase characteristics. The impulse response can later be shifted forward to begin at $n=0$, producing a causal filter with pure linear-phase. As we showed at the start of this chapter, the frequency response of a zero-phase nonrecursive filter takes the general form (see equation (5.5)):

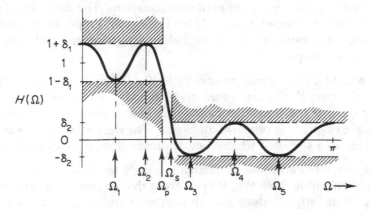

Figure 5.21 Specifying an equiripple low-pass filter.

$$H(\Omega) = \sum_{k=-M}^{M} b_k \exp(-jk\Omega) = b_0 + 2 \sum_{k=1}^{M} b_k \cos k\Omega \tag{5.30}$$

where the coefficients b_k equal the terms of the filter's impulse response $h[n]$. Hence we may also write:

$$H(\Omega) = h[0] + 2 \sum_{k=1}^{M} h[k] \cos k\Omega \tag{5.31}$$

Now a term $\cos k\Omega$ can always be expressed as a sum of powers of $\cos \Omega$. Therefore equation (5.31) can be recast as:

$$H(\Omega) = \sum_{k=0}^{M} c_k (\cos \Omega)^k \tag{5.32}$$

where the coefficients c_k are related to the impulse response values. This shows that the frequency response $H(\Omega)$ of such a filter can be written as an Mth-order trigonometric polynomial. We conclude that it can display up to $(M-1)$ local extrema within the range $0 < \Omega < \pi$, corresponding to ripple peaks and troughs. Furthermore, differentiating equation (5.32) with respect to Ω we obtain:

$$H'(\Omega) = \frac{dH(\Omega)}{d\Omega} = -\sin\Omega \sum_{k=1}^{M} kc_k (\cos \Omega)^{k-1} \tag{5.33}$$

Since $\sin\Omega$ is zero when $\Omega = 0$ and π, there must be a maximum, or minimum, at both these frequencies. Hence there are, at most, $(M+1)$ extrema over the range $0 \le \Omega \le \pi$. In general the extrema are not divided equally between passband and stopband.

Not surprisingly the design parameters M, δ_1, δ_2, Ω_p and Ω_s interact, and cannot all be independently specified. Two main approaches were developed in the early 1970s:

Hermann and Schuessler specified the parameters M, δ_1, and δ_2, allowing Ω_p and Ω_s to vary. They showed that the equiripple behavior of Figure 5.21 could be expressed by a set of nonlinear equations. The difficulty of solving the equations for large values of M led Hofstetter, Oppenheim, and Siegel to develop an iterative algorithm for finding a trigonometric polynomial with the required properties.

Parks and McClellan chose to specify M, Ω_p, Ω_s, and the ripple *ratio* δ_1/δ_2, while allowing the actual value of δ_1 to vary. Their approach has the advantage that the transition bandwidth — an important feature of most practical designs — is properly controlled. The design problem was shown to reduce to a so-called Chebyshev approximation over disjoint sets.

The approach of Parks and McClellan is widely used, so we will say a few more words about it. Basically, they showed that equiripple approximations giving optimum error performance in passband and stopband must display either $(M+2)$ or $(M+3)$ *alternations* of the error function over the range $0 \le \Omega \le \pi$.

By 'alternations' we mean successive reversals of the peak error (between desired and actual responses), at the maximum permitted level.

Two typical examples are shown in Figure 5.22, for the case $M=7$ (a filter with $(2M+1)=15$ impulse response terms). The alternations are marked with dots. In part (a) of the figure there are $M+3=10$ alternations — the maximum number possible. Note that the alternations at Ω_p and Ω_s are *not* local extrema of $H(\Omega)$. The other eight alternations, one of which occurs at $\Omega=0$ and one at $\Omega=\pi$, are local extrema. They correspond to the $(M+1)$ possible extrema of $H(\Omega)$. This form of response was named *extraripple* by Parks and McClellan, because it has one more alternation than the $(M+2)$ demanded for optimum error performance. The resulting filter turns out to be the same as that derived using the technique of Hofstetter *et al.*

Part (b) of the figure illustrates another case, this time with just $(M+2)$ alternations. As usual, $H(\Omega)$ displays local extrema at $\Omega=0$ and $\Omega=\pi$ — but the former is not an alternation, because it does not reach the permitted ripple level. Two other cases with $(M+2)$ alternations, not illustrated here, are also possible. In one, the ripple peak at $\Omega=0$ is an alternation, but that at $\Omega=\pi$ is not; in the other case, the ripple peaks at $\Omega=0$ and π are both alternations, but there is one less intermediate ripple in the response.

We therefore see that there are several minor variations of Parks–McClellan filters, all of which offer optimum equiripple performance. Such filters give the sharpest passband–stopband transition for specified ripple levels and filter order. It is perhaps unlikely that you will ever have to design one in detail yourself; but you are quite likely to meet them as part of a standard DSP software package.

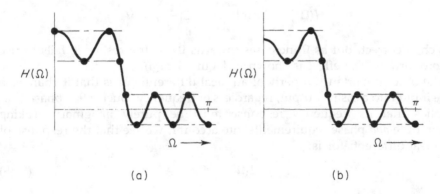

(a)   (b)

Figure 5.22 Parks–McClellan equiripple filters.

5.5 DIGITAL DIFFERENTIATORS

Differentiation of a signal gives a measure of its instantaneous rate-of-change, or slope. It is a fairly common signal processing requirement. To take a simple example, we might have a signal representing the position of a moving object or vehicle. By differentiating, we could assess its velocity; by differentiating again, its acceleration.

Various techniques, nonrecursive and recursive, have been devised for differentiating digital signals. In this section we concentrate on a valuable nonrecursive technique based on the Fourier Transform method and the use of window functions. Apart from its intrinsic value, the technique emphasizes that the desired frequency response of an LTI processor is not necessarily of the 'rectangular' form assumed in previous sections.

We should first make some remarks about the simplest possible approach to digital differentiation. An LTI system which forms the *first-order difference (FOD)* of an input signal:

$$\{x[n] - x[n-1]\} \tag{5.34}$$

may be thought of as a 'differentiator'. It gives a simple measure of slope. However, we can readily show that accurate differentiation is only achieved over the lower part of the frequency range $0 \le \Omega \le \pi$. The corresponding frequency response is:

$$H(\Omega) = 1 - \exp(-j\Omega) = 1 - \cos\Omega + j \sin\Omega \tag{5.35}$$

giving a magnitude function:

$$|H(\Omega)| = \{(1-\cos\Omega)^2 + \sin^2\Omega\}^{1/2} = 2 \sin\left(\frac{\Omega}{2}\right) \tag{5.36}$$

Now an ideal differentiator would have a magnitude response proportional to Ω. (This is because differentiation of a signal $\sin n\Omega$ with respect to n gives the signal $\Omega\cos n\Omega$.) For small values of Ω the FOD system gives:

$$|H(\Omega)| = 2 \sin\left(\frac{\Omega}{2}\right) \approx 2\left(\frac{\Omega}{2}\right) = \Omega \tag{5.37}$$

which is correct. But as Ω increases towards $\Omega = \pi$, the response fails to rise in proportion. The effect is shown in Figure 5.23(a).

The other important property of an ideal differentiator is that it changes a sine input into a cosine output, regardless of frequency. Such a 90° phase shift implies that its frequency response must be purely imaginary. Taking magnitude *and* phase requirements into account, we see that the response of an ideal differentiator is:

$$H(\Omega) = j\Omega \tag{5.38}$$

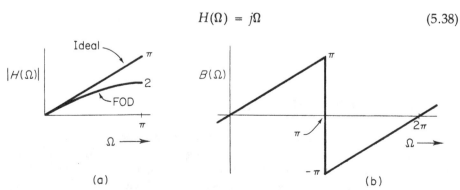

(a)   (b)

Figure 5.23 Frequency responses of digital differentiators.

How do we go about designing a digital processor with this form of frequency response? To answer the question, we need to consider the nature of the magnitude and phase responses of an LTI system. First, we should note that the frequency response $H(\Omega)$ corresponding to any real impulse response $h[n]$ can be written in the form:

$$H(\Omega) = A(\Omega) + jB(\Omega) \qquad (5.39)$$

Here $A(\Omega)$ is an *even* function of Ω representing cosines in $h[n]$, and $B(\Omega)$ is an *odd* function representing sines. So far in this chapter we have always assumed that a desired frequency response is an even, zero-phase, function. That is, $H(\Omega) = A(\Omega)$. The corresponding impulse response is then even, and symmetrical about $n=0$.

However it is not the only possible approach. In principle we can just as easily treat $H(\Omega)$ as an odd, purely imaginary, function $jB(\Omega)$. In the case of a digital differentiator this is the natural thing to do because, as we have shown, we need an imaginary frequency response. The corresponding impulse response will turn out to be odd — that is, antisymmetrical about $n=0$.

We may now use the Fourier Transform method to specify the coefficients of an ideal, nonrecursive, differentiating filter. Figure 5.23(b) shows the required form for $B(\Omega)$. It is odd; and, as with any digital LTI processor, it is periodic in Ω.

Example 5.2 Find the impulse response $h[n]$ corresponding to the frequency response $H(\Omega) = jB(\Omega)$, where $B(\Omega)$ is as shown in Figure 5.23(b). Sketch the form of $h[n]$ for a causal differentiating filter truncated to 21 terms.

Solution We have:

$$H(\Omega) = jB(\Omega) = j\Omega, \quad -\pi < \Omega < \pi$$

The inverse Fourier Transform (see equation (5.10)) is:

$$h[n] = \frac{1}{2\pi} \int_{2\pi} H(\Omega) \exp(j\Omega n) \, d\Omega$$

giving:

$$h[n] = \frac{1}{2\pi} \int_{-\pi}^{\pi} j\Omega \exp(j\Omega n) \, d\Omega \qquad (5.40)$$

Integrating by parts, we obtain:

$$h[n] = \frac{1}{2\pi} \left\{ \left[j\Omega \, \frac{\exp(j\Omega n)}{jn} \right]_{-\pi}^{\pi} - \int_{-\pi}^{\pi} \frac{\exp(j\Omega n)}{n} \, d\Omega \right\}$$

$$= \frac{1}{2\pi} \left[\exp(j\Omega n) \left\{ \frac{\Omega}{n} - \frac{1}{jn^2} \right\} \right]_{-\pi}^{\pi}$$

$$= \frac{1}{2\pi} \left\{ \exp(jn\pi) \left\{ \frac{\pi}{n} + \frac{j}{n^2} \right\} - \exp(-jn\pi) \left\{ \frac{-\pi}{n} + \frac{j}{n^2} \right\} \right\}$$

If n is odd, $\exp(jn\pi) = \exp(-jn\pi) = -1$. If n is even, $\exp(jn\pi) = \exp(-jn\pi) = 1$. We obtain:

$$h[n] = -\frac{1}{n}, \quad n = \pm 1, 3, 5 \ldots$$

and

$$h[n] = \frac{1}{n}, \quad n = \pm 2, 4, 6$$

The case $n=0$ is a little awkward because of the denominators n and n^2 in the above expressions. However if we put $n=0$ in equation (5.40), we readily find that $h[0] = 0$.

These results show that $h[n]$ is antisymmetrical about $n=0$, with its central value $h[0]$ equal to zero. Although the values decay to either side of $n=0$, in theory the 'tails' are infinitely long.

Figure 5.24 shows the impulse response truncated to 21 terms, and shifted to begin at $n=0$. (Strictly, perhaps we should say '20 terms', because the middle one is zero!) This causal version of the differentiator will give a best-fit approximation to the desired frequency response in the least-squares sense, and will introduce a pure delay of ten sampling intervals.

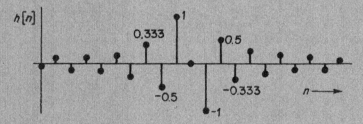

Figure 5.24

Unfortunately, truncation of the impulse response produces a familiar effect: the desired frequency response is degraded by ripples, particularly near any sharp discontinuity. We could compress the ripples into a smaller frequency range by increasing the number of impulse response coefficients. But the differentiator would be less economic, with a longer delay.

The best approach is to use a tapered window, just as in previous sections. Program no. 19 in Appendix A1 computes and plots the frequency response of a differentiator, truncated to $(2M+1)$ terms using either a rectangular or a Hamming window. Plots for $M=10$ are reproduced in Figure 5.25. The Hamming window gives a far superior performance over most of the frequency range, at the expense of a wider transition in the region of $\Omega = \pi$. Of course, the performance of either type of window improves if the value of M is increased.

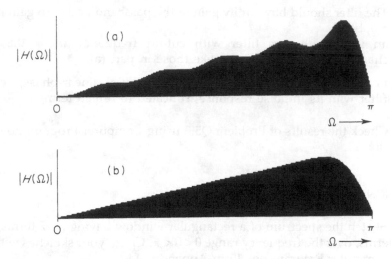

Figure 5.25 Frequency responses of two nonrecursive differentiators (*abscissa: 320 samples*) based on (a) a rectangular window, and (b) a hamming window

As we have said before, there are other possible approaches to digital differentiation. But the one we have developed here is valuable, and ties in well with the other material of this chapter.

PROBLEMS

SECTION 5.2

Q5.1 Predict the form of the frequency response of the following low-pass moving-average filters, and check your answers by running Computer Program no. 12 in Appendix A1:

- (a) 9-term impulse response, all terms equal to $\frac{1}{9}$
- (b) 19-term impulse response, all terms equal to $\frac{1}{19}$.

Also sketch the z-plane zero configurations of the two filters.

Q5.2 Modify Computer Program no. 12 in Appendix A1 to accept a 'cosine-modulated' impulse response, as described in the main text. Use the new version of the program:

- (a) to check the bandpass characteristic shown in Figure 5.6(b), and
- (b) to compute and plot the frequency response of a 21-term high-pass filter.

SECTION 5.3.1

Q5.3 Use the inverse Fourier Transform to find the impulse response $h[n]$ of:

- (a) an ideal, zero-phase, low-pass filter with cut-off frequency $\Omega_1 = 0.4\pi$.

The filter should have unity gain in the passband and zero gain in the stopband.

(b) an ideal high-pass filter with cut-off frequency $\Omega_1 = 0.8\pi$. Its characteristics are otherwise like those in part (a).

In each case, sketch the impulse response of a causal, linear-phase, version of the filter with its impulse response truncated to fifteen terms.

Q5.4 Check the results of Problem Q5.3 using Computer Program no. 13 in Appendix A1.

SECTION 5.3.2

Q5.5 Sketch the spectrum of a rectangular window having (a) 7 terms, and (b) 15 terms, over the frequency range $0 < \Omega < \pi$. Check your sketches with the aid of Computer Program no. 15 in Appendix A1.

Q5.6 Computer Program no. 15 in Appendix A1 gives a decibel plot of the spectrum of a rectangular window. Modify the program to plot the spectrum of a triangular window (see equation (5.20)). Hence check the form of Figure 5.14 in the main text.

SECTION 5.3.3

Q5.7 Tabulate the sample values of:

(a) a von Hann window with 11 terms,
(b) a Hamming window with 13 terms (you need only estimate 'one half' of each window).

Q5.8 Use Computer Program no. 16 in Appendix A1 to check the results of Problem Q5.7, and to plot the window spectra.

Q5.9 The Hamming window function is usually quoted as:

$$w[n] = 0.54 + 0.46 \cos\left(\frac{n\pi}{M}\right), \qquad -M \le n \le M$$

$$= 0 \text{ elsewhere.}$$

Using Computer Program no. 16, investigate the effects of small changes in the relative sizes of the window's DC and cosine components (nominally 0.54 and 0.46 respectively), on the sidelobe levels of a 25-term window. Keep the *sum* of the two components constant at 1.0.

Q5.10 Use Computer Program no. 16 to find the decibel levels of the first three sidelobes of:

(a) a von Hann window with 31 terms,
(b) a Hamming window with 17 terms.

Why are window sidelobe levels more or less independent of window length?

SECTION 5.3.4

Q5.11 A linear-phase low-pass nonrecursive filter is to be designed using the Kaiser window. A ripple level $\delta = 0.002$, and a transition width $\Delta = 0.1\pi$, are acceptable. Find the value of the Kaiser parameter α, and the required window length $(2M+1)$.

Q5.12 Repeat Q5.11 for a high-pass filter with ripple level $\delta = 0.005$ and transition width $\Delta = 0.15\pi$.

Q5.13 Using Computer Program no. 18 in Appendix A1, check the Kaiser filter plots shown in Figure 5.20 of the main text.

SECTION 5.5

Q5.14 The simplest LTI processor which approximates a digital differentiator has the difference equation:

$$y[n] = x[n] - x[n-1]$$

A less-widely used alternative is to estimate the *central difference*, using the equation:

$$y[n] = 0.5\{x[n] - x[n-2]\}$$

Sketch the magnitude responses of the two approximations on the same diagram, over the range $0 < \Omega < \pi$. Contrast their performance with that of an ideal differentiator. By how many dB is each response lower than that of the ideal differentiator at the frequency $\Omega = 0.2\pi$?

Q5.15 Figure 5.23(b) in the main text shows the frequency response of an ideal

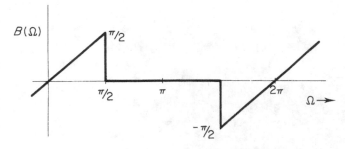

Figure Q5.15

differentiator. However in some practical situations it may be better to provide differentiation over a reduced frequency range. A good example is when a signal is contaminated by high-frequency 'noise', which tends to be accentuated by 'ideal' differentiation.

Use the inverse Fourier Transform to find the impulse response of the 'reduced-range' differentiator $H(\Omega) = jB(\Omega)$, where $B(\Omega)$ is as shown in Figure Q5.15.

Q5.16 Use Computer Program no. 19 in Appendix A1, suitably modified, to check the frequency response of the 'reduced-range' differentiator designed in Q5.15. Try truncating the impulse response to 51 terms with a Hamming window.

CHAPTER 6
Design of Recursive Digital Filters

6.1 INTRODUCTION

The output from a recursive digital filter depends on one or more previous output values, as well as on inputs. In other words it involves *feedback*. From the DSP point of view, its great advantage is computational economy. A filter characteristic requiring (say) 100 or more coefficients in a nonrecursive realization can often be obtained using just a few recursive coefficients. However, there are two potential disadvantages. First, a recursive filter may become unstable if its feedback coefficients are chosen badly. Secondly, recursive designs cannot generally provide the linear-phase responses so readily achieved by nonrecursive methods.

In most cases a recursive filter has an *infinite impulse response (IIR)*. Although the impulse response $h[n]$ decays towards zero as $n \rightarrow \infty$, it theoretically continues forever. Assuming the filter is causal ($h[n] = 0$ for $n < 0$) this means that the impulse response cannot be symmetrical in form. Therefore the filter cannot display a pure linear-phase characteristic.

In the previous chapter we noted that a nonrecursive digital filter possesses only z-plane zeros (apart from any poles at the origin). By contrast, a recursive filter has one or more strategically placed z-plane poles. In general we may write its transfer function and difference equation as:

$$H(z) = \frac{\displaystyle\sum_{k=0}^{M} b_k z^{-k}}{\displaystyle\sum_{k=0}^{N} a_k z^{-k}} \tag{6.1}$$

and:

$$a_0 y[n] + a_1 y[n-1] + \ldots a_N y[n-N] = b_0 x[n] + b_1 x[n-1] + \ldots b_M x[n-M] \tag{6.2}$$

where $N > 0$ and $M \geq 0$. Factorizing the numerator and denominator polynomials of equation (6.1), we obtain the pole-zero description of the filter (see also equation (4.21)):

$$H(z) = \frac{K(z-z_1)(z-z_2)(z-z_3) \ldots}{(z-p_1)(z-p_2)(z-p_3) \ldots} \tag{6.3}$$

The art of designing recursive filters is to approximate a desired performance — usually specified in terms of a frequency response characteristic — with as few poles and zeros as possible. Many techniques have been devised for doing this, and in this chapter we introduce a few of the best known ones.

The frequency response corresponding to equation (6.3) is:

$$H(\Omega) \;=\; \frac{K(\exp(j\Omega) - z_1)(\exp(j\Omega) - z_2)(\exp(j\Omega) - z_3) \;\cdots}{(\exp(j\Omega) - p_1)(\exp(j\Omega) - p_2)(\exp(j\Omega) - p_3) \;\cdots} \tag{6.4}$$

This may look complicated, but it can readily be evaluated by computer. Therefore if we specify the poles and zeros of a recursive filter we can estimate and plot its frequency response characteristic. The basic reason why recursive techniques are so powerful is that we have separate control over the numerator and denominator of $H(\Omega)$. In particular, we can produce sharp response peaks by arranging that the magnitude of the denominator becomes small at the appropriate frequencies.

We should note in passing that the Fourier Transform method used for nonrecursive filter design in the previous chapter is not helpful here. Although it gives the impulse response corresponding to a desired frequency response, we cannot in general realize the impulse response with a recursive algorithm. Other approaches are needed, and we shall find our previous work on the z-transform extremely useful in this chapter.

6.2   SIMPLE DESIGNS BASED ON z-PLANE POLES AND ZEROS

One possible approach to recursive filter design is to choose the z-plane poles and zeros intuitively. We can then find the filter's difference equation, and compute its frequency response. The method is rather 'hit-or-miss', and only suitable for undemanding applications. Nevertheless some useful filters can be rapidly designed in this way.

The approach relies upon ideas discussed in Section 4.3.2. We can visualize the frequency response of an LTI processor by drawing vectors from the various poles and zeros to points on the unit circle. A pole close to the unit circle gives rise to a well-defined response peak; a zero close to (or on) the unit circle produces a trough (or null).

In Section 4.3.3 we developed these ideas in relation to first and second-order systems. In particular, we showed how placing a single pole on the real axis in the z-plane can give an elementary low-pass or high-pass filter; and how a complex conjugate pole-pair can give a bandpass filter. A series of impulse and frequency response plots for second-order systems were illustrated in Figure 4.11. If you are unfamiliar with this work, we suggest you review it before proceeding.

Our aim in this section is to show how z-plane poles and zeros can be positioned to give a variety of simple, but useful, recursive filters. Checking the designs will be much easier if we have a computer program which accepts pole-zero locations as input data, and plots the frequency response on the screen.

Referring back to equation (6.4) we see that a frequency response $H(\Omega)$ can be expressed as the product of numerator factors of the form $\{\exp(j\Omega) - z_n\}$, divided by the product of denominator factors of the form $\{\exp(j\Omega) - p_n\}$. (We will ignore the gain factor K at present, since it only affects the scale of $H(\Omega)$ — not the shape.) Therefore we can build up an overall response by assessing the contributions of individual poles or pole-pairs, and zeros or zero-pairs, in turn. This is equivalent to synthesizing the system as a series of cascaded first and second-order subsystems. Such a realization is often referred to as the *cascade canonic form* in the DSP literature.

In most practical situations we are more interested in the magnitude of $H(\Omega)$ than the phase. So we will aim to produce a decibel plot of $|H(\Omega)|$ against frequency — just like Programs 15–18 in Chapter 5.

Suppose we specify a real pole at $z = \alpha$. It contributes the following factor to the *denominator* of $H(\Omega)$:

$$F_1(\Omega) = \{\exp(j\Omega) - \alpha\} = (\cos\Omega - \alpha) + j\sin\Omega \qquad (6.5)$$

Its magnitude contribution is therefore:

$$|F_1(\Omega)| = \{(\cos\Omega - \alpha)^2 + \sin^2\Omega\}^{1/2} = \{1 - 2\alpha\cos\Omega + \alpha^2\}^{1/2} \qquad (6.6)$$

A real zero at $z = \alpha$ would give an identical contribution, but to the *numerator* of $|H(\Omega)|$. If we specify an nth-order pole, or zero, the contribution is simply raised to power n.

Let us now consider complex-conjugate pole-pairs, or zero-pairs. A pole-pair with polar coordinatets $(r, \pm\theta)$ makes a contribution to the denominator of $H(\Omega)$:

$$\begin{aligned} F_2(\Omega) &= \{\exp(j\Omega) - r\exp(j\theta)\}\{\exp(j\Omega) - r\exp(-j\theta)\} \\ &= \exp(2j\Omega) - 2r\cos\theta\,\exp(j\Omega) + r^2 \\ &= (\cos 2\Omega - 2r\cos\theta\cos\Omega + r^2) + j(\sin 2\Omega - 2r\cos\theta\sin\Omega) \end{aligned} \qquad (6.7)$$

The magnitude is:

$$|F_2(\Omega)| = \{(\cos 2\Omega - 2r\cos\theta\cos\Omega + r^2)^2 + (\sin 2\Omega - 2r\cos\theta\sin\Omega)^2\}^{1/2} \qquad (6.8)$$

A complex conjugate zero-pair would make a similar contribution, but to the numerator.

Program no. 20 in Appendix A1 uses equations (6.6) and (6.8) to compute the magnitude response of any filter specified as a set of poles and zeros. It starts by requesting information on any real poles or zeros; then on any complex-conjugate pairs. The computation proceeds by assessing magnitude contributions one-by-one, modifying the value of $|H(\Omega)|$ accordingly. The process is repeated for 320 values of frequency in the range $0 < \Omega < \pi$. The peak gain value corresponding to $K = 1$ in equation (6.4) is printed out. After normalization by this value, a decibel plot of $|H(\Omega)|$ is produced.

Let us start with some representative examples which illustrate the effects of poles or zeros on their own. (We ignore poles or zeros at the origin which may be needed to produce a causal, or minimum delay, filter. They do not affect the spectral magnitude function). Figure 6.1(a) shows the plot produced by a single real pole placed at $z = 0.9$; in part (b) we see the effect of a real, second-order, zero placed at $z = -0.8$. Part (c) of the figure refers to a complex pole-pair at radius $r = 0.975$ and angles $\theta = \pm 150°$ in the z-plane (similar plots were drawn to linear scales in Figure 4.11). And part (d) shows the response of a complex zero-pair placed on the unit circle at $\theta = \pm 50°$. The general form of these plots is as expected, based on our work on pole and zero vectors in Chapter 4.

Such characteristics may be useful as they stand for simple digital filtering tasks. In other cases we need to build up a response by specifying a mixture of poles and zeros, both real and complex. The overall response is simply the product of individual responses; or, plotted to logarithmic scales, their decibel values are additive.

We can demonstrate the last point by specifying a single filter with *all* the poles and zeros used in parts (a) to (d) of Figure 6.1. That is, we use the following input data for Computer Program no. 20:

> a real pole at $z = 0.9$, of order 1
> a real zero at $z = -0.8$, of order 2
> a complex pole-pair with $r = 0.975$, $\theta = 150°$
> a complex zero-pair with $r = 1$, $\theta = 50°$

The pole-zero configuration is shown in Figure 6.2(a); the screen plot of $|H(\Omega)|$ is reproduced in part (b). The four responses of Figure 6.1 have been combined additively on a logarithmic scale, with their principal features preserved.

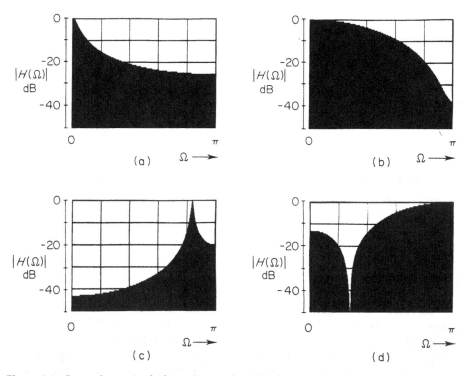

Figure 6.1 Spectral magnitude functions produced by (a) a single real pole at $z = 0.9$; (b) a second-order zero at $z = -0.8$; (c) a complex conjugate pole pair at $r = 0.975$, $\theta = \pm 150°$; and (d) a complex conjugate zero pair on the unit circle at $\theta = \pm 50°$ *(each abscissa: 320 samples)*.

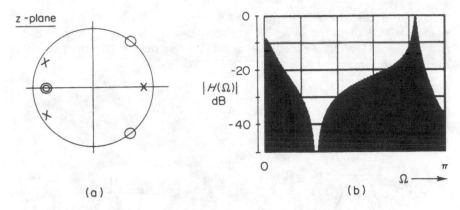

Figure 6.2 (a) A pole–zero configuration, and (b) the equivalent spectral magnitude function, normalized to a peak value of 0 dB (*abscissa: 320 samples*).

Example 6.1 Design a recursive digital bandpass filter with the following characteristics:

 (a) A passband centered at $\Omega = \pi/2$, with a bandwidth of $\pi/40$ between −3 dB points, and a peak gain of unity.
 (b) Steady-state rejection of components at $\Omega = 0$ and $\Omega = \pi$.

Use Program no. 20 in Appendix A1 to produce a plot of the magnitude characteristic, and specify the filter's difference equation.

Solution To meet part (a) of the specification we place a complex-conjugate pole pair at $\theta = \pm 90°$. Its radius r must be chosen to give the desired bandwidth. The bandwidth requirement is illustrated by Figure 6.3(a), which shows a small portion of the magnitude characteristic $|H(\Omega)|$. Note first that the '−3dB points' are where the response falls to $1/\sqrt{2}$ of its peak value $(20 \log_{10} (1/\sqrt{2}) = -3dB)$. The interval between −3 dB points is widely used by filter designers as a convenient measure of bandwidth. In this design it should equal $\pi/40$ radians, or 4.5°.

Part (b) of the figure explains the relationship between the bandwidth and the pole radius r. It shows a small portion of the unit circle in the region of one of the poles, to a greatly expanded scale. If we consider a vector drawn from the pole to a succession of points on the unit circle, we see that its minimum length must be $(1-r)$. This corresponds to the response peak. Points B and C give a pole vector $\sqrt{2}$ times as long, and therefore represent the −3 dB points. If we assume that the unit circle approximates a straight line in this region, then PAB is a right-angle triangle and the distance d also equals $(1-r)$. The assumption is reasonable if the pole is close to the unit circle (say $r > 0.9$).

The other assumption we need to make is that the response peak is entirely due to the pole — not its complex conjugate, nor any other poles or zeros. Again, this is near to the truth if the pole is more-or-less 'on its own', and close to the unit circle.

Accepting these assumptions, we see that the -3 dB bandwidth corresponds to a distance $2d = 2(1-r)$ around the unit circle, and hence to a change in Ω of $2(1-r)$ radians, or $360(1-r)/\pi$ degrees. In the present example we have:

$$2(1-r) = \pi/40, \text{ giving } r = 0.961$$

We will use this value in our design.

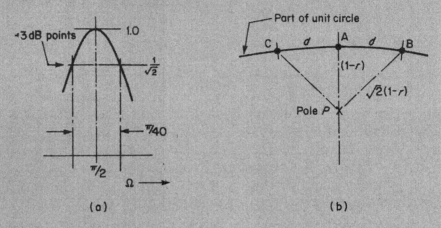

Figure 6.3 (a) Measuring the -3 dB bandwidth; (b) relationship between bandwidth and the radius of a z-plane pole.

The specification also calls for rejection of components at $\Omega = 0$ and $\Omega = \pi$. We therefore place one zero on the unit circle at $z=1$, and another at $z = -1$. The complete pole-zero configuration is shown in Figure 6.4(a).

Given these pole and zero locations, Program no. 20 produces a plot of the form shown in part (b) of the figure. It confirms the -3 dB bandwidth of about $\pi/40$. The program also estimates the maximum gain of the filter as 26.15 (or 28.35 dB). Since we require a maximum gain of unity, the constant K in equation (6.3) must be set to $(26.15)^{-1}$, or 0.03824. The filter's transfer function becomes:

$$H(z) = \frac{Y(z)}{X(z)} = \frac{0.03824(z-1)(z+1)}{\left\{z - 0.961 \exp\left(j\frac{\pi}{2}\right)\right\}\left\{z - 0.961 \exp\left(-j\frac{\pi}{2}\right)\right\}}$$

$$= \frac{0.03824\,(z^2 - 1)}{z^2 + 0.9235}$$

The corresponding difference equation is:

$$y[n+2] + 0.9235\,y[n] = 0.03824\{x[n+2] - x[n]\}$$

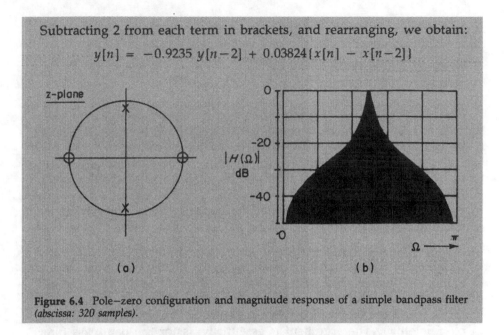

Subtracting 2 from each term in brackets, and rearranging, we obtain:

$$y[n] = -0.9235\, y[n-2] + 0.03824\{x[n] - x[n-2]\}$$

(a)   (b)

Figure 6.4 Pole–zero configuration and magnitude response of a simple bandpass filter (*abscissa: 320 samples*).

We have already admitted that this approach to design is rather 'hit-or-miss', and only suitable when the desired performance is not tightly specified. It relies largely on the designer's intuition, and is not optimized in any clear sense. Nevertheless, there is at least one other type of recursive filter which can readily be specified in this way.

Suppose we require a simple bandstop filter for rejecting a narrow band of unwanted frequencies. The obvious approach is to place a pair of complex-conjugate zeros on the unit circle at the appropriate points. However, this does not give a sharp rejection 'notch', and components to either side of the center-frequency are substantially affected (see Figure 6.1(b)). A much better scheme is to use the pole-zero configuration of Figure 6.5(a). Over most of the frequency range $0 < \Omega < \pi$, adjacent pole and zero vectors drawn to the unit circle are

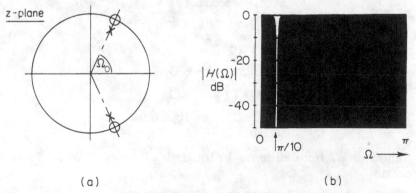

(a)   (b)

Figure 6.5 (a) Poles and zeros of a 'notch' filter; (b) response of a notch design for rejecting mains-supply interference from an EKG signal (*abscissa: 320 samples*).

almost identical in length, and the response is close to unity. However, in the immediate vicinity of $\Omega = \Omega_0$, the zero vector becomes much shorter than the adjacent pole vector, producing a narrow notch.

Example 6.2   An EKG (electrocardiogram) signal, representing the electrical activity of the heart, is sampled at 1.2 kHz. It has been contaminated by mains-supply interference at 60 Hz during recording.

Design a digital bandstop filter for rejecting the unwanted interference, based on the pole-zero configuration of Figure 6.5(a). The rejection notch should be 10 Hz wide at the -3 dB points. Specify the filter's difference equation, and use Computer Program no. 20 to plot its frequency response magnitude characteristic.

Solution   We first note that the sampling rate of 1.2 kHz gives 'adequate' sampling of analog signal components up to 600 Hz, according to the Sampling Theorem. 600 Hz corresponds to $\Omega = \pi$. Therefore the required center-frequency of 60 Hz corresponds to $\Omega_0 = 0.1\pi$.

The width of the rejection notch depends on the radius r of the complex pole-pair. The argument is essentially as for the bandpass filter illustrated in Figure 6.4. As we approach $\Omega = \Omega_0$, the zero vector reduces more rapidly than the adjacent pole vector. At a distance $(1-r)$ along the unit circle from the center-frequency, the ratio of their lengths is approximately $1/\sqrt{2}$. Hence the -3 dB points are about $2(1-r)$ radians apart. In this case π radians corresponds to 600 Hz. For a notch 10 Hz wide we therefore require that:

$$\frac{2(1-r)}{\pi} = \frac{10}{600}, \text{ giving } r = 1 - \frac{\pi}{120} = 0.97382$$

We can now specify the filter's transfer function as:

$$H(z) = \frac{Y(z)}{X(z)} = \frac{\{z - \exp(j0.1\pi)\}\{z - \exp(-j0.1\pi)\}}{\{z - 0.97382\exp(j0.1\pi)\}\{z - 0.97382\exp(-j0.1\pi)\}}$$

$$= \frac{z^2 - 2z\cos(0.1\pi) + 1}{z^2 - 1.9476\cos(0.1\pi)z + 0.94833}$$

$$= \frac{z^2 - 1.9021z + 1}{z^2 - 1.8523z + 0.94833}$$

The corresponding difference equation is:

$$y[n+2] - 1.8523\,y[n+1] + 0.94833\,y[n]$$
$$= x[n+2] - 1.9021\,x[n+1] + x[n]$$

Subtracting 2 from all terms in square brackets, and rearranging, we obtain:

$$y[n] = 1.8523\,y[n-1] - 0.94833\,y[n-2] +$$
$$x[n] - 1.9021\,x[n-1] + x[n-2]$$

The frequency response characteristic, produced by Computer Program no. 20 in Appendix A1, is shown in Figure 6.5(b).

We have met this particular filter, and difference equation, before. Its action on a typical EKG waveform, contaminated by supply frequency interference, was illustrated in Figure 1.4. We also showed the transients it produces, in response to a switched sinusoid at its centre-frequency, in Figure 2.14. The frequency response was plotted to linear scales in Figure 3.13. And you now know how to design it!

6.3 FILTERS DERIVED FROM ANALOG DESIGNS

The history of analog filter design goes back more than sixty years. It was therefore natural that, in the comparatively recent development of DSP, designers should look for ways of converting successful analog filters into digital equivalents.

Our approach in this book is essentially digital, and we do not wish to discuss analog filters in any detail. Instead, we will try to summarize their links with recursive digital filters, setting the scene with some general remarks. A more rigorous treatment can be found in several of the references listed in the bibliography.

If you have studied analog signal and systems theory, you will know that the *Laplace transform* plays a similar role to that of the z-transform in the digital case. For example, an analog LTI filter can always be described by a frequency-domain transfer function of the general form:

$$H(s) = \frac{K(s-z_1)(s-z_2)(s-z_3) \cdots}{(s-p_1)(s-p_2)(s-p_3) \cdots} \tag{6.9}$$

where s is the *Laplace variable* and K is a constant, or gain, factor. Apart from this factor, the filter is characterized by its poles (p_1, p_2, p_3 ...) and zeros (z_1, z_2, z_3 ...), which can be plotted in the complex *s-plane*.

Although the form of equation (6.9) is identical to that describing the transfer function $H(z)$ of a digital processor (see equation (6.3)), the variable s is *not* the same as the variable z. For example we know that the frequency response of a digital processor can be found by making the substitution $z \rightarrow \exp(j\Omega)$; but the equivalent substitution in the analog case is $s \rightarrow j\omega$, where ω is the angular frequency in radians per second. It follows that the imaginary axis in the s-plane ($s = j\omega$) corresponds to the unit circle in the z-plane, and that the interpretation of pole/zero locations is different in the two cases. Another essential difference is that the frequency response of an analog filter is not a periodic function. Any conversion from an analog design into a digital one must clearly take these factors into account.

To summarize, we need to be able to convert a transfer function $H(s)$ into a transfer function $H(z)$, so that the frequency response of the digital filter over the range $0 < \Omega < \pi$ approximates, in an acceptable manner, that of the analog filter over the range $0 < \omega < \infty$. In the following sections we look at two methods of achieving this.

6.3.1   THE BILINEAR TRANSFORMATION: BUTTERWORTH AND CHEBYSHEV FILTERS

One of the most effective ways of converting an analog filter into a digital equivalent is by means of a *bilinear transformation*. In this section we apply the method to *Butterworth* and *Chebyshev* filters, which are among the best known of all analog filter families. Our approach has three main parts:

A summary of the characteristics of analog Butterworth and Chebyshev filters.

An introduction to the bilinear transformation.

Use of a computer program to specify the z-plane poles and zeros of digital Butterworth and Chebyshev filters with low-pass, high-pass, or bandpass characteristics.

Butterworth and Chebyshev filters offer two different ways of approximating an ideal, 'rectangular', response characteristic. We illustrate this for an analog low-pass filter in Figure 6.6. The ideal response has unity transmission in the passband, zero transmission in the stopband. A Butterworth filter approximates it with a so-called *maximally-flat* passband characteristic: the gain falls off gradually towards the passband edge, passing through $1/\sqrt{2}\,(-3\text{ dB})$ at the nominal cut-off frequency ω_1. If the order (and hence complexity) of the filter is increased, its passband and stopband performances improve, and the transition from passband to stopband becomes sharper.

The Chebyshev approximation gives an *equiripple* performance in the passband. The response oscillates between 1.0 and $(1+\epsilon^2)^{-1/2}$, where ϵ is a ripple parameter controlled by the designer. The number of passband ripples increases with the order of the filter. For a given order, the stopband performance is superior to the Butterworth design, with a sharper transition. Therefore a Chebyshev filter is generally preferred when some passband ripple is acceptable. The greater the ripple, the better the stopband performance becomes.

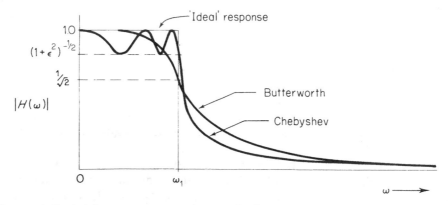

Figure 6.6 Typical frequency response (magnitude) functions of Butterworth and Chebyshev analog low-pass filters.

The magnitude functions of these two important types of analog filter are given by:

$$|H(\omega)| = \frac{1}{\left\{1 + \left(\dfrac{\omega}{\omega_1}\right)^{2n}\right\}^{1/2}} \qquad \text{(Butterworth)} \qquad (6.10)$$

and:

$$|H(\omega)| = \frac{1}{\left\{1 + \epsilon^2 C_n^2\left(\dfrac{\omega}{\omega_1}\right)\right\}^{1/2}} \qquad \text{(Chebyshev)} \qquad (6.11)$$

where n is the filter order and ω_1 is the nominal cut-off frequency. C_n is the so-called Chebyshev polynomial of nth order. It oscillates between 0 and 1 in the passband (for any value of $n > 0$), rising to large values in the stopband. The amount of passband ripple δ is related to the parameter ϵ by the expression:

$$\delta = 1 - (1 + \epsilon^2)^{-1/2} \qquad (6.12)$$

The zero-order and first-order Chebyshev polynomials are:

$$C_0(x) = 1 \quad \text{and} \quad C_1(x) = x \qquad (6.13)$$

Second and higher-order polynomials may be successively generated using the recursive relationship:

$$C_n(x) = 2xC_{n-1}(x) - C_{n-2}(x) \qquad (6.14)$$

It is possible to derive equivalent high-pass and bandpass filters from a low-pass *prototype*. The techniques are part of the stock-in-trade of the analog filter designer. We shall see later that such flexibility is also possible in the case of digital filters.

Although the magnitude characteristics of Butterworth and Chebyshev filters are good, their phase responses are less impressive. They depart considerably from the ideal linear-phase characteristic, especially towards the cut-off point. Also, the phase response of a Chebyshev filter displays unwanted ripples. In this respect Butterworth and Chebyshev filters are inferior to the nonrecursive designs described in Chapter 5.

We now introduce the bilinear transformation, showing how it can be used to design equivalent digital filters. Let us first consider the function:

$$F(z) = \frac{z - 1}{z + 1} \qquad (6.15)$$

which is 'bilinear' in the sense that its numerator and denominator are both linear in z. This is not the only possible bilinear function, but it is the most widely used. To explain its value in the present context, we need to find its spectrum. Thus:

$$F(\Omega) = \frac{\exp(j\Omega) - 1}{\exp(j\Omega) + 1}$$

$$= \frac{\exp(j\Omega/2)\{\exp(j\Omega/2) - \exp(-j\Omega/2)\}}{\exp(j\Omega/2)\{\exp(j\Omega/2) + \exp(-j\Omega/2)\}}$$

$$= \frac{2j\,\sin(\Omega/2)}{2\,\cos(\Omega/2)} = j\,\tan\left(\frac{\Omega}{2}\right) \tag{6.16}$$

$F(\Omega)$ is purely imaginary, and periodic. Its magnitude varies between 0 and ∞ as Ω varies between 0 and π.

Next, suppose that we know the transfer function of a 'desirable' analog filter, expressed in the general form given by equation (6.9). As already noted, its frequency response is found by substituting $j\omega$ for s:

$$H(\omega) = H(s)\Big|_{s=j\omega} = \frac{K(j\omega - z_1)(j\omega - z_2)(j\omega - z_3)\,\cdots}{(j\omega - p_1)(j\omega - p_2)(j\omega - p_3)\,\cdots} \tag{6.17}$$

The complete response is clearly generated as ω varies from 0 to ∞. If we substitute $F(\Omega) = j\tan(\Omega/2)$ for $j\omega$, exactly the same values must be produced as Ω varies between 0 and π. In other words we obtain a function $H(\Omega)$ in which the complete frequency response of the analog filter is *compressed* into the range $0 < \Omega < \pi$.

To summarize, the bilinear transformation gives a digital filter whose response over the range $0 < \Omega < \pi$ reproduces that of the analog filter over the range $0 < \omega < \infty$. However the resulting compression of the frequency scale is nonlinear. The shape of the tan function means that the compression, or 'warping', effect is very small near $\Omega = 0$; but it increases dramatically as we approach $\Omega = \pi$. Actually, as DSP designers we do not need to concern ourselves unduly with this effect. Butterworth and Chebyshev filters designed by this method are valid designs in their own right, and we can define their properties without constantly referring back to the analog prototypes.

How do we derive the transfer function $H(z)$ of such a filter? The foregoing discussion shows that replacing $j\omega$ by $F(\Omega)$ in equation (6.17) is equivalent to substituting $F(z)$ for s in equation (6.9). Therefore if we know, or can find, the transfer function $H(s)$ of a 'desirable' analog filter — Butterworth, Chebyshev, or otherwise — we can readily generate the transfer function $H(z)$ of its digital counterpart. Fortunately, in the case of well-known filters such as the Butterworth and Chebyshev families the work has already been done. We can use known formulae to specify $H(z)$, or the z-plane poles and zeros, directly.

There are several reasons why the bilinear transformation is so valuable for deriving digital Butterworth and Chebyshev filters:

The 'maximally flat', or 'equiripple', amplitude properties of the filters are preserved when the frequency axis is compressed.

There is no aliasing of the analog frequency response. Thus the response of a low-pass filter falls to zero at $\Omega = \pi$. This is a useful feature in many practical applications.

The method yields a recursive filter which is computationally efficient.

It is now time to define the properties of these filters more closely. As we have seen, the bilinear transformation effectively replaces $j\omega$ by $j\tan(\Omega/2)$ in a frequency response expression. It follows from equations (6.10) and (6.11) that the magnitude responses of low-pass Butterworth and Chebyshev digital filters are given by:

$$|H(\Omega)| = \frac{1}{\left\{1 + \left[\dfrac{\tan(\Omega/2)}{\tan(\Omega_1/2)}\right]^{2n}\right\}^{1/2}} \qquad \text{(Butterworth)} \qquad (6.18)$$

and:

$$|H(\Omega)| = \frac{1}{\left\{1 + \epsilon^2 C_n^2\left[\dfrac{\tan(\Omega/2)}{\tan(\Omega_1/2)}\right]\right\}^{1/2}} \qquad \text{(Chebyshev)} \qquad (6.19)$$

For completeness we also give formulae for the z-plane pole-zero locations. You need not concern yourself with their details, because we have incorporated them in a computer program which we will describe, and use, a little later.

A Butterworth low-pass digital filter of nth order has n poles arranged on a circular locus in the z-plane, and an nth-order real zero at $z = -1$. The poles are given by the values of P_m falling inside the unit circle, where the real and imaginary parts of P_m are respectively:

$$PR_m = \left\{1 - \tan^2\left(\frac{\Omega_1}{2}\right)\right\}/d$$

$$PI_m = 2\tan\left(\frac{\Omega_1}{2}\right)\sin\left(\frac{m\pi}{n}\right)/d, \qquad m = 0, 1, \ldots (2n-1) \quad (6.20)$$

and:

$$d = 1 - 2\tan\left(\frac{\Omega_1}{2}\right)\cos\left(\frac{m\pi}{n}\right) + \tan^2\left(\frac{\Omega_1}{2}\right) \qquad (6.21)$$

If n is even, the terms $(m\pi/n)$ are replaced by $(2m+1)\pi/2n$.

In the case of a Chebyshev low-pass filter, the corresponding formulae are:

$$PR_m = 2\left\{1 - a\tan\left(\frac{\Omega_1}{2}\right)\cos\phi\right\}/d - 1 \qquad (6.22)$$

$$PI_m = 2b\tan\left(\frac{\Omega_1}{2}\right)\sin\phi \qquad (6.23)$$

$$d = \left\{1 - a\tan\left(\frac{\Omega_1}{2}\right)\cos\phi\right\}^2 + b^2\tan^2\left(\frac{\Omega_1}{2}\right)\sin^2\phi \qquad (6.24)$$

where $\phi = m\pi/n$ and $m = 0, 1, \ldots (2n-1)$. The parameters a and b are given by:

$$a = 0.5(c^{1/n} - c^{-1/n}); \quad b = 0.5(c^{1/n} + c^{-1/n})$$

where:

$$c = (1 + \epsilon^{-1} + \epsilon^{-2})^{1/2} \tag{6.25}$$

If n is even, $\phi = m\pi/n$ is again replaced by $(2m+1)\pi/2n$. The poles of a Chebyshev filter do not lie on a circular locus; its shape is known as a 'cardioid'. In addition to the poles, there is an nth-order zero at $z = -1$.

Computer Program no. 21 in Appendix A1 is based on these formulae. It calculates and prints out the pole/zero locations of any desired Butterworth or Chebyshev low-pass filter. (It also copes with high-pass and bandpass designs — a matter to be discussed later.) Knowing the poles and zeros, we can specify the filter's transfer function and difference equation. We may also use Computer Program no. 20 to produce a decibel plot of its frequency response characteristic.

Example 6.3 A Butterworth low-pass filter is required with a cut-off frequency $\Omega_1 = 0.2\pi$. Its response should be at least '30 dB down' at $\Omega = 0.4\pi$.

  (a) Estimate the minimum order of filter required.
  (b) Use Computer Program no. 21 to find the z-plane poles and zeros, and sketch them.
  (c) Derive the filter's difference equation.
  (d) Use Program no. 20 to produce a decibel plot of its frequency response.

Solution

  (a) For a cut-off frequency of 0.2π, equation (6.18) gives the response magnitude at $\Omega = 0.4\pi$ as:

$$|H(0.4\pi)| = \cfrac{1}{\left\{1 + \left[\cfrac{\tan 0.2\pi}{\tan 0.1\pi}\right]^{2n}\right\}^{1/2}} = \cfrac{1}{\{1 + 2.236^{2n}\}^{1/2}}$$

Now -30 dB corresponds to a response ratio of $\log_{10}^{-1}\left(\cfrac{-30}{20}\right)$ $= 0.03162$. Hence we require that:

$$\cfrac{1}{\{1 + 2.236^{2n}\}^{1/2}} \leq 0.03162$$

giving:

$$1 + 2.236^{2n} \geq 1000, \quad \text{or} \quad n \geq 4.29$$

Since the filter order must be an integer, we choose $n = 5$.

(b) By running Program no. 21, and selecting its Butterworth and low-pass options, we find that a 5th-order filter with a cut-off at 36° (0.2π) has a real zero, of order 5, at $z = -1$. The pole locations, given in terms of radius(r) and angle(θ) are:

r	$\theta°$
0.50953	0
0.83221	34.644
0.59619	23.125

The first of these is interpreted as a first-order real pole at $z = 0.50953$. The others denote complex-conjugate pairs at r, $\pm \theta$. The filter's pole-zero configuration is sketched in Figure 6.7(a). The five poles lie on a circular locus in the z-plane. The locus intersects the unit circle at points corresponding to the cut-off frequency Ω_1.

(c) Knowing the poles and zeros, we could multiply out the factors of $H(z)$ to obtain a numerator and denominator polynomial, and hence derive the filter's difference equation. However this would involve a lot of multiplication of coefficients (a problem which worsens as the filter order increases). A convenient alternative is to treat the overall filter as a cascaded set of first and second-order subfilters, as shown in Figure 6.7(b). The first-order subfilter has a single real pole and zero; each second-order subfilter comprises a complex pole-pair and a second-order zero. In this way the total complement of five poles and a 5th-order zero is built up (of course, if we design an even-order filter, we only need second-order sub-filters). Note that intermediate outputs are labeled $v[n]$ and $w[n]$.

The transfer function of the first-order subfilter takes the form:

$$\frac{V(z)}{X(z)} = \frac{(z+1)}{(z-\alpha)}$$

giving the difference equation:

$$v[n] = \alpha v[n-1] + x[n] + x[n-1]$$

Each second-order subfilter has a transfer function of the form:

$$\frac{W(z)}{V(z)} = \frac{(z+1)^2}{\{z-r\exp(j\theta)\}\{z-r\exp(-j\theta)\}}$$

$$= \frac{z^2 + 2z + 1}{z^2 - 2r\cos\theta\, z + r^2}$$

yielding a difference equation:

$$w[n] = 2r\cos\theta\, w[n-1] - r^2 w[n-2] + x[n] + 2x[n-1] + x[n-2]$$

Inserting the pole values found in part (b), we obtain the following set of difference equations:

$$v[n] = 0.50953v[n-1] + x[n] + x[n-1]$$
$$w[n] = 1.3693w[n-1] - 0.69257w[n-2]$$
$$+ v[n] + 2v[n-1] + v[n-2]$$
$$y[n] = 1.0966y[n-1] - 0.35544\,y[n-2]$$
$$+ w[n] + 2w[n-1] + w[n-2]$$

The three equations are used *together*, the output of one feeding the input of the next. As already noted, an alternative would be to derive a single high-order difference equation involving just x and y.

(d) Computer Program no. 20 is fed with the following data:

No. of separate real poles:	1
value, order, of pole:	0.50953, 1
No. of separate real zeros:	1
value, order, of zero:	$-1, 5$
No. of complex pole-pairs:	2
radius, angle, of each:	0.83221, 34.644
	0.59619, 23.125
No. of complex zero-pairs:	0

The program produces the screen plot shown in Figure 6.7(c). Note that the -3 dB point occurs at $\Omega = 0.2\pi$, and that the response is more than 30 dB down at $\Omega = 0.4\pi$, as required.

The program also prints out the maximum gain of the filter. In

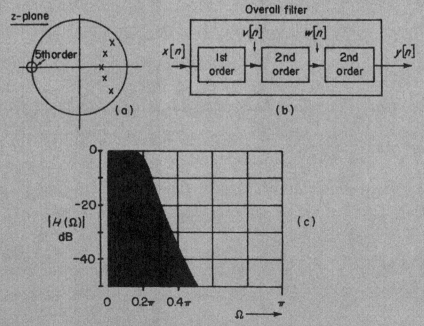

Figure 6.7 A 5th-order Butterworth low-pass digital filter *(abscissa of part (c): 320 samples)*

this case it occurs at $\Omega = 0$, and equals 780 (57.8 dB). This means that, for signals in the passband, the difference equations in part (c) will give output sample values $y[n]$ which are much greater than the input samples $x[n]$. If the scaling effect is unacceptable, it can be offset by multiplying all input samples by $\frac{1}{780}$ — or by sharing this factor between the three difference equations.

If we can accept some passband ripple, a Chebyshev design should allow us to meet the same stopband specification with a filter of lower order. We illustrate this in Worked Example 6.4 below.

Another point concerns the derivation of an equivalent high-pass filter. Referring back to Figure 6.7(a), we see that the low-pass filter's poles are clustered around $z=1$, with a 5th-order zero at $z = -1$. If we change the sign of the real parts of all poles and zeros, we get a 'mirror-image' pattern with high-pass characteristics. All poles are to the left of the imaginary axis, and the zero is at $z=1$. Program no. 21 designs high-pass filters in this way; having found the poles and zeros of the low-pass prototype, it changes the signs of all their real parts.

Example 6.4
  (a) Use Computer Programs 20 and 21 to plot the frequency response of a 3rd-order Chebyshev low-pass filter with 3 dB passband ripple and a cut-off frequency of 0.2π (36°). Does the filter meet the cut-off specification of Example 6.3?
  (b) Plot the pole-zero configuration, and frequency response magnitude function, of a 6th-order Butterworth high-pass filter with a cut-off frequency of 0.7π.

Solution
  (a) 3 dB passband ripple corresponds to a *fractional* ripple of 0.2929. Computer Program no. 21 specifies the filter as follows:

> a real zero, of order 3, at $z = -1$
> poles at:

r	$\theta°$
0.82343	0
0.91467	32.794

Given this information, Program no. 20 produces the decibel plot shown in Figure 6.8(a). We see that the filter just meets the 30 dB cut-off at $\Omega = 0.4\pi$ required in Example 6.3. Since it is of 3rd-order (rather than the 5th-order Butterworth filter discussed in Example 6.3), it involves considerably less computation.
  (b) Choosing the Butterworth and high-pass options, Program no. 21 specifies a 6th-order zero at $z=1$, and three complex pole-pairs as follows:

r	$\theta°$
0.80853	126.95
0.52174	135.78
0.35026	160.39

They are plotted in Figure 6.8(b). The frequency response characteristic, produced by Program no. 20, is shown in part (c) of the figure. This filter could be implemented by three 2nd-order difference equations, as discussed in Worked Example 6.3.

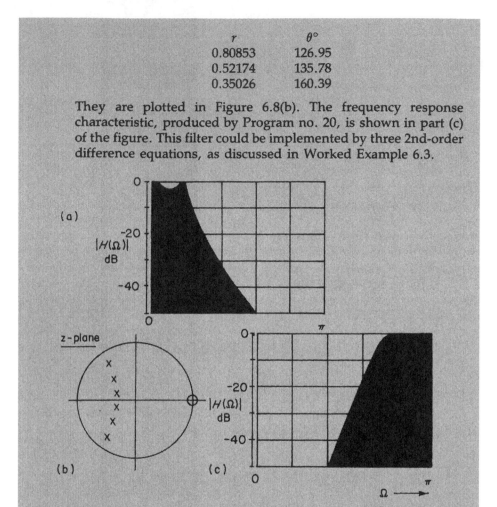

Figure 6.8 (a) Frequency response of a 3rd-order Chebyshev low-pass filter; (b) pole-zero configuration of a 6th-order Butterworth high-pass filter, and (c) its frequency response (*abscissa: 320 samples*).

We have not yet exhausted the possibilities offered by Computer Program no. 21! It can also be used to design Butterworth and Chebyshev *bandpass* filters. Having found the poles and zeros of the low-pass prototype, the program converts them using a *low-pass to bandpass transformation*. The resulting bandpass filter has twice as many poles and zeros as the prototype, so its order is always even.

The low-pass to bandpass transformation may be summarized as follows. Suppose we require a bandpass filter with a lower cut-off frequency Ω_2 and an upper cut-off frequency Ω_3. The program first finds the poles and zeros of a low-pass prototype with cut-off frequency $\Omega_1 = (\Omega_3 - \Omega_2)$. A pole (or zero) located at $z = \alpha$ then gives two poles (or zeros) in the bandpass design, at locations:

$$z = 0.5A(1+\alpha) \pm \{0.25A^2(1+\alpha)^2 - \alpha\}^{1/2} \qquad (6.26)$$

where:

$$A = \cos\left(\frac{\Omega_3 + \Omega_2}{2}\right)\bigg/\cos\left(\frac{\Omega_3 - \Omega_2}{2}\right)$$

When α is complex, equation (6.26) demands complex arithmetic. The program allows for this by separate manipulation of real and imaginary parts.

The zeros of a Butterworth or Chebyshev low-pass prototype always occur at $z = -1$. Equation (6.26) converts them into an equal number of zeros at $z = 1$, and at $z = -1$. For example a 10th-order bandpass filter, derived from a 5th-order prototype, has 5th-order zeros at $z = 1$ and $z = -1$. The frequency response therefore displays true nulls at $\Omega = 0$ and $\Omega = \pi$.

Figure 6.9(a) shows the z-plane poles and zeros of such a Chebyshev filter, with cut-off frequencies of 0.2778π (50°) and 0.5222π (94°), and 2 dB passband ripple. You may like to check the pole/zero configuration for yourself, using Program no. 21. The decibel response plot produced by Program no. 20 is reproduced in part (b) of the figure. Note how the cluster of five poles gives an 'equiripple' passband with five humps. A convenient way of implementing the filter would be as five cascaded subfilters, each involving a zero at $z = 1$, a zero at $z = -1$, and a complex-conjugate pole-pair.

In our discussion of the bilinear transformation, we have concentrated on Butterworth and Chebyshev filters because they are relatively easy to understand, and are widely used by analog and digital designers. However, other types of analog filter can also be converted into digital equivalents using the bilinear transformation. For example, there is an alternative type of Chebyshev design with ripples in the stopband, rather than the passband. More valuable is the class of *elliptic filters*, which display equiripple characteristics in both passband and stopband. Like the nonrecursive equiripple designs introduced in Section 5.4, digital elliptic filters offer the sharpest passband-stopband transition of any recursive filter with the same order and ripple specification. This makes them very useful in applications requiring a narrow transition band, when a certain amount of passband and stopband ripple can be tolerated.

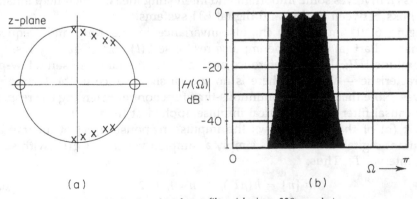

Figure 6.9 A 10th-order Chebyshev bandpass filter *(abscissa: 320 samples).*

The frequency response magnitude function of an analog low-pass elliptic filter takes the form:

$$|H(\Omega)| = \frac{1}{\left\{1 + \epsilon^2 R_n^2\left(\frac{\omega}{\omega_1}, L\right)\right\}^{1/2}} \tag{6.27}$$

where R_n is the *Chebyshev rational function* of nth-order. This is not the same as the Chebyshev polynomial C_n in equation (6.11). However, the parameters ϵ and ω_1 play the same role as in the normal type of Chebyshev filter. That is, ϵ controls the amount of passband ripple, and ω_1 is the nominal cut-off frequency. The additional parameter L controls the transition width and stopband ripple, and interacts with ω_1 to affect the actual cut-off frequency.

We have already mentioned the frequency compression, or 'warping', effect of the bilinear transformation. Compression does not destroy an equiripple characteristic — it merely alters the frequencies at which the ripples occur. Therefore application of the transformation to elliptic filters preserves their valuable equiripple properties.

Elliptic functions are complicated, and it is unlikely that you will ever have to design digital elliptic filters from first principles. However, you may well meet them as part of a standard DSP software package.

6.3.2 IMPULSE-INVARIANT FILTERS

Another method of deriving a digital filter from an analog filter is known as *impulse-invariance*. In this case the design criterion is that the impulse response of the digital filter should be a sampled version of that of the reference analog filter. We shall see that the criterion produces a different relationship between the analog and digital filters in the frequency domain, compared with the bilinear transformation described in the previous section.

The impulse-invariant technique can be applied to a wide range of filters, including Butterworth and Chebyshev types. In practice it tends to be less effective, and more awkward to use, than the bilinear transformation. Nevertheless it involves some important and interesting ideas, which help illustrate the links between analog and digital LTI systems.

Figure 6.10 summarizes impulse-invariance in the time and frequency domains. Part (a) shows the impulse response $h(t)$ and frequency response magnitude $|H(\omega)|$ of the reference analog filter. We have chosen a low-pass characteristic — although there is no special significance in the form of the curves. Note that $h(t)$ is a continuous-time function representing the response of a causal filter to an electrical impulse applied at time $t=0$.

Part (b) of the figure shows the impulse response $h_1[n]$ of an impulse-invariant digital filter. $h_1[n]$ is simply a sampled version of $h(t)$, with sampling interval T_1. Thus:

$$h_1[n] = h(nT_1), \qquad n=0, 1, 2, \ldots \tag{6.28}$$

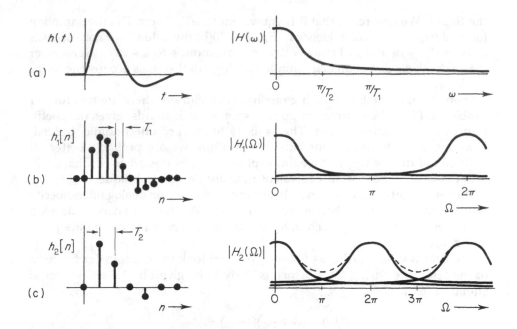

Figure 6.10   The idea of impulse-invariance.

In Section 1.2 we described how sampling an analog signal causes repetition of its spectrum at multiples of the sampling frequency. Although we are now sampling an impulse response rather than a signal, the same ideas apply. The frequency response of our impulse-invariant filter will be a repeating version of that of the analog filter. It follows that if we sample at too low a rate, there will be significant spectral overlap, leading to aliasing. The frequency response of the digital filter will then be a poor replica of the characteristic we are trying to copy.

It is important to realize that, in general, the frequency response of an analog low-pass filter only tends to zero as $\omega \to \infty$. It follows that a certain amount of aliasing is inherent in the impulse-invariant approach. However, this can be made small by ensuring that the samples of $h(t)$ are closely spaced — as in part (b) of the figure.

Figure 6.10(c) shows the effects of reducing the sampling rate to half its previous value. We have now defined an alternative digital filter (also impulse-invariant) for which:

$$h_2[n] = h(nT_2), \qquad n=0, 1, 2, \ldots \qquad (6.29)$$

The aliasing effect, shown on the right-hand side, is much more serious. The frequency response $|H_2(\Omega)|$, sketched as a dotted line, is a poor replica of $|H(\omega)|$. We conclude that the effectiveness of the impulse-invariant technique depends on an adequate sampling rate, and on choosing an analog reference filter with a limited bandwidth.

It is important to be clear about the frequency scales in the various parts of the figure. We first recall that Ω is equivalent to ωT, where T is the sampling interval (equation (1.28)). Hence in Figure 6.10(b) the value $\Omega = \pi$ corresponds to $\omega = \pi/T_1$; whereas in Figure 6.10(c) it corresponds to $\omega = \pi/T_2$. Remember that, in both cases, adequately-sampled analog input signals occupy the range $0 < \Omega < \pi$.

Even when a suitable sampling rate has been chosen, there are two further problems. First, the sampled impulse response, as it stands, gives the coefficients for a *nonrecursive* filter. This is likely to be very uneconomic. Indeed, if we are going to use a nonrecursive algorithm, we are probably better off with one of the highly effective linear-phase designs described in Chapter 5. What is needed here is an equivalent *recursive* design technique.

The second problem is that most books and references on analog filters specify transfer functions rather than impulse responses. We therefore need to develop a frequency-domain approach involving transfer functions, or z-plane poles and zeros.

The best way forward is as follows. We first look up the transfer function of the reference analog filter, which is likely to be given in the *series* form of equation (6.9):

$$H(s) = \frac{K(s-z_1)(s-z_2)(s-z_3) \cdots}{(s-p_1)(s-p_2)(s-p_3) \cdots} \qquad (6.30)$$

Assuming there are no repeated poles (that is, no two poles have the same value), we may use a partial fraction expansion to express $H(s)$ in the following *parallel* form:

$$H(s) = \frac{K_1}{(s-p_1)} + \frac{K_2}{(s-p_2)} + \frac{K_3}{(s-p_3)} + \cdots \qquad (6.31)$$

In effect we are *decomposing* the analog filter into a set of single-pole subfilters, whose outputs are added together. This is illustrated by Figure 6.11(a).

It is quite straightforward to specify a digital, impulse-invariant, version of

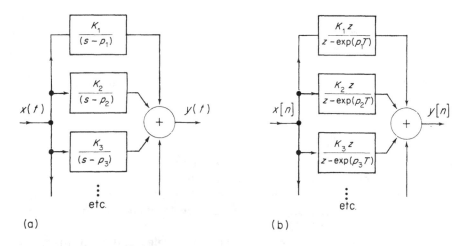

(a)   (b)

Figure 6.11 Designing an impulse-invariant filter by parallel decomposition.

each subfilter. This is because the impulse response of each analog subfilter takes a simple exponential form. Thus, for the ith subfilter:

$$h_i(t) = K_i \exp(p_i t), \quad t \geq 0$$
$$= 0, \quad t < 0 \tag{6.32}$$

The impulse response of the impulse-invariant digital subfilter is therefore:

$$h_i[n] = h_i(nT) \tag{6.33}$$

where T is the chosen sampling interval. Using equation (6.32) we have:

$$h_i[n] = K_i \exp(np_i T), \quad n \geq 0$$
$$= 0, \quad n < 0 \tag{6.34}$$

The transfer function of the digital subfilter equals the z-transform of its impulse response. Fortunately this is easily found. Thus:

$$H_i(z) = \sum_{n=0}^{\infty} K_i \exp(np_i T) z^{-n} = \sum_{n=0}^{\infty} K_i \{\exp(p_i T) z^{-1}\}^n$$

$$= \frac{K_i}{1 - \exp(p_i T) z^{-1}} = \frac{K_i z}{z - \exp(p_i T)} \tag{6.35}$$

The overall digital filter is now built up as a parallel set of subfilters, as shown in Figure 6.11(b). Like the analog case, the overall impulse response is simply equal to the sum of the individual impulse responses. And since each of these is impulse-invariant, so is the complete filter.

Equation (6.35) shows that each digital subfilter has a zero at the origin of the z-plane, and a pole at $z = \exp(p_i T)$. Actually the zero is not essential, because it does not affect the shape of the frequency response. However, it does ensure that the filter is minimum-delay.

We now apply the above design process to two examples — the first simple, the second rather more challenging.

Example 6.5 One of the most basic types of analog LTI processor is a first-order low-pass filter with the transfer function:

$$H(s) = \frac{1}{1 + s\tau}$$

where τ is the so-called *time constant*. (Such a filter may be constructed using a resistor-capacitor combination.)

(a) Sketch the frequency response magnitude $|H(\omega)|$ and impulse response $h(t)$ of such a filter, assuming $\tau = 1$ second.
(b) Find the transfer function and difference equation of the impulse-invariant digital filter having a sampling interval of 0.05 s.
(c) Specify the difference equation if the sampling interval is changed to 0.5 s.
(d) Use Computer Program no. 20 in Appendix A1 to plot the

magnitude responses of the two digital filters, and compare with $|H(\omega)|$ in (a) above.

Solution

(a) Putting $\tau=1$, and replacing s by $j\omega$, we have:

$$H(\omega) = \frac{1}{1+j\omega}, \qquad \text{giving } |H(\omega)| = \frac{1}{(1+\omega^2)^{1/2}}$$

Since this is a single-pole filter there is no need for a parallel decomposition. Equation (6.32) gives directly:

$$h(t) = \exp(-t), \qquad t \geq 0$$
$$= 0, \qquad\qquad t < 0$$

$|H(\omega)|$ and $h(t)$ are sketched in Figure 6.12. Note that the -3 dB point ($|H(\omega)| = 1/\sqrt{2}$) occurs when $\omega=1$.

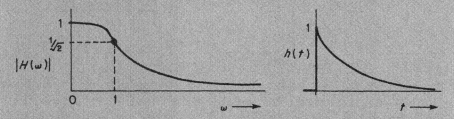

Figure 6.12 Characteristics of a first-order analog low-pass filter.

(b) Using equation (6.35), the transfer function is:

$$H(z) = \frac{Y(z)}{X(z)} = \frac{1}{1-\exp(-T)z^{-1}} = \frac{1}{1-0.9512z^{-1}}$$

The difference equation is therefore:

$$y[n] = 0.9512\, y[n-1] + x[n]$$

(c) Substituting the value $T=0.5$ we obtain the difference equation:

$$y[n] = 0.6065\, y[n-1] + x[n]$$

(d) We may use Computer Program no. 20, which assumes a *series* (cascade) realization, because the filter is first-order. The filter in part (b) above has a single pole at $z = 0.9512$; that in part (c) has its pole at $z = 0.6065$. Using these pole locations as data, we obtain the two plots shown in Figure 6.13.

   Note that when $T = 0.05$ s, the frequency $\Omega=\pi$ corresponds to $\omega=\pi/T = 62.8$ radians/second. When $T = 0.5$ s, $\Omega=\pi$ corresponds to $\omega = 6.28$ radians/second. These are marked on the figure. We see that in each case the -3 dB cut-off occurs close to 1 radian/

second, as in the analog reference filter. However, the aliasing effect is much more serious with the lower sampling rate.

We do not often require a filter with a cut-off frequency as low as 1 radian/second. However the above results are also relevant to other cut-off frequencies, providing the sampling rate is scaled in proportion. For example they apply to a cut-off frequency of 1000 radians/second, with $T = 5 \times 10^{-5}$ s and $T = 5 \times 10^{-4}$ s respectively.

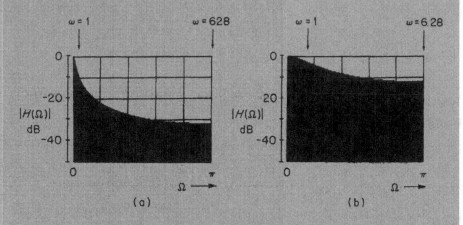

Figure 6.13 Responses of first-order, impulse-invariant, digital low-pass filters *(abscissa: 320 samples).*

Of course we hardly need the technique of impulse invariance to design a first-order filter! We have included the foregoing example to clarify the method, and to illustrate the relationship between the analog and digital designs. In particular it shows how important the sampling rate is in determining the amount of aliasing of the analog characteristic.

Example 6.6 The transfer function of an analog, 3rd-order, Butterworth low-pass filter with a cut-off frequency of 1 radian/second is:

$$H(s) = \frac{1}{(s+1)(s + 0.5 + j0.866)(s + 0.5 - j0.866)}$$

(a) Design an impulse-invariant digital equivalent, based on a sampling interval of 0.5 s. Specify its difference equation.
(b) Use Computer Programs 20 and 21 in Appendix A1 to compare its frequency response with that of a 3rd-order Butterworth filter designed by the bilinear transformation method.
(c) Sketch the pole-zero pattern of the impulse-invariant filter, and

use Computer Program no. 4 to produce a screen plot of its impulse response.

Solution
(a) We first use a partial fraction expansion to express $H(s)$ in parallel form:

$$H(s) = \frac{K_1}{(s+1)} + \frac{K_2}{(s + 0.5 + j0.866)} + \frac{K_3}{(s + 0.5 - j0.866)}$$

Care is needed with the algebra, because it involves complex numbers. We obtain:

$$K_1 = 1; \; K_2 = -0.5 + j0.2887; \; K_3 = -0.5 - j0.2887$$

Given $T = 0.5$, we may use equation (6.35) to find the component parts of the digital transfer function. Thus:

$$H(z) = \frac{z}{(z - 0.6065)} + \frac{(-0.5 + j0.2887)z}{z - (0.7788 \exp(-j0.433))}$$

$$+ \frac{(-0.5 - j0.2887)z}{z - (0.7788 \exp(j0.433))}$$

Note that the three poles are at $z = 0.6065$, and $z = r \exp(\pm j\theta)$ where $r = 0.7788$ and $\theta = 0.433$ radians (24.8°). We may combine the two complex expressions to give:

$$H(z) = \frac{z}{(z - 0.6065)} + \frac{z(0.8956 - z)}{(z^2 - 1.414z + 0.6065)}$$

The filter can either be implemented in this form, as *paralleled* first and second-order subsystems; or we can convert into the *series* form. The latter is required here, because we wish to use Computer Program no. 20 to check the frequency response — and the program assumes the series form. Thus:

$$H(z) = \frac{z(z^2 - 1.414z + 0.6065) + z(0.8956 - z)(z - 0.6065)}{(z - 0.6065)(z^2 - 1.4138z + 0.6065)}$$

which reduces to:

$$H(z) = \frac{0.08701z(z + 0.7315)}{(z^3 - 2.0203z^2 + 1.464z - 0.3678)}$$

The difference equation of the impulse-variant filter is therefore:

$$y[n] = 2.0203 \, y[n-1] - 1.464y[n-2] + 0.3678y[n-3]$$
$$+ 0.08701x[n-1] + 0.06365x[n-2]$$

(b) The pole/zero locations are now used as data for Computer Program no. 20. We ignore the zero at $z=0$, since it does not affect the frequency response magnitude. The required information is therefore:

No. of separate real poles: 1
    value, order, of pole: 0.6065, 1

No. of separate real zeros: 1
    value, order, of zero: $-0.7315, 1$

No. of complex pole pairs: 1
    radius and angle: 0.7788, 24.8°

No. of complex zero pairs: 0

The program produces the screen plot shown in Figure 6.14(a). In this case $\Omega = \pi$ corresponds to $\omega = \pi/T = 6.28$ radians/second. We see that the -3 dB cut-off point occurs at $\omega = 1$, as required by the specification.

To compare with the response of a 3rd-order Butterworth filter designed by the bilinear transformation method, we run Program no. 21. The required cut-off frequency is $\Omega = \omega T = 0.5$ radian, or 28.65°. The program specifies a third-order zero at $z = -1$; a real pole at $z = 0.5932$; and a complex pole pair at radius 0.7831 and angles $\pm 25.32°$. This information, used as input data for Program no. 20, produces the plot shown in Figure 6.14(b).

The two plots are very similar at low frequencies. However, the cut-off slope of the impulse-invariant filter is less steep at higher frequencies, due to some aliasing of the analog filter characteristic. This disadvantage is partly offset by the fact that the impulse-invariant filter has fewer z-plane zeros and requires slightly less computation.

(c) The poles and zeros of the impulse-invariant filter are sketched in Figure 6.15(a). Note that only one zero contributes to shaping

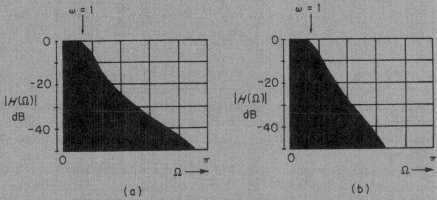

(a)   (b)

Figure 6.14 Responses of 3rd-order Butterworth low-pass filters designed by (a) impulse-invariance, and (b) the bilinear transformation (*abscissa: 320 samples*).

the frequency response characteristic, compared with the three zeros of the equivalent bilinear transformation design.

The impulse response is plotted by using the filter's multiplier coefficients as data for Computer Program no. 4. The difference equation found in part (a) above shows that the coefficients are:

recursive: 2.0203, −1.464, 0.3678
nonrecursive: 0, 0.08701, 0.06365

The resulting plot is shown in Figure 6.15(b). This must be a sampled version (with $T = 0.5$ second) of the impulse response of the reference analog filter.

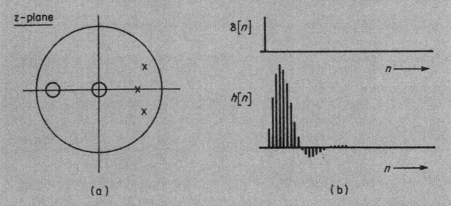

Figure 6.15 (a) Pole–zero configuration, and (b) impulse response of an impulse-invariant Butterworth low-pass filter (*abscissa: 50 samples*).

You may feel that the method of impulse invariance is rather cumbersome, involving some awkward algebra. Not surprisingly, most serious users of the technique rely on substantial help from a digital computer! Although our own approach has been strictly introductory, we hope it has given a good idea of the basis of the method.

6.4 FREQUENCY SAMPLING FILTERS

We next consider a recursive design method which has no direct counterpart in the theory of analog filters. The *frequency-sampling* method is a good example of a DSP technique developed from first principles. It gives great flexibility over the choice of filter magnitude characteristic. It also produces filters with finite impulse responses (FIRs), which offer the advantage of a linear-phase response.

It is sometimes implied that *all* FIR filters are non-recursive, but this is not so. It is true that they *may* always be implemented nonrecursively, but in some cases an equivalent recursive operation is not only possible, but much more economic. The frequency-sampling filter is an important example of this idea.

To understand frequency-sampling, we need to develop some background. We ask you to be patient if the first part of the discussion seems a little abstract. All should become clear in due course!

It is helpful to start by considering a *digital resonator* having a complex conjugate pole-pair on the unit circle in the z-plane, and a second-order zero at the origin. Its transfer function is:

$$H(z) = \frac{Y(z)}{X(z)} = \frac{z^2}{\{z-\exp(j\theta)\}\{z-\exp(-j\theta)\}} = \frac{z^2}{z^2 - 2\cos\theta\, z + 1} \tag{6.36}$$

giving the difference equation:

$$y[n] = 2\cos\theta\, y[n-1] - y[n-2] + x[n] \tag{6.37}$$

As a simple example, let $\theta = 60°$, so that $\cos\theta = 0.5$. Then:

$$y[n] = y[n-1] - y[n-2] + x[n] \tag{6.38}$$

We refer to such a system as a resonator because its poles are actually *on* the unit circle, representing the margin of stability. Its impulse response continues forever, neither increasing nor decreasing in amplitude. This is confirmed by the impulse response of Figure 6.16(a), which takes the form of a continuing oscillation at a frequency corresponding to the pole positions.

As it stands, such a resonator is not a useful processor, because it is unstable. However, its impulse response can be made finite by cascading with a very simple form of nonrecursive filter, known as a *comb filter*. The combination of comb filter and resonator provides the basic building block for a complete frequency-sampling filter.

The operation of a comb filter–resonator combination can be visualized from the block diagram of Figure 6.16(b). An impulse input to the comb filter causes it to produce one positive and one negative output impulse, separated by m sampling intervals. The first of these 'excites' the resonator, which starts to generate its characteristic oscillation. The second impulse brings the oscillation to a halt m sampling intervals later. The combination of comb filter and resonator therefore produces a simple, recursive, FIR filter.

It is also instructive to consider pole-zero locations. The comb filter has the transfer function:

$$H(z) = 1 - z^{-m} = \frac{(z^m - 1)}{z^m} \tag{6.39}$$

giving m zeros uniformly spaced around the unit circle. These produce a 'comb' frequency response, illustrated in Figure 6.16(c) for the case when $m = 24$. Since the resonator contributes a complex-conjugate pole-pair, the overall pole-zero configuration is as shown in part (d) of the figure. We see that the poles of the resonator are *exactly cancelled* by two of the comb filter's zeros, giving an overall filter which has, in effect, only z-plane zeros. This must of course be the case, because its impulse response is finite.

At first sight it may seem rather pointless to use a resonator, only to have its poles cancelled. However, the essential point is that it provides a more

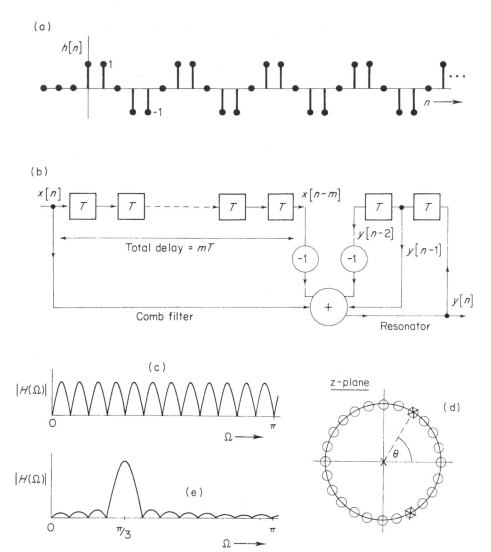

Figure 6.16 Basis of the frequency-sampling technique.

economic algorithm. Assuming $m = 24$, the recursive difference equation of the block diagram in part (b) of the figure is:

$$y[n] = y[n-1] - y[n-2] + x[n] - x[n-24] \qquad (6.40)$$

This requires only three additions/subtractions. But since the impulse response continues for 24 sampling intervals, with eight values of ± 1 and eight of -1, a nonrecursive realization would need many more additions and subtractions. The advantage of recursive operation becomes even greater if m is increased.

Figure 6.16(e) shows that inclusion of the resonator converts the comb filter response into an elementary bandpass characteristic. The center of the pass-band corresponds to the resonator pole locations. If the parameter m is increased, the width of the main passband reduces, and the characteristic tends to a $\sin x/x$, or sinc, function. Such a filter can be valuable in its own right for simple applications; more importantly in the present context, it forms the basis for a complete frequency-sampling filter.

We are finally in a position to explain the frequency-sampling concept. Suppose we require a digital filter with the response magnitude characteristic of Figure 6.17(a). We first sample it, as in part (b) of the figure. (For illustrative purposes we show just five samples; more would generally be used in a practical design.) The required response is now built up by superposing a set of sinc functions, each weighted by one of the sample values a_1 to a_5, and arranged around it. This is illustrated to an expanded horizontal scale in part

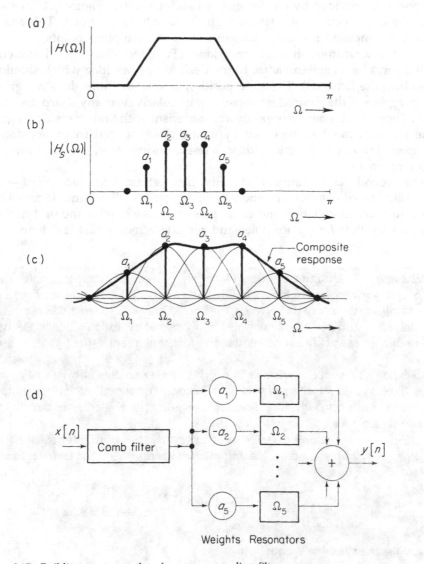

Figure 6.17 Building up a complete frequency-sampling filter.

(c). Each of the sinc functions is provided by a comb filter—resonator combination of the type already discussed.

Fortunately the complete frequency-sampling filter is made much more economic by using a single comb filter, which feeds all the resonators in parallel. The overall filter structure is illustrated by Figure 6.17(d). Note that alternate weights must be inverted, because there is a phase reversal between the outputs of adjacent resonators.

The idea of reconstituting a continuous function from its sampled version, by arranging a weighted sinc function around each of the sample values, is very important in signal theory. The reconstitution of an analog signal from its time-domain samples can be viewed in this way. Of course we are here sampling a frequency function, rather than a time function, and each sinc response is provided by an elementary bandpass filter. Figure 6.17(c) shows that the sinc functions all pass through zero at *adjacent* samples. This ensures that the composite response is correct at all the sampling points.

Two further matters should be mentioned. First, the actual filter characteristic will always be an *approximation* to the desired one (an idea which should by now be quite familiar!). The superposition of sinc functions does not give an exact replica of the desired response — particularly near any sharp discontinuity. Therefore it is better to specify a characteristic with finite transition regions. The problem can also be eased by taking closer-spaced frequency-domain samples; but of course this leads to a less economic filter, because there are more resonators.

The second matter concerns stability. If we attempt to place poles — and cancelling zeros — exactly on the unit circle, very small arithmetic errors may prevent exact cancellation and cause poles to move outside the unit circle. In practice we therefore place poles and zeros at a radius just less than unity.

Example 6.7 Design a frequency-sampling filter which approximates the frequency response shown in Figure 6.18(a). Take frequency-domain samples at 3° intervals between 75° and 93° (Ω = 1.309 to 1.623 radians) inclusive, and place z-plane poles and zeros at radius r = 0.999. Derive a suitable set of difference equations for implementing the filter.

Solution The specification calls for the seven samples shown in Figure 6.18(b). We therefore need seven resonators. The signals at various points in the frequency-sampling filter are conveniently labeled as in part (c) of the figure.

Since we are taking samples at 3° intervals, the comb filter must have 360/3 = 120 z-plane zeros. Placing them at radius 0.999, the transfer function is:

$$H(z) = \frac{W(z)}{X(z)} = \frac{z^{120} - (0.999)^{120}}{z^{120}} = 1 - 0.886867z^{-120}$$

giving the difference equation:

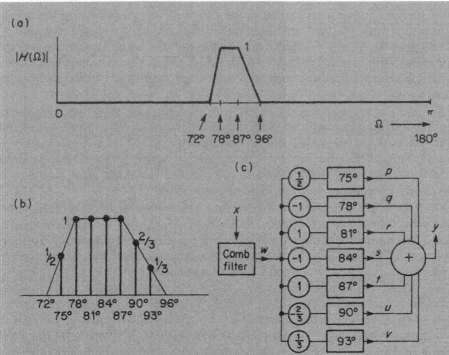

Figure 6.18 Design of a frequency-sampling bandpass filter.

$$w[n] = x[n] - 0.886867x[n-120]$$

The resonator equation (equation (6.37)), modified for a pole radius less than unity, takes the general form:

$$y[n] = 2r\cos\theta\, y[n-1] - r^2\, y[n-2] + x[n]$$

Each of the seven resonators required here has its own angle θ. Also, the sample weighting can be incorporated as a gain factor applied to the input (we must remember to invert alternate weights, as previously explained). Using $r = 0.999$, and the signal symbols given in Figure 6.18(c), the seven resonator equations are readily found:

$$
\begin{aligned}
p[n] &= 0.517121p[n-1] - 0.998001p[n-2] + 0.5w[n]\\
q[n] &= 0.415408q[n-1] - 0.998001q[n-2] - w[n]\\
r[n] &= 0.312556r[n-1] - 0.998001r[n-2] + w[n]\\
s[n] &= 0.208848s[n-1] - 0.998001s[n-2] - w[n]\\
t[n] &= 0.104567t[n-1] - 0.998001t[n-2] + w[n]\\
u[n] &= -0.998001u[n-2] - 0.666667w[n]\\
v[n] &= -0.104567v[n-1] - 0.998001v[n-2] + 0.333333w[n]
\end{aligned}
$$

To realize the complete filter, we precede these equations with the comb filter equation already given; and we follow them with a final equation which superposes all resonator outputs:

$$y[n] = p[n] + q[n] + r[n] + s[n] + t[n] + u[n] + v[n]$$

The total set of nine equations may be incorporated in a program loop.

You would probably like some proof that these frequency-sampling equations really work! We have therefore included them in Computer Program no. 22 in Appendix A1, which calculates and plots the filter's impulse and frequency responses. The impulse response is found by loading a unit sample into the input signal array. The frequency response is then calculated by discrete Fourier transformation.

The plots are reproduced in Figure 6.19. The impulse response $h[n]$ starts at $n=0$, and ends abruptly after 120 sampling intervals. The filter is therefore FIR. The frequency response magnitude $|H(\Omega)|$, plotted to linear scales, compares quite well with the desired characteristic of Figure 6.18(a). The unwanted sidelobes and ripple could, of course, be reduced by taking more samples of the desired response, and using more resonators.

Since the impulse response has 120 terms, a nonrecursive version of the same filter would need 120 multipliers. This compares with the 17 multiplications, plus a few additions/subtractions, specified by our set of frequency-sampling equations. The advantage of recursive operation would increase even further if the frequency-domain samples were closer-spaced.

We stated earlier that frequency-sampling filters display linear-phase characteristics, implying that their impulse responses must be symmetrical in form. Actually if you look closely at $h[n]$ in Figure 6.19, you will see that it is *antisymmetrical*. The effect is to add a phase shift of $\pi/2$ (90°) at all frequencies — in addition to the pure linear-phase term. If this is a disadvantge for a particular application, the design approach can be modified slightly to give pure linear-phase. Readers interested in this possibility will find a problem on it at the end of the chapter.

Before leaving the topic of frequency-sampling filters, we should consider the value of an individual comb filter-resonator as a digital processor. We have already mentioned that such a combination may be useful in its own right.

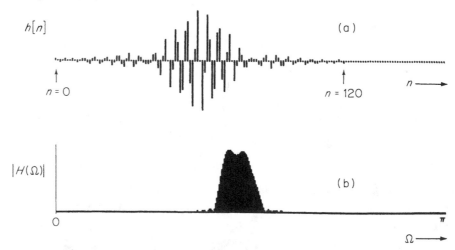

Figure 6.19 Impulse and frequency responses of a frequency-sampling bandpass filter (*abscissae: 160 and 320 samples*).

For example, equation (6.40) specifies a simple bandpass filter with an approximately sinc form of response — see also Figure 6.16(e). From a practical point of view, the most interesting filters of this type are ones with *integer multipliers*.

No distinction has so far been made in this book between integer and decimal arithmetic. The digital filters we have described are specified by decimal multipliers — typically given with 4 or 5 figure accuracy. However, a DSP algorithm in which signal samples and multipliers are represented by integers will generally be faster in operation. High-level languages often make special provision for integer arithmetic. And if you are programming in machine code, you will probably find that integer operations are relatively fast and easy.

What restrictions must be placed on a comb filter–resonator combination for its difference equation to have integer multipliers? A resonator with its poles on the unit circle is defined by equation (6.37):

$$y[n] = 2\cos\theta\ y[n-1] - y[n-2] + x[n] \qquad (6.41)$$

Cascaded with a comb filter having m sampling delays, the difference equation becomes:

$$y[n] = 2\cos\theta\ y[n-1] - y[n-2] + x[n] - x[n-m] \qquad (6.42)$$

The criterion for integer arithmetic is therefore met if $2\cos\theta$ is an integer. There are three possibilities, corresponding to bandpass functions with different center-frequencies, as follows:

θ	$2\cos\theta$	Center-frequency
60°	1	$\Omega = \pi/3$
90°	0	$\Omega = \pi/2$
120°	−1	$\Omega = 2\pi/3$

Alternatively, we can place a single cancelling pole at $\theta = 0°$ or $180°$, giving a first-order low-pass or high-pass function respectively:

$$y[n] = y[n-1] + x[n] - x[n-m] \qquad \text{(low-pass)} \qquad (6.43)$$
$$y[n] = -y[n-1] + x[n] - x[n-m] \qquad \text{(high-pass)} \qquad (6.44)$$

The first of these is a recursive version of our old friend the moving-average filter (see Sections 1.1.2 and 5.2). Note that when we work in integer arithmetic, we can place poles *exactly on* the unit circle, because cancellation by coincident zeros is perfect.

It is possible to cascade filters of this type, modifying the passband characteristics and reducing sidelobe levels. Even so, it must be admitted that their frequency responses are far from ideal. For this reason they are often omitted from books on digital filtering, especially those with a communication engineering bias. However they can be of genuine value in DSP applications where programming simplicity and operating speed are paramount. If you are interested in this type of design, you will find two relevant problems included at the end of the chapter, and further information given in reference no. 17 in the bibliography.

6.5   DIGITAL INTEGRATORS

Integration, like differentiation, is a common signal processing operation which can be performed digitally. If we assume that the signal samples represent an underlying analog waveform, then we expect integration to give a measure of the area under the curve. There are several well-known integration algorithms, including the *running sum*, the *trapezoid rule*, and *Simpson's rule*. In this section we consider such schemes as digital filtering operations, and compare their properties in the time and frequency domains.

Digital integration and differentiation can be approximated by a variety of algorithms, recursive and nonrecursive. However, whereas the normal approach to differentiation is nonrecursive (see Section 5.5), integration is naturally a recursive operation. It requires a running estimation of elements of area. The currently-available output from the integrator is updated to take account of each new input sample.

The simplest integration algorithm is the *running sum*. Indeed, if our requirement is merely to add up all the samples of a digital signal, this is all we shall ever need! The difference equation, first given as equation (1.42), is:

$$y[n] = y[n-1] + x[n] \tag{6.45}$$

The corresponding transfer function and frequency response are respectively:

$$H[z] = \frac{Y(z)}{X[z]} = \frac{1}{(1-z^{-1})} = \frac{z}{(z-1)} \tag{6.46}$$

and:

$$H(\Omega) = \frac{1}{1-\exp(-j\Omega)} = \frac{\exp(j\Omega)}{\exp(j\Omega)-1} \tag{6.47}$$

We shall discuss the implications of equations (6.46) and (6.47) a little later.

An alternative is to use the *trapezoid rule*. In this case the difference equation, widely quoted in books on numerical analysis, takes the form:

$$y[n] = y[n-1] + \tfrac{1}{2}\{x[n] + x[n-1]\} \tag{6.48}$$

In other words the output is updated by the *average* of two adjacent input values. The transfer function and frequency response are:

$$H(z) = \frac{\tfrac{1}{2}(1 + z^{-1})}{(1 - z^{-1})} = \frac{z + 1}{2(z-1)} \tag{6.49}$$

and:

$$H(\Omega) = \frac{\exp(j\Omega) + 1}{2\{\exp(j\Omega)-1\}} \tag{6.50}$$

A third well-known method of digital integration is *Simpson's rule*. The corresponding equations are:

$$y[n] = y[n-2] + \tfrac{1}{3}\{x[n] + 4x[n-1] + x[n-2]\} \tag{6.51}$$

$$H(z) = \frac{\frac{1}{3}(1 + 4z^{-1} + z^{-2})}{(1 - z^{-2})} = \frac{z^2 + 4z + 1}{3(z^2 - 1)} \tag{6.52}$$

and:

$$H(\Omega) = \frac{\exp(2j\Omega) + 4\exp(j\Omega) + 1}{\{3\exp(2j\Omega) - 1\}} \tag{6.53}$$

Note that we have specified causal versions of all three types of integrator, in which the output depends on present and previous inputs.

How can we assess the relative merits of the three algorithms? It is helpful to begin by considering their performance in the time domain. Assuming that the aim is to measure the area under the equivalent analog curve, then the approximations involved are summarized by Figure 6.20. Part (a) represents the running sum integrator. For convenience we take the sampling interval as one unit of time ($T = 1$). The total area is assessed as a series of small rectangular elements, such as the one shown shaded. The area of each element is just equal to the sample value it surrounds. In effect the underlying analog signal is being approximated by a *zero-order-hold* waveform.

Part (b) of the figure illustrates the trapezoid rule, equation (6.48). Each area element is now a trapezium. We see that the integral is assessed by representing the analog signal as a set of straight-line sections — often referred to as a *first-order-hold* approximation.

Simpson's rule is illustrated in part (c) of the figure. It may be shown that equation (6.51) is equivalent to fitting a *quadratic* to three adjacent sample values. Each area element now covers two sampling intervals.

We may look on each of these methods as giving a *polynomial approximation* to the underlying analog signal. The running sum uses a zero-order polynomial: that is, it assumes each sample value is held for one complete sampling interval. The trapezoid rule uses a *first-order* polynomial — a straight line joining two adjacent samples. And Simpson's rule is based on a quadratic or *second-order* polynomial, joining three adjacent samples. Higher-order polynomials may in principle be used to give improved approximations, fitted to a larger number of sample values. Indeed, such polynomial curve-fitting is often used for reasons other than the design of digital integrators.

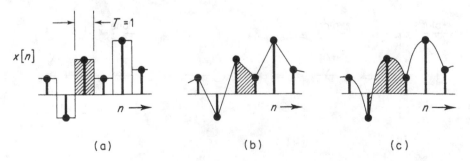

Figure 6.20 Digital integration: (a) the running-sum technique; (b) the trapezoid rule; (c) Simpson's rule.

From the integration point of view, there is little to choose between the various orders of polynomial when we are dealing with slowly-varying signals. However, with faster rates of change, it is clear that the different algorithms are likely to provide rather different estimates of the integral. To take this idea further, we need to consider performance in the frequency domain.

Let us first note that an ideal *analog* integrator would have a magnitude response inversely proportional to frequency, with a phase shift of $\pi/2$ radians. This is because:

$$\int \cos(\omega t)\,\mathrm{d}t = \frac{1}{\omega}\sin(\omega t) \qquad (6.54)$$

A digital integrator should offer a similar performance. However since the frequency response of any digital processor is repetitive, we can only hope to approximate the desired characteristic over the range $0 \le \Omega \le \pi$ — the range occupied by an adequately-sampled analog signal.

We have already given expressions for the transfer functions of our three types of digital integrator (equations (6.46), (6.49), and (6.52)). Figure 6.21 shows the corresponding pole-zero configurations, and we can infer the general form of their frequency responses using concepts developed in Section 4.3.2.

Part (a) of the figure refers to the running-sum integrator. Its frequency response is shaped by the real pole at $z = 1$, giving a magnitude inversely proportional to Ω at low frequencies. This is because the pole vector (which contributes to the transfer function *denominator*) initially increases linearly with frequency. However, as $\Omega \to \pi$ the increase is less marked, so we may expect the integrator to overemphasize high frequencies somewhat.

The opposite effect can be expected from the pole-zero plot of the trapezoidal integrator in part (b), because of the zero at $z = -1$.

Figure 6.21(c) shows the poles and zeros of the Simpson's-rule algorithm. We now have a pole at $z = -1$, as well as $z = 1$, and must therefore expect exaggeration of high frequencies — even though the effect will be counteracted to some extent by the two zeros lying on the negative real axis.

We can investigate these ideas quantitatively with the help of Computer Program no. 20 in Appendix A1. Already used several times in this chapter, the program accepts pole and zero locations as data, and produces a decibel plot of the corresponding magnitude response, normalized to a peak value of 0 dB.

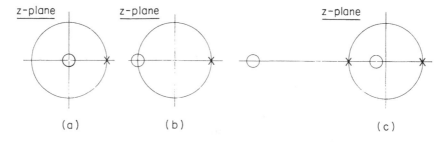

Figure 6.21 Pole-zero configurations of the running sum, trapezoid, and Simpson integrators.

We have to be a little careful here, because our integrators all have a pole *on* the unit circle at $z=1$, giving an infinite response at $\Omega=0$ (the Simpson algorithm will also have an infinite response at $\Omega=\pi$). However it is simple to modify the program to plot responses between (say) $\Omega = 0.05\pi$ and $\Omega = 0.95\pi$, accepting that they will be normalized to a peak of 0 dB over this slightly reduced range.

The results are shown in Figure 6.22. In part (a) we have sketched the magnitude response of an ideal integrator, for ease of reference. The curve is normalized to 0 dB at $\Omega = 0.05\pi$. Part (b) shows the screen plot for the running-sum algorithm. The response is close to ideal up to about $\Omega = 0.4\pi$, but becomes slightly too high as $\Omega \rightarrow \pi$. Part (c), representing the trapezoid rule, is also good up to about $\Omega = 0.4\pi$, but falls away strongly as $\Omega \rightarrow \pi$. And part (d) — the Simpson's-rule integrator — behaves well up to slightly higher frequencies (say $\Omega = 0.5\pi$) than the other two, but overaccentuates frequencies towards $\Omega = \pi$. These findings confirm our previous discussion.

To summarize, we see that there is little to choose between the three algorithms up to about $\Omega = 0.4\pi$. Above this frequency they display marked differences, and the choice must depend on the application. For example, if we need to integrate a signal which is contaminated by random fluctuations or 'noise' (much of which is generally high-frequency), it may be best to use the trapezoid rule. In this way the effects of the noise will be reduced. Whichever algorithm is used, it is clearly important to appreciate the form of its frequency response, and make due allowance. This is particularly true if the signal, after integration, is to be subjected to spectral analysis.

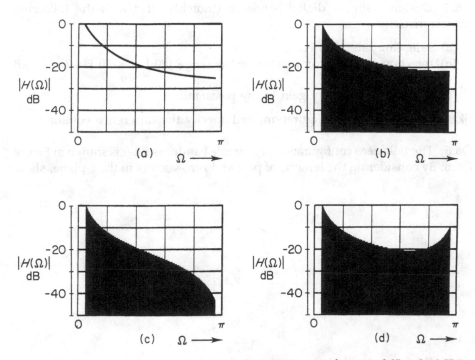

Figure 6.22 Frequency responses of four digital integrators over the range $0.05\pi < \Omega < 0.95\pi$. (a) Ideal; (b) running sum; (c) trapezoid; (d) Simpson *(each abscissa: 320 samples)*.

PROBLEMS

SECTION 6.2

Q6.1 Sketch, to linear scales, the frequency response (magnitude) of recursive filters specified by the following z-plane pole-zero locations: (a) pole at $z = -0.9$, zero at $z=1$; (b) poles at $z = \pm j\, 0.95$, zeros at $z = \pm 1$; (c) second-order pole at $z = 0.95$, zeros at $z = -0.5 \pm j\, 0.8$.

Q6.2 Confirm the results of Q6.1 using Computer Program no. 20 in Appendix A1. (Bear in mind that the computer plots have logarithmic vertical scales.)

Q6.3 Convert the following gain values to decibels: (a) 10; (b) 1; (c) 0.001; (d) 0.0004; (e) 0

Q6.4 Choose a suitable pole-zero configuration for a recursive digital filter with the following properties:

(a) a passband centered at $\Omega = 0.3\pi$, with a -3 dB bandwidth of 0.03π;
(b) Steady-state rejection of components at $\Omega = 0.8\pi$.

Sketch the pole-zero configuration, and derive the filter's difference equation.
   Finally, check your design by using the pole-zero locations as input data for Program no. 20 in Appendix A1.

Q6.5 Design a simple digital bandstop ('notch') filter with the following characteristics:

(a) sampling rate 1000 Hz,
(b) rejection of signal components in the range 100 Hz $\pm$ 10 Hz (to -3 dB points),
(c) approximately unity gain in the passband.

Sketch the pole-zero configuration, and specify the difference equation.

Q6.6 The pole-zero configuration of a simple bandpass filter is shown in Figure Q6.6. By considering the lengths of pole and zero vectors in the z-plane, show

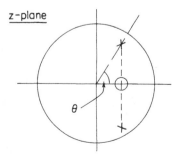

z-plane

Figure Q6.6

that the peak response is more or less independent of the center-frequency θ, provided the pole radius is close to unity.

SECTION 6.3.1

Q6.7 Make accurate sketches, to both linear and logarithmic (dB) vertical scales, of the following analog low-pass filters:

(a) Butterworth 5th order,
(b) Chebyshev 5th-order, 10 per cent ripple ($\delta = 0.1$).

Assume the cut-off frequency is 1 kHz in each case, and sketch the responses between 0 and 4 kHz.

Q6.8 Make an accurate sketch of the frequency response (in dB) of a 3rd-order Butterworth digital low-pass filter based on the bilinear transformation, with cut-off frequency $\Omega_1 = 0.2\pi$.

Q6.9 Using Computer Program no. 21 in Appendix A1, find and sketch the z-plane poles and zeros of the following digital filters based on the bilinear transformation:

(a) Butterworth low-pass, 7th-order, cut-off frequency $\Omega_1 = 0.3\pi$.
(b) Chebyshev low-pass, 6th-order, 20 per cent ripple, cut-off frequency $\Omega_1 = 0.4\pi$.
(c) Butterworth high-pass, 4th-order, cut-off frequency 0.8π.
(d) Chebyshev bandpass, 8th-order, 3 dB ripple, passband edges at 0.2π and 0.5π.

Make sure you can relate the pole-zero locations to the expected form of the frequency response in each case.

Q6.10 You are asked to design a Butterworth digital high-pass filter with a cut-off frequency $\Omega_1 = 0.7\pi$. Its response should be at least 30 dB down at $\Omega = 0.5\pi$, and at least 50 dB down at $\Omega = 0.3\pi$. What minimum order of filter is required?

Q6.11 Using Computer Program no. 21, find the poles and zeros of a Chebyshev 3rd-order high-pass filter with cut-off frequency $\Omega_1 = 0.9\pi$ and 3 dB passband ripple. Then use Program no. 20 to plot the filter's frequency response. What cut-off (in dB) does the filter achieve at $\Omega = 0.8\pi$?
   Specify the filter's difference equation, corrected to give a peak gain of unity in the passband.

Q6.12 Recast the difference equation of Q6.11 in the form of two separate equations, representing cascaded first- and second-order subfilters. Include the gain-correction term in the first-order subfilter.

SECTION 6.3.2

Q6.13 An analog bandpass filter has the transfer function:

$$H(s) = \frac{s}{(s + 1)(s + 2)}$$

where s is the Laplace variable. Design a recursive impulse-invariant filter based on a sampling interval of 0.1 s, giving your answer in terms of a single difference equation relating input and output.

Use Computer Program no. 20 in Appendix A1 to plot the frequency response of the filter, checking that it has a bandpass characteristic. (*Note*: the program does not require details of any poles or zeros at the origin, because they have no effect on the response magnitude function.)

Q6.14 Repeat Q6.13 for a sampling interval of 0.3 s, checking that the aliasing effect in the region of $\Omega = \pi$ is now more pronounced.

Q6.15 A Butterworth analog low-pass filter of 4th-order, with a cut-off frequency of 1 radian/s, has the transfer function:

$$H(s) = \frac{1}{(s^2 + 1.8478\ s + 1)(s^2 + 0.7654\ s + 1)}$$

Design a recursive, impulse-invariant, digital filter based on a sampling interval of 0.25 s. Specify difference equations for a parallel-form realization.

Q6.16 Recast the difference equations of Q6.15 as a single equation representing a series-form realization of the filter. Hence use Computer Program no. 20 in Appendix A1 to plot its frequency response.

SECTION 6.4

Q6.17 Design a frequency-sampling filter which approximates the frequency response of Figure Q6.17, taking samples at 4° intervals between 40° and 60° ($\Omega = 0.2222\pi$ to 0.333π) as shown. Place poles and zeros at a radius of 0.998 in the z-plane. Specify a suitable set of difference equations for implementing the filter.

Modify Computer Program no. 22 in Appendix A1 to calculate and plot the

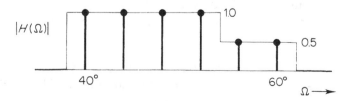

Figure Q6.17

filter's impulse and frequency responses. Why is the frequency response a rather poor approximation to the desired shape, and how might it be improved?

Q6.18 As mentioned in the main text, the impulse response of the frequency-sampling filter derived in Worked Example 6.7 is antisymmetrical in form (see Figure 6.19). This gives a phase response with an extra 90° phase shift at all frequencies. If the additional phase term is unacceptable, the impulse response must be made symmetrical.

(a) Show that the impulse response can only be symmetrical if each resonator executes an *odd* number of half-cycles of oscillation.

(b) Given condition (a), convince yourself that the comb filter must be redesigned to supply two *positive* impulses in response to an input impulse (rather than one positive, the other negative), and that it cannot have a z-plane zero at $z = 1$.

Using the above information, redesign the frequency-sampling filter of Figure 6.18 to give a pure linear-phase response. Use a comb filter with 120 zeros as before, but shift the desired frequency response forward by 1.5° (and take samples between 73.5° and 91.5° inclusive). Specify a modified set of difference equations.

Alter Computer Program no. 22 in Appendix A1 to accept the new equations. Check that the impulse response is symmetrical and that the frequency response is as expected.

Q6.19 If the poles of a digital resonator are restricted to certain points on the unit circle, then a comb filter—resonator combination provides a simple recursive filter which can be implemented using integer arithmetic.

The poles and zeros of a low-pass and a bandpass filter of this type are illustrated in Figure Q6.19. In each case:

(a) Sketch the frequency response magnitude characteristic.
(b) Specify the filter's difference equation.
(c) Find, and sketch, its impulse response.

Do the filters have pure linear-phase characteristics, and if so, why?

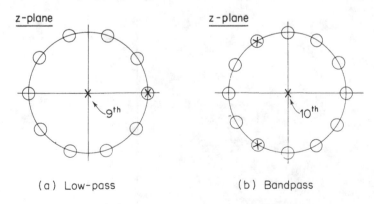

(a) Low-pass   (b) Bandpass

Figure Q6.19

Q6.20 A high-pass filter has the transfer function:

$$H(z) = \frac{(z^{12} - 1)^2(z - 1)}{z^{23}(z + 1)^2}$$

(a) Sketch its pole-zero configuration and frequency response (magnitude).
(b) Find its difference equation and show that the filter can be implemented using integer arithmetic.
(c) Find and sketch its impulse response.

What computational advantage does the filter offer, over an equivalent nonrecursive design?

SECTION 6.5

Q6.21 Show that:

(a) the trapezoid rule for integration is equivalent to fitting a first-order polynomial to two adjacent sample values, and
(b) Simpson's rule is equivalent to fitting a second-order polynomial (quadratic) to three adjacent sample values.

Q6.22 Estimate the relative responses (in dB) of the following three types of digital integrator, *compared with the ideal integrator*, at the frequencies $\Omega = 0.2\pi$, 0.5π, and 0.9π:

running sum,
trapezoid rule,
Simpson's rule.

CHAPTER 7

The Discrete and Fast Fourier Transforms

7.1 INTRODUCTION

It would be hard to exaggerate the importance of the *Discrete Fourier Transform (DFT)* in digital signal processing. When implemented using a *Fast Fourier Transform (FFT)* algorithm, the DFT offers rapid frequency-domain analysis and processing of digital signals, and investigation of LTI systems. The development of FFT algorithms from the mid-1960s onwards gave a huge impetus to DSP, making practicable a range of valuable techniques.

We have already described two Fourier representations for digital signals in Chapter 3:

A discrete-time version of the *Fourier Series*, applicable to periodic signals. We interpreted the spectral coefficients a_k as harmonics of the series, producing a line spectrum.

A discrete-time version of the *Fourier Transform*, applicable to aperiodic signals and LTI processors, and giving rise to continuous functions of the variable Ω.

These two representations are close cousins of their continuous-time counterparts in classical Fourier analysis. There is, however, one major difference: they yield spectral functions which are periodic. This is an inevitable consequence of working with sampled signals and data.

The DFT may be regarded as a third Fourier representation, applicable to aperiodic digital signals of finite length. It is closely related to our work in Chapter 3 — particularly to the discrete Fourier Series. Therefore if you are unfamiliar with this material, we suggest you review it before proceeding.

We have two main aims in this chapter. First, we wish to explain the basis of the DFT, and its relationship with the other Fourier representations. We will then discuss the computational problems of implementing the DFT directly, and explain how it may be speeded up using a variety of FFT algorithms.

7.2 THE DISCRETE FOURIER TRANSFORM (DFT)

7.2.1 BASIS OF THE DFT

Truly periodic signals are rarely encountered in practical DSP. Aperiodic signals and data with a finite number of nonzero sample values are far more common — for example, the dollar price of gold (Figure 1.3) or the midday temperature record of Figure 2.11(a). The Discrete Fourier Transform (DFT) of such a signal $x[n]$, defined over the range $0 \le n \le (N-1)$, is given by:

$$X[k] = \sum_{n=0}^{N-1} x[n] \exp(-j2\pi kn/N)$$

$$= \sum_{n=0}^{N-1} x[n]\, W_N^{kn} \tag{7.1}$$

where $W_N = \exp(-j2\pi/N)$, and the spectral coefficients $X[k]$ are evaluated for $0 \le k \le (N-1)$. The inverse DFT, or IDFT, which allows us to recover the signal from its spectrum, is given by:

$$x[n] = \frac{1}{N} \sum_{k=0}^{N-1} X[k]\, W_N^{-kn} \tag{7.2}$$

where the values of $x[n]$ are evaluated for $0 \le n \le (N-1)$.

If we use equation (7.1) to calculate additional values of $X[k]$, outside the range $0 \le k \le (N-1)$, we find that they form a periodic spectral sequence. Likewise, using equation (7.2) to calculate additional values of $x[n]$ outside the range $0 \le n \le (N-1)$ yields a periodic version of the signal. We therefore see that the DFT and IDFT both represent a finite-length sequence as one period of a periodic sequence. In effect the DFT considers an aperiodic signal $x[n]$ to be periodic for the purposes of computation.

Note that the only difference between the DFT and the IDFT is the scaling factor of $(1/N)$, and a sign change in the exponent. Therefore if we have an algorithm for computing the DFT, it is a simple matter to modify it to compute the IDFT. This is a direct consequence of the symmetry between time and frequency domains.

Equations (7.1) and (7.2) are essentially the same as the analysis and synthesis equations of the discrete Fourier Series (equations (3.1) and (3.2)). The only difference is that we have now incorporated the scaling factor $1/N$ in the synthesis equation, rather than the analysis equation (in line with the usual definition of the DFT). Also, we have labeled the spectral coefficients in equation (7.1) as $X[k]$, rather than a_k, to emphasize the transform-pair relationship between a signal and its spectrum. Otherwise the DFT and discrete Fourier Series equations are identical. This suggests that equation (7.1) may be viewed in two ways. If the signal $x[n]$ is truly periodic, the equation gives a form of discrete Fourier Series; but if $x[n]$ is basically aperiodic, and only being treated as periodic for the purposes of computation, the equation represents the DFT. Of course, we must not get too concerned about this; a periodic signal of period N and an aperiodic signal of length N are both completely defined by N values!

You may feel that all this is unnecessarily complicated. After all, we developed a discrete-time version of the Fourier Transform for aperiodic signals and LTI systems in Section 3.3. Can we not use it to describe the corresponding spectral functions, without making any artificial assumptions about periodicity? The answer is both 'yes' and 'no'. On the one hand we can certainly find an expression for a continuous spectral function $X(\Omega)$ or $H(\Omega)$. However, difficulties arise when we wish to estimate, plot, or process it. A computer or digital hardware cannot work with a continuous function. All it can do is estimate the function for a discrete set of Ω values — in other words, produce a *sampled version*. This raises the obvious question: how many samples should we take? Too few will give an inadequate representation; whereas too many will involve us in unnecessary computation.

Back in Section 3.3.2 we computed sample values of some continuous spectral functions. For example, Figure 3.12 shows 320 samples of the response $H(\Omega)$ of a bandpass filter over the range $0 \le \Omega \le \pi$. The criterion for choosing 320 samples was vague; about all we can say is that they gave a good clear screen plot, without taking too long to compute! Clearly, this is not good enough if we wish to transform, and inverse transform, digital sequences as quickly and efficiently as possible. Rather than rely on intuition, we must establish how many samples are theoretically *necessary*.

The answer to this important question is provided by a frequency-domain version of the famous Sampling Theorem, first discussed in Section 1.2. It turns out to be very helpful for understanding the nature of the DFT, and its relationship with the Fourier Series and Fourier Transform representations introduced in Chapter 3.

The Sampling Theorem tells us that a continuous (analog) signal of limited bandwidth may be completely represented by a set of time-domain samples, provided they are spaced close enough together. Nor surprisingly, the essential symmetry of Fourier transformation means that an equivalent result holds in the frequency domain. It may be stated as follows:

> The continuous spectrum of a signal with limited duration T_0 seconds may be completely represented by regularly-spaced frequency-domain samples, provided the samples are spaced not more than $1/T_0$ Hz apart.

Now an aperiodic digital signal with N finite sample values has duration $T_0 = NT$ seconds, where T is the sampling interval in the time domain. Hence its spectrum can be completely represented by frequency-domain samples spaced $1/NT$ Hz, or $2\pi/NT$ radians per second, apart. Since the variable Ω is equal to ωT, where ω is the radian frequency (see equation (1.28)), we conclude that we must sample at intervals in Ω of $2\pi/N$ (or less).

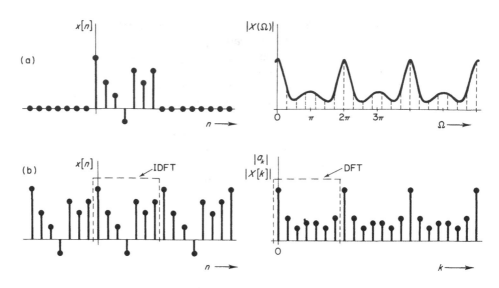

Figure 7.1 Relationships between the DFT, Fourier Transform and discrete Fourier Series.

The other point to remember is that the spectrum of a digital signal (or frequency response of a digital processor) is always periodic in Ω, with period 2π. This is a consequence of sampling. Since one period is clearly sufficient to define it, sampling at the minimum interval of $2\pi/N$ gives us $2\pi \div 2\pi/N = N$ frequency-domain samples.

Returning to equations (7.1) and (7.2), we see that the DFT does exactly this: it provides us with N distinct spectral coefficients $X[k]$ for a signal with N distinct sample values. Conversely, the IDFT regenerates the N signal values from the N spectral coefficients. The values of $X[k]$ may be regarded as frequency-domain samples of the 'underlying' Fourier Transform $X(\Omega)$. And the DFT, by just obeying the Sampling Theorem in the frequency domain, gives us the most economic spectral representation possible.

The DFT copes with complex, as well as real, signals. If a signal has N complex sample values (each with a real and imaginary part), then it has $2N$ degrees of freedom in the time domain. The N values of its spectrum are also complex (real and imaginary part, or magnitude and phase), giving $2N$ degrees of freedom in the frequency domain. Conversely, if the signal is real, its N sample values give just N degrees of freedom in the time domain. In this case it turns out that half the spectral coefficients are sufficient to define the spectrum, because the other half displays a 'mirror-image' pattern — a point already made when discussing the discrete Fourier Series in Section 3.2.1. We therefore see that whether a signal is complex or real, the DFT gives the same number of degrees of freedom in the two domains. This attractive result stems from the symmetry of Fourier transformation.

Figure 7.1 helps summarize the relationships between the DFT, the Fourier

Transform, and the discrete Fourier Series. Part (a) shows a real aperiodic signal $x[n]$ and, on the right-hand side, its spectrum (only the magnitude of $X(\Omega)$ is given, for convenience). The spectrum is continuous and periodic, with period 2π. In this case we are assuming a real time function, so the portion of $|X(\Omega)|$ between 0 and π is a mirror-image of that between π and 2π. The spectrum may therefore be defined by its fluctuations over any frequency interval equal to π. Part (b) of the figure shows a periodic version of the same signal, and, on the right, its discrete Fourier Series (which is also periodic). The spectral coefficients may be regarded as samples of $X(\Omega)$. Finally, the figure shows that the DFT estimates a single period of the repetitive spectrum, and the IDFT a single period of the repetitive signal — as indicated by the dotted lines.

7.2.2 PROPERTIES OF THE DFT

In this section we summarize a number of important properties of the DFT — including linearity, time-shifting, convolution and modulation. The properties are closely related to those of the discrete Fourier Series listed in Table 3.1, and discussed in Section 3.2.2. (You may find it helpful to refer back to the general points made in that section.) We also describe the effects on the DFT when the signal being transformed is real or complex, even or odd.

We again use a double-headed arrow to denote the relationship between a signal and its spectrum. Thus $x[n] \leftrightarrow X[k]$ signifies that the aperiodic signal $x[n]$, defined for $0 \le n \le (N-1)$, has DFT coefficients $X[k]$, defined for $0 \le k \le (N-1)$. The sequence $x[n]$ transforms into the sequence $X[k]$; conversely, $X[k]$ inverse transforms into $x[n]$. We recall that if the DFT of $x[n]$ is estimated for values of n outside this range, it is found to be periodic. The same is true of the IDFT of $X[k]$. Thus:

$$x[n] = x[n+N] \qquad \text{for all } n$$

and

$$X[k] = X[k+N] \qquad \text{for all } k \qquad (7.3)$$

The property can also be interpreted by evaluating the indices modulo-N. A detailed account of the periodic, or *circular*, nature of the DFT may be found in various more advanced references — for example, the book by Oppenheim and Schafer listed in the bibliography. We shall say more about it later in this chapter, and again in Chapter 8.

Let us now turn to the important DFT properties mentioned at the start of this section. They may be stated as follows:

Linearity

If
$$x_1[n] \leftrightarrow X_1[k] \quad \text{and} \quad x_2[n] \leftrightarrow X_2[k]$$

Then
$$Ax_1[n] + Bx_2[n] \leftrightarrow AX_1[k] + BX_2[k] \qquad (7.4)$$

Time-shifting

If

$$x[n] \leftrightarrow X[k]$$

Then

$$x[n - n_0] \leftrightarrow X[k] \exp(-j2\pi k n_0/N) = X[k] \, W_n^{k n_0} \tag{7.5}$$

Convolution

If

$$x_1[n] \leftrightarrow X_1[k] \quad \text{and} \quad x_2[n] \leftrightarrow X_2[k]$$

Then

$$\sum_{m=0}^{N-1} x_1[n] x_2[m - n] \leftrightarrow X_1[k] X_2[k] \tag{7.6}$$

Modulation

If

$$x_1[n] \leftrightarrow X_1[k] \quad \text{and} \quad x_2[n] \leftrightarrow X_2[k]$$

Then

$$x_1[n] x_2[n] \leftrightarrow \frac{1}{N} \sum_{m=0}^{N-1} X_1[m] X_2[k - m] \tag{7.7}$$

Properties (7.6) and (7.7) show that time-domain convolution is equivalent to frequency-domain multiplication, and vice versa. Once again, we must remember that all shifts involved in these properties are interpreted as circular, or modulo-N.

We should next consider what happens to the DFT if the signal $x[n]$ being transformed is real, or complex. We have generally simplified our account of DSP in this book by considering only real signals. However, the DFT applies equally well to complex ones, and many practical DFT algorithms cater for them. It is therefore important to appreciate how the DFT coefficients are affected. The definition of the DFT, given by equation (7.1), is:

$$X[k] = \sum_{n=0}^{N-1} x[n] \, W_N^{kn}, \quad \text{where } W_N = \exp(-j2\pi/N) \tag{7.8}$$

$$= \sum_{n=0}^{N-1} x[n] \{ \cos(2\pi kn/N) - j \, \sin(2\pi kn/N) \} \tag{7.9}$$

Now cosines are even functions, and sines are odd functions. Hence if $x[n]$ is real, the real part of $X[k]$ must be even, and the imaginary part must be odd. It also follows that the magnitude and phase of $X[k]$ must be even and odd respectively, and that $X[k] = X^*[-k]$, where the asterisk denotes the complex conjugate. This spectral symmetry means that the DFT can be completely defined by just half the total set of spectral coefficients. Thus the DFT of a real, N-valued, signal yields two real coefficients, $X[0]$ and $X[N/2]$, together with $N/2 - 1$ distinct complex coefficients (the other $N/2 - 1$ complex coefficients are their complex conjugates). We therefore need two real and $N/2 - 1$ complex coefficients to define the spectrum, giving N degrees of freedom. These correspond to the N degrees of freedom on the time-domain — a point already made in the previous section.

If $x[n]$ is a complex signal, with sample values having real and imaginary parts, then the above spectral symmetry is lost and all coefficients are needed to define the DFT.

When $x[n]$ is real, and also even such that $x[n] = x[-n]$, then its spectrum contains only cosines terms. Therefore the imaginary part of $X[k]$ is zero. Conversely, when $x[n]$ is both real and odd, its spectrum contains only sine terms and the real part of $X[k]$ is zero. In both cases the number of degrees of freedom is further reduced by a factor of 2. We summarize these additional properties in the following table, using symbols $\mathcal{R}$ and $\mathcal{I}$ to denote real and imaginary parts respectively.

$x[n]$		$X[k]$	
$\mathcal{R}$	$\mathcal{I}$	$\mathcal{R}$	$\mathcal{I}$
even	0	even	0
odd	0	0	odd
0	even	0	even
0	odd	odd	0

The practical significance of the table is that it can be used to improve the efficiency of DFT algorithms designed to work on restricted types of data.

For ease of reference the main properties of the DFT are listed in Table 7.1 at the end of the chapter.

7.2.3 COMPUTING THE DFT

A major practical consideration when computing the DFT is its speed. This is a complicated matter which depends not only on the algorithm and programming language used, but also on the hardware. In this book we are mainly interested in implementing DSP algorithms in high-level languages on general-purpose computers. Multiplications are usually the most time-consuming operations in such situations, so we will pay special attention to them in the following discussion. However, it is important to realize that special-purpose DSP hardware is often designed to achieve fast multiplication, so that other aspects of programming may become equally, or more, significant.

The most obvious approach to computing the DFT and IDFT is to implement equations (7.1) and (7.2) directly. As we shall see, the method is straightforward, but relatively slow. Expressing the imaginary exponentials in terms of sines and cosines, we have, for the DFT:

$$X[k] = \sum_{n=0}^{N-1} x[n] \exp(-j2\pi kn/N)$$

$$= \sum_{n=0}^{N-1} x[n]\{\cos(2\pi kn/N) - j\sin(2\pi kn/N)\} \qquad (7.10)$$

and for the IDFT:

$$x[n] = \frac{1}{N} \sum_{k=0}^{N-1} X[k] \exp(j2\pi kn/N)$$

$$= \frac{1}{N} \sum_{k=0}^{N-1} X[k]\{\cos(2\pi kn/N) + j\sin(2\pi kn/N)\} \tag{7.11}$$

Apart from the $1/N$ scaling factor and sign change in the IDFT, the computations required in the two cases are identical. We will therefore concentrate here on the DFT, equation (7.10).

Let us start by assuming that the signal $x[n]$ is real. The real and imaginary parts of the spectrum are then given by:

$$\Re(X[k]) = \sum_{n=0}^{N-1} x[n] \cos(2\pi kn/N) \tag{7.12}$$

and

$$\Im(X[k]) = -\sum_{n=0}^{N-1} x[n] \sin(2\pi kn/N) \tag{7.13}$$

These equations may readily be recast in terms of magnitude and phase, using:

$$|X[k]| = \{\Re(X[k])^2 + \Im(X[k])^2\}^{1/2} \tag{7.14}$$

and

$$\phi_k = \arctan \frac{\Im(X[k])}{\Re(X[k])} \tag{7.15}$$

If the signal $x[n]$ is complex, the computations are a little more involved. Let us write $x[n] = r[n] + ji[n]$, where $r[n]$ is the real part and $i[n]$ is the imaginary part. Also, substituting Ω_0 in place of $2\pi/N$ for convenience, the DFT becomes:

$$X[k] = \sum_{n=0}^{N-1} (r[n] + ji[n])\{\cos(kn\Omega_0) - j\sin(kn\Omega_0)\} \tag{7.16}$$

giving:

$$\Re(X[k]) = \sum_{n=0}^{N-1} r[n] \cos(kn\Omega_0) + i[n] \sin(kn\Omega_0) \tag{7.17}$$

and

$$\Im(X[k]) = \sum_{n=0}^{N-1} i[n] \cos(kn\Omega_0) - r[n] \sin(kn\Omega_0) \tag{7.18}$$

   Unfortunately, direct implementation of these equations is costly in terms of the number of multiplications involved. Equations (7.17) and (7.18), as they stand, require a total of $4N^2$ floating-point multiplications; equations (7.12) and (7.13) (for real data) require $2N^2$; and if the signal is even or odd, as well as real, we still need N^2. In each case there are similar numbers of integer multiplications and floating-point additions/subtractions to be performed. In general, we may expect the computation time to be roughly proportional to N^2.

   You may like to return briefly to Section 3.2.1, where we used two programs (listed as Programs 7 and 8 in Appendix A1) to compute the discrete Fourier Series of some real periodic signals. Since the DFT uses similar equations, the programs may be regarded as DFTs — although the interpretation of results is slightly different. Program No. 8, which transforms a 64-point signal, involves a modest amount of computation. But since the number of multiplications is proportional to N^2 in a direct implementation, transformation of 2048-point or 4096-point signals (a fairly common requirement) is quite a different matter. When truly high-speed processing is required, the computational cost of such lengthy transforms can easily become a limiting factor.

   The good news is that many of the multiplications turn out to be redundant. Basically this is because the exponentials (or sines and cosines) specified in the DFT and IDFT equations are calculated many times as k and n vary. This is particularly true with lengthy transforms. Highly efficient FFT algorithms are available for reducing the redundancy, and are described in the next section.

   Another problem with direct implementation of the DFT is the number of trigonometric evaluations required. Our equations show that a sine or cosine is required for each floating-point product, and this can be very time-consuming. An effective solution is to use a *table lookup* — that is, we calculate and store the sine and cosine functions just once at the start of the program. To avoid redundancy in the table, we store only N values of each and subsequently evaluate the product (kn) modulo-N.

   Suppose, for example, we wish to compute the DFT of a real, 64-point, signal. Equations (7.12) and (7.13) show that we need values of $\cos(2\pi kn/64)$ and $\sin(2\pi kn/64)$, with k and n both varied over the range 0 to 63 inclusive. Each trigonometric function has only 64 values per period, and these are first calculated and stored in the table lookup. We subsequently evaluate (kn) modulo-64, using the result to indicate which values are to be taken from the table.

   Program no. 23 in Appendix A1 computes such a DFT — both with and without a table lookup. A suitable signal $x[n]$ is first generated. Then, on the first pass ($P=1$), the DFT equations are implemented directly without the benefit of a table. This section of the program, plus the subsequent conversion to magnitude and phase and the screen plot, are essentially the same as in Program no. 8. (The only differences are that we have renamed the real and imaginary DFT arrays XXR and XXI respectively, and have omitted the $1/N$

scaling factor). On the second pass ($P=2$), a table lookup is generated by storing the cosines and sines in arrays C and S respectively. The product (kn) is evaluated modulo-64, and the DFT is recalculated and plotted on the screen. Typical time-savings due to the use of the table lookup are of the order of 50 per cent.

The screen plot produced by the program (on both passes) is shown in Figure 7.2. We have specified the same signal $x[n]$ as in Program no. 8, so the plot is also similar (see Figure 3.3). It is only the interpretation which differs slightly. Thus we are now regarding $x[n]$ as aperiodic, and the DFT coefficients as samples of one period of the underlying continuous spectrum $X(\Omega)$. We could in principle recover $X(\Omega)$ from the samples if we wished; however in practice it is rarely necessary to do so. The DFT is an entirely 'adequate' representation of the spectrum, according to the frequency-domain version of the Sampling Theorem.

Various other modifications and improvements can be used to speed up direct implementation of the DFT. For example, some of the values of $x[n]$ may be zero, and can be omitted from the computation. Sometimes all N values of $X[k]$ are not needed. Another factor is the cost of test and branch instructions in the program; if this is high, it may be helpful to incorporate portions of *straight-line code*. These and other matters are discussed in the book by Burrus and Parks (see Bibliography).

Although we cannot go into detail here, it is important to realize that there are various alternative approaches to computing the DFT. The *Goertzel* algorithm views the DFT as the output of a recursive digital filter which makes use of the periodicity of W_N^{kn} (see equation (7.1)) to reduce computation. *Rader's method* converts the DFT into a form of cyclic convolution — a technique which has been developed and refined for prime-length DFTs by Winograd. The *Chirp*

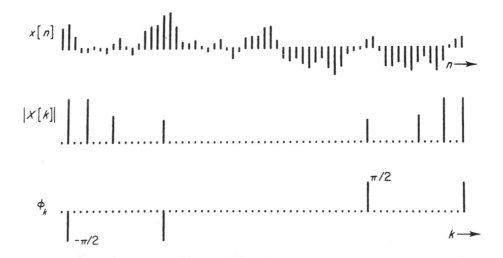

Figure 7.2 A direct DFT computation (*abscissa: 64 samples*).

z-Transform method involves multiplying the signal by a frequency-modulated (chirp) function, and can be used to evaluate the spectrum at unequal intervals. Further information is given in the books by Burrus and Parks, and Oppenheim and Schafer (see Bibliography).

Now that we have introduced the main aspects of the DFT, it is time to turn our attention to the most widely-used method of reducing computational redundancy — the Fast Fourier Transform.

7.3 THE FAST FOURIER TRANSFORM (FFT)

7.3.1 BASIS OF THE FFT

Highly efficient algorithms for computing the DFT were first developed in the 1960s. Collectively known as Fast Fourier Transforms (FFTs), they all rely upon the fact that the standard DFT involves redundant calculation. The DFT of an N-length signal (equation (7.1)) is defined by:

$$X[k] = \sum_{n=0}^{N-1} x[n] \exp(-j2\pi kn/N) = \sum_{n=0}^{N-1} x[n] \, W_N^{kn} \tag{7.19}$$

where $W_N = \exp(-j2\pi/N)$ and $X[k]$ is evaluated for $0 \le k \le (N-1)$. It turns out that the same values of $x[n] \, W_N^{kn}$ are calculated many times as the computation proceeds — particularly if the transform is lengthy. This is because W_N^{kn} is a periodic function with a limited number of distinct values. The same is true of the IDFT, equation (7.2). It is the aim of FFT algorithms to eliminate the redundancy.

Strictly speaking, there is no such thing as 'the FFT'. Rather, there is a collection of algorithms with different features, advantages, and limitations. An algorithm which is suitable for programming in high-level language on a general-purpose computer may not be the best for special-purpose DSP hardware. What the different algorithms have in common is their general approach — the *decomposition* of the DFT into a number of successively shorter, and simpler, DFTs.

Before getting down to detail, you may find it helpful if we demonstrate the periodic nature of W_N^{kn}. For simplicity let us consider a short DFT — say $N=8$. In this case the complete computation involves varying both k and n over the range 0 to 7 inclusive, giving a product (kn) which varies between 0 and 49. However, there are only eight distinct values of W_8^{kn}, tabulated in Figure 7.3. (Since half the values are the negative of the other half, we might say that there are only four distinct values.) The table forms a matrix with diagonal symmetry. Furthermore, many of the values are ± 1, implying simple additions and subtractions. All these features — and especially the periodicity and symmetry of W_N^{kn} — contribute to the redundancy of the standard DFT.

There are various ways of explaining FFT decomposition. Our approach in this section may be summarized as follows. First, we show that a DFT can be

		Value of n						
		0	1	2	3	4	5	6	7
Value of k	0	1	1	1	1	1	1	1	1
	1	1	$\dfrac{(1-j)}{\sqrt{2}}$	$-j$	$\dfrac{-(1+j)}{\sqrt{2}}$	-1	$\dfrac{-(1-j)}{\sqrt{2}}$	j	$\dfrac{(1+j)}{\sqrt{2}}$
	2	1	$-j$	-1	j	1	$-j$	-1	j
	3	1	$\dfrac{-(1+j)}{\sqrt{2}}$	j	$\dfrac{(1-j)}{\sqrt{2}}$	-1	$\dfrac{(1+j)}{\sqrt{2}}$	$-j$	$\dfrac{-(1-j)}{\sqrt{2}}$
	4	1	-1	1	-1	1	-1	1	-1
	5	1	$\dfrac{-(1-j)}{\sqrt{2}}$	$-j$	$\dfrac{(1+j)}{\sqrt{2}}$	-1	$\dfrac{(1-j)}{\sqrt{2}}$	j	$\dfrac{-(1+j)}{\sqrt{2}}$
	6	1	j	-1	$-j$	1	j	-1	$-j$
	7	1	$\dfrac{(1+j)}{\sqrt{2}}$	j	$\dfrac{-(1-j)}{\sqrt{2}}$	-1	$\dfrac{-(1+j)}{\sqrt{2}}$	$-j$	$\dfrac{(1-j)}{\sqrt{2}}$

Figure 7.3 Tabulation of W_8^{kn}.

expressed in terms of shorter, simpler, DFT's by dividing the signal $x[n]$ into subsequences. The method, which is widely described in the DSP literature, will be referred to as *conventional decomposition*. Having introduced you to some of the basic ideas in this way, we will go on to develop various types of FFT algorithm using an alternative approach known as *index-mapping*.

It is important to understand that conventional decomposition and index-mapping are two ways of looking at the same problem. There is no essential difference between them. We choose to develop index-mapping later in this section for two main reasons. First, it offers a compact and convenient notation. Secondly, unlike conventional decomposition, it may readily be applied to DFT's of any length which is not prime. Although we restrict ourselves in this book to transform lengths $N=2^i$, where i is a positive integer, other lengths of FFT are sometimes required. Those of you who go on to design or use such algorithms will very likely meet index-mapping again.

Conventional decomposition may be introduced by considering a signal $x[n]$ broken down into shorter, interleaved, subsequences. The process is referred to as *decimation-in-time*, and gives rise to one of the major classes of FFT algorithm. Suppose we have a signal with N sample values, where N is an integer power of 2. We first separate $x[n]$ into two subsequences, each with $N/2$ samples. The first subsequence consists of the even-numbered points in $x[n]$, and the second consists of the odd-numbered points. Writing $n=2r$ when n is even, and $n=2r+1$ when n is odd, the DFT may be recast as:

$$X[k] = \sum_{n=0}^{N-1} x[n] \, W_N^{kn}, \qquad 0 \le k \le (N-1)$$

$$= \sum_{r=0}^{N/2-1} x[2r] \, W_N^{2rk} + \sum_{r=0}^{N/2-1} x[2r+1] \, W_N^{(2r+1)k}$$

$$= \sum_{r=0}^{N/2-1} x[2r] \, (W_N^2)^{rk} + W_N^k \sum_{r=0}^{N/2-1} x[2r+1] \, (W_N^2)^{rk} \qquad (7.20)$$

We note that:

$$W_N^2 = \exp(-2j2\pi/N) = W_{N/2} \qquad (7.21)$$

and we may therefore write:

$$X[k] = \sum_{r=0}^{N/2-1} x[2r] \, W_{N/2}^{rk} + W_N^k \sum_{r=0}^{N/2-1} x[2r+1] \, W_{N/2}^{rk}$$

$$= G[k] + W_N^k H[k] \qquad (7.22)$$

We have now expressed the original N-point DFT in terms of two $N/2$-point DFTs, $G[k]$ and $H[k]$. $G[k]$ is the transform of the even-numbered points in $x[n]$, and $H[k]$ is the transform of the odd-numbered points. Note that (unfortunately!) we must multiply $H[k]$ by an additional term W_N^k before adding it to $G[k]$. The reason for this is not hard to understand. The subsequences into which we have decomposed $x[n]$ both contain $N/2$ sample values, but they are displaced from one another in time-origin by one sampling interval. The first value, $x[0]$, in the even-numbered subsequence occurs one sampling interval before the first value, $x[1]$, in the odd-numbered subsequence — and so on. Referring back to the time shifting property of the DFT, equation (7.5), we see that a time shift of one sampling interval is indeed equivalent to multiplying the corresponding spectrum by W_N^k.

If we assume that the transform length N is an integer power of 2, it follows that $N/2$ is even. Therefore we can take the decomposition further, by breaking each $N/2$-point subsequence down into two shorter, $N/4$-point subsequences. The process can continue until, in the limit, we are left with a series of 2-point subsequences, each of which requires a very simple 2-point DFT. A complete decomposition of this type gives rise to one of the commonly-used radix-2, *decimation-in-time*, FFT algorithms.

We can demonstrate the decimation-in-time process quite easily in the case of a short transform — for example, $N=8$. The signal $x[n]$ is defined for:

$$n = \{0 \quad 1 \quad 2 \quad 3 \quad 4 \quad 5 \quad 6 \quad 7\}$$

We start by forming two subsequences of the even-numbered and odd-numbered points:

$$n = \{0 \quad 2 \quad 4 \quad 6\} \quad \text{and} \quad \{1 \quad 3 \quad 5 \quad 7\}$$

Each subsequence is further decimated, giving:

$$n = \{0 \quad 4\} \quad \text{and} \quad \{2 \quad 6\} \quad \text{and} \quad \{1 \quad 5\} \quad \text{and} \quad \{3 \quad 7\}$$

The 8-point DFT is now realized as four 2-point DFTs, performed on pairs of sample values separated by half the transform length. Thus $x[0]$ is paired with $x[4]$, $x[2]$ is paired with $x[6]$, and so on.

Although the original DFT has been recast in terms of a series of 2-point DFTs, we must be careful not to assume that the resulting computation is trivial. It is perfectly true that each 2-point DFT involves very simple arithmetic — indeed, we shall see later that it requires just one addition and one subtraction. However, we have already showed that the transform of a sequence cannot be found by merely adding the transforms of its subsequences. Additional factors are involved (W_N^k in equation (7.22)), because the subsequences are displaced from one another in time-origin. If we decompose a lengthy DFT into successively shorter DFTs, such factors arise at every stage of the decomposition. Furthermore (just to be awkward!), they are different at each stage, because the time shifts and intervals between samples are different.

Part of the challenge in designing effective FFT algorithms lies in incorporating these unwelcome (but entirely necessary) factors in such a way that the amount of computation is minimized. It is quite possible to develop the ideas and techniques further in terms of conventional decomposition (see, for example, the book by Oppenheim and Schafer listed in the Bibliography). However, for reasons noted earlier, we prefer to turn our attention now to the alternative approach known as *index-mapping*. When once the basis of index-mapping has been explained, we will give a more detailed account of decomposition into 2-point transforms, and apply it to a variety of FFT algorithms.

The index-mapping approach may be introduced as follows. We start by expressing the transform length N as the product of two factors N_1 and N_2:

$$N = N_1 N_2 \tag{7.23}$$

We also define two new indices:

$$n_1 = 0, 1, 2 \ldots (N_1 - 1)$$

and

$$n_2 = 0, 1, 2 \ldots (N_2 - 1) \tag{7.24}$$

The following linear equation may now be used to *map* the values of n_1 and n_2 into the time-index n of the DFT:

$$n = (M_1 n_1 + M_2 n_2)_N \tag{7.25}$$

where M_1 and M_2 are constants, and the brackets and subscript N denote modulo-N. The mapping can be made one-to-one, in the sense that all the required values of n ($0 \le n \le N - 1$) are generated once (and only once) as n_1 and n_2 vary over their specified ranges. In a similar way the frequency-index k of the DFT can be mapped using:

$$k = (J_1 k_1 + J_2 k_2)_N \tag{7.26}$$

In general it is the form of equations (7.25) and (7.26), and the choice of the various constants, which define the FFT decomposition.

Let us begin with a simple example — a 4-point DFT which we will decompose into 2-point DFTs. We can use it to illustrate many important features of FFT algorithms, and to introduce further terminology. In this case $N=4$ and $N_1=N_2=2$. For reasons which will become clear as we proceed, we choose $M_1=2$, $M_2=1$, $J_1=1$, and $J_2=2$. Hence:

$$n = 2n_1 + n_2 \quad \text{and} \quad k = k_1 + 2k_2 \tag{7.27}$$

with n_1, n_2, k_1 and k_2 all taken over the range 0 to 1. The mappings of n and k are shown in Figure 7.4. Note that each of the required values of n and k is represented once, and once only.

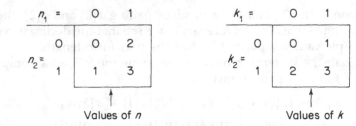

Figure 7.4 Index maps for a 4-point FFT decomposition.

The DFT (equation (7.19)) for $N=4$ is given by:

$$X[k] = \sum_{n=0}^{N-1} x[n]\, W_N^{kn} = \sum_{n=0}^{3} x[n]\, W_4^{kn} \tag{7.28}$$

It is helpful initially to substitute for n (but not for k), using equation (7.27):

$$X[k] = \sum_{n_2=0}^{1} \sum_{n_1=0}^{1} x[n_1, n_2]\, W_4^{k(2n_1+n_2)} \tag{7.29}$$

For simplicity, let us write $X[k]$ as X and $x[n_1, n_2]$ as x. Then:

$$X = \sum_{n_2=0}^{1} \sum_{n_1=0}^{1} x W_4^{2kn_1}\, W_4^{kn_2} \tag{7.30}$$

Since $W_4^{kn_2}$ is not a function of n_1, we have:

$$X = \sum_{n_2=0}^{1} W_4^{kn_2} \sum_{n_1=0}^{1} x W_4^{2kn_1} \tag{7.31}$$

giving:

$$X = 1 \sum_{n_1=0}^{1} x W_4^{2kn_1} + W_4^{k} \sum_{n_1=0}^{1} x W_4^{2kn_1} \tag{7.32}$$

At this point it is worth referring back to our earlier discussion of conventional decomposition. In equation (7.22), we expressed an N-point DFT in terms of two $N/2$-point DFTs, together with an additional multiplier W_N^k. This was the result of decimating the original signal $x[n]$ into two subsequences. We now see from equation (7.32) that index-mapping has achieved the same result

— indeed equation (7.22) becomes the same as equation (7.32) if we put $N=4$. This confirms that the index-mappings shown in Figure 7.4 are equivalent to a decimation-in-time of the original 4-point sequence into two 2-point subsequences.

Let us examine (7.32) a little more carefully. Referring to the index map for n, we see that if $n_2=0$ then $n=0$, 2 when $n_1=0$, 1; and if $n_2=1$ then $n=1$, 3 when $n_1=0$, 1. Hence:

$$X = \{x[0] + x[2]\,W_4^{2k}\} + W_4^k\,\{x[1] + x[3]\,W_4^{2k}\} \tag{7.33}$$

Therefore one of the 2-point DFTs involves pairing $x[0]$ and $x[2]$; the other involves pairing $x[1]$ and $x[3]$. Once again, we see that the decimation creates pairs of sample values separated by half the transform length.

If we evaluate the W_4 terms modulo-4, we obtain the following complete set of equations for the 4-point transform:

$$X[0] = \{x[0] + x[2]W_4^0\} + W_4^0\{x[1] + x[3]W_4^0\}$$
$$X[1] = \{x[0] + x[2]W_4^2\} + W_4^1\{x[1] + x[3]W_4^2\}$$
$$X[2] = \{x[0] + x[2]W_4^0\} + W_4^2\{x[1] + x[3]W_4^0\}$$
$$X[3] = \{x[0] + x[2]W_4^2\} + W_4^3\{x[1] + x[3]W_4^2\}$$

$$\tag{7.34}$$

This way of representing a 4-point DFT may be visualized more easily with the help of the *signal flow graph* in Figure 7.5(a). Starting on the left-hand side with the data values, we go through two stages of processing and end up on the right-hand side with the DFT coefficients $X[0]$ to $X[3]$. The first stage of processing involves forming two 2-point DFTs, by pairing $x[0]$ with $x[2]$ and $x[1]$ with $x[3]$. The results are combined in the second stage of processing, which incorporates the additional factor W_4^k. This stage also takes the form of two 2-point DFTs. In effect we have decomposed the original 4-point DFT into two sets of 2-point DFTs, defining a radix-2 FFT algorithm.

The convention used in the figure is that branches entering a node are summed to give the variable at that node. Weightings, or *transmittances*, are represented by circular symbols. The number inside each symbol denotes a power of W_4 — for example, a '2' denotes W_4^2. If there is no symbol, the branch transmittance is unity. By following paths through the graph we can express any of the outputs (X) in terms of the inputs (x). You may like to check equations (7.34) against the graph in this way.

The flow graph structure, and the branch transmittances, are directly related to equations (7.31) thru (7.34). In particular, it is worth noting that the inner sum over n_1 in equation (7.31) gives us the transmittance values in the first processing stage; the outer sum over n_2 gives the transmittances in the second stage. Note how each node has two input and/or output branches. In the first stage, one of each pair of branches is horizontal, the other 'jumps' one step vertically. In the second stage, one of each pair is horizontal and the other jumps two steps vertically. This reflects the increasing vertical span of the individual 2-point DFTs as we move from left to right across the graph.

Several of the symbols in Figure 7.5(a) denote W_4^0, which equals unity.

However, by keeping the symbols in the graph we emphasize its structure and symmetry. If the '0' symbols are omitted, we get the equivalent signal flow graph of figure 7.5(b).

It is important to note that, whereas the output data is arranged in *natural-order*, the input data array has been reordered or *shuffled*. This is because the initial 2-point DFTs require $x[0]$ to be paired with $x[2]$, and $x[1]$ to be paired with $x[3]$. The correct input sequence can be deduced from the index maps of Figure 7.4. Comparing values of k and n in the two 'boxes', we see that if the output index k is taken through the natural-order sequence $0-1-2-3$, the corresponding locations in the 'n-box' give the shuffled sequence $0-2-1-3$.

As we pointed out in our discussion of conventional decomposition, an N-point DFT can be decomposed into two $N/2$-point DFTs by decimating the input sequence. Each $N/2$-point DFT can in turn be decomposed into $N/4$-point DFTs, and so on. In the limit we are left with a series of 2-point transforms, and the FFT decomposition is complete. Such complete decomposition is not essential, but it is the usual approach — and the one we adopt in this chapter.

Note how simple the above FFT decomposition is. Part (b) of the figure shows that most of the branches have unity transmission, representing additions. Furthermore:

$$W_4^1 = \exp(-j2\pi/4) = -j$$
$$W_4^2 = (W_4^1)^2 = j^2 = -1$$

and
$$W_4^3 = W_4^1 W_4^2 = j \tag{7.35}$$

Hence the symbols '1' and '3' denote weighting by $\pm j$. They are simply conversions between real and imaginary parts — not multiplications. The symbol '2' represents weighting by -1, or subtraction. We see that a length-4 FFT requires no floating-point multiplications at all. The arithmetic values of the various transmittances are entered in part (c) of the figure, giving a third version of the signal flow graph.

The graph can be developed a little further. If (as we claim!) the problem has been reduced to 2-point transforms, it should be possible to express the branch transmittances in terms of $W_2 = \exp(-j2\pi/2) = -1$. Now:

$$W_4^0 = 1 = W_2^0; \; W_4^2 = W_2^1 = -1; \quad \text{and} \quad W_4^3 = W_4^2 W_4^1 = W_2^1 W_4^1$$
$$\tag{7.36}$$

Making these substitutions, we get the signal flow graph in Figure 7.5(d). Although most of the transmittances have been recast as powers of W_2 (and therefore have values ± 1), we cannot eliminate the terms W_4^1. They are referred to as *twiddle factors*. The twiddle factors represent additional arithmetic over and above that required by the 'straight' 2-point transforms in the first stage of processing. Note that, in this case, the two twiddle factors could easily be combined, because they occur in branches coming from the same node — a matter we return to a little later.

You may recall our earlier discussion of the additional multiplier W_N^k which arose when we decimated a signal into subsequences, and expressed its transform in terms of the transforms of the subsequences (equations (7.22) and

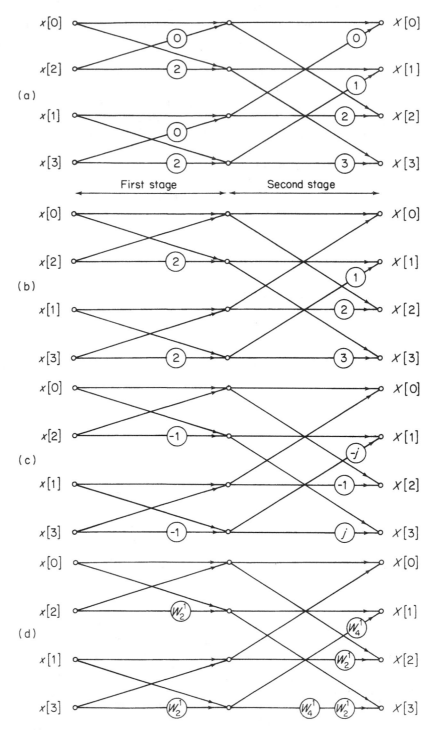

Figure 7.5 Signal flow graphs for a 4-point, in-place decimation-in-time FFT algorithm with shuffled input and natural-order output. (a) Branch transmittances expressed as powers of W_4; (b) unity transmittances omitted; (c) transmittances expressed as arithmetic values; (d) recast in terms of W_2 and twiddle factors.

(7.32)). Twiddle factors arise in the same way. They represent the additional computations required, after the complete DFT has been recast (as far as possible) in terms of elementary 2-point transforms.

Figure 7.5 shows there are several alternative ways of labeling branch transmittances on a signal flow graph. One difficulty when reading books and references on the FFT is that they tend to adopt different labeling schemes and conventions. However, if you start with the scheme in Figure 7.5(a) — using powers of W_N to denote transmittances for an overall DFT of length N — then the alternatives in parts (b), (c), and (d) of the figure are easily deduced. The scheme in part (a) also has the advantage of displaying a clear pattern.

It is quite possible to derive the transmittances and twiddle factors of Figure 7.5(d) by re-examining equation (7.31). Substituting for k as well as n (using equation (7.27)), we have:

$$X = \sum_{n_2=0}^{1} W_4^{(k_1+2k_2)n_2} \sum_{n_1=0}^{1} x W_4^{2(k_1+2k_2)n_1}$$

$$= \sum_{n_2=0}^{1} W_4^{k_1 n_2} W_4^{2k_2 n_2} \sum_{n_1=0}^{1} x W_4^{2k_1 n_1} W_4^{4k_2 n_1} \qquad (7.37)$$

Now $W_4^{4k_2 n_1} = W_4^0 = 1$ for all values of k_2 and n_1, so it does not contribute. Furthermore $W_4^2 = W_2$. Therefore:

$$X = \sum_{n_2=0}^{1} W_4^{k_1 n_2} W_2^{k_2 n_2} \sum_{n_1=0}^{1} x W_2^{k_1 n_1} \qquad (7.38)$$

Comparing with Figure 7.5(d) we see that the term $W_2^{k_1 n_1}$ in the inner sum gives the branch transmittances in the first processing stage; $W_2^{k_2 n_2}$ corresponds to transmittances in the second stage, together with twiddle factors $W_4^{k_1 n_2}$. The non-contribution of $W_4^{4k_2 n_1}$ simplifies the computation and has the effect of *uncoupling* the inner sum, which depends only on k_1. Such uncoupling results from an appropriate choice of constants in equation (7.27).

It is important to realize that we have developed all the above ideas in relation to one simple radix-2 FFT decomposition. We must be careful not to jump to too many general conclusions! The basic ideas will be expanded in the next section, where we design several types of FFT algorithm. In the meantime there are a number of additional features which must be mentioned.

The first of these concerns the important notion of an *FFT butterfly*. We recall that the 2-point transforms on the left-hand side of Figure 7.5 each involve one addition and one subtraction. This can be viewed as the basic computational element of the FFT, and is represented by Figure 7.6(a). Note that we have now indicated the subtraction by placing '-1' against the appropriate branch. Because of its shape, the figure is known as a 'butterfly'. In some books and references the shape is simplified by omitting the top and bottom horizontal lines, giving the form shown in part (b) of the figure. Even without arrows or labels, it has the same interpretation.

The basic FFT butterfly is also relevant to the second (and any subsequent)

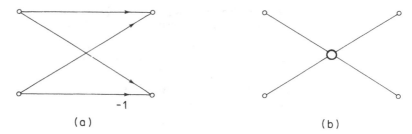

Figure 7.6 (a) Basic FFT butterfly; (b) alternative butterfly symbol.

processing stages, provided we include appropriate twiddle factors. Let us take the lower right-hand 2-point DFT in Figure 7.5(d) as an example. As already noted, its two twiddle factors W_4^1 can be combined, giving the flow graph of Figure 7.7(a). This is now in the form of a single twiddle factor followed by the basic FFT butterfly — as shown in part (b). The same perspective can be applied to all the 2-point transforms in a signal flow graph, giving an FFT decomposition which is expressed entirely in terms of butterflies and twiddle factors. From the point of view of computational efficiency, this is extremely valuable.

Closely related to twiddle factors and FFT butterflies is the idea of *in-place computation*. The butterflies in Figures 7.6 and 7.7 can be computed using a few elementary program instructions, storing the results *back in the original data locations*. This is because the data, once used, is not needed again. The idea can be extended to a complete signal flow graph, using the same (complex) data array throughout the computation, and updating it stage-by-stage. Storage requirements are thereby minimized. However, it is important to realize that such in-place computation requires either input or output data shuffling, to keep the nodes of each butterfly at the right horizontal level in the flow graph.

Most FFT decompositions involve more than the two processing stages shown in Figure 7.5. If we decompose an N-point DFT right down to 2-point DFTs, V processing stages are needed where $V = \log_2 N$. Let us denote the complex data array resulting from the mth stage as $X_m(l)$, where $l = 0,1,2 \ldots N-1$,

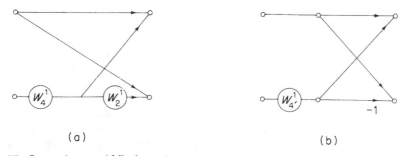

Figure 7.7 Separating a twiddle factor from a basic FFT butterfly.

and $m = 1,2 \ldots V$. We can think of the input data (after shuffling) as being loaded initially into the array and denoted by $X_0(l)$. The general form of butterfly for the $(m+1)^{\text{th}}$ stage of computation is then as shown in Figure 7.8, where T_F denotes the appropriate twiddle factor. $X_{m+1}(p)$ overwrites $X_m(p)$, and so on. The figure gives the following equations, which are implemented over and over again:

$$X_{m+1}(p) = X_m(p) + T_F X_m(q)$$
$$X_{m+1}(q) = X_m(p) - T_F X_m(q) \tag{7.39}$$

In many practical implementations the twiddle factor multiplications are considered as part of the butterfly and absorbed within it.

A common form of FFT program structure is based on three nested loops. The outer loop steps through the processing stages one at a time; the inner loops implement the butterfly and twiddle factor calculations. As with the standard DFT, various refinements and modifications are possible — including table lookups, and algorithms designed for real data. The book by Burrus and Parks

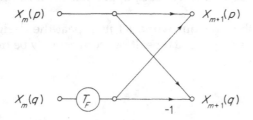

Figure 7.8 General form of butterfly and twiddle factor for a decimation-in-time FFT.

(see Bibliography) gives a useful account, and lists a number of detailed FFT programs.

To end this section we should consider the speed advantage of the FFT over a direct implementation of the DFT. We noted in the previous section that the number of complex multiplications in a direct implementation is of the order N^2. If N is an integer power of 2, and the FFT decomposition proceeds right down to 2-point transforms, there are $\log_2 N$ stages of FFT computation. Nominally there are N complex multiplications per stage, giving a total of $N \log_2 N$. Hence the speed advantage is expected to be approximately:

$$\frac{N^2}{N \log_2 N} = \frac{N}{\log_2 N} \tag{7.40}$$

This value is widely quoted in the DSP literature. Note that the advantage increases dramatically with transform length. For example if $N=8$, it is less than 3; but if $N = 2048 = 2^{11}$, it is nearly 200.

Expression (7.40) is based simply on the nominal amount of complex multiplication required. When programming FFTs in high-level languages on general-purpose computers, multiplications may indeed take up much of the execution time. However, special-purpose DSP hardware is often specifically designed to achieve fast multiplication, and this may render expression (7.40) inaccurate.

7.3.2   FFT ALGORITHMS

We have already pointed out that there is no such thing as 'the FFT'. Rather, there are many types of FFT algorithm based on the same general approach — decomposition of a DFT into successively shorter DFTs. Our aim in this section is to describe some of the most popular types of algorithm, using them to consolidate and extend ideas introduced in the previous section.

The simple FFT decomposition already examined (Figure 7.5) was valuable for introducing many of the basic ideas. It may be summarized as a *radix-2, decimation-in-time, in-place*, FFT algorithm, with *input data-shuffling* and a *natural-order output*. We hope that you already appreciate the meaning of these various terms!

To consolidate the previous work, let us repeat the design process for an 8-point FFT of the same general type. It can conveniently be treated as a worked example.

Example 7.1   An 8-point, radix-2, decimation-in-time, in-place, FFT may be defined by the following index-map equations (compare with equations (7.25) thru (7.27):

$$n = 4n_1 + 2n_2 + n_3 \quad \text{and} \quad k = k_1 + 2k_2 + 4k_3 \qquad (7.41)$$

with all six independent variables taken over the range 0 to 1.

(a) Construct index maps for n and k. Assuming a natural-order output, specify the shuffled input sequence.
(b) Draw a signal flow graph for the FFT, expressing all branch transmittances as powers of W_8.
(c) Redraw the flow graph in terms of basic (2-point) FFT butterflies and twiddle factors. How many complex multiplications are required by the FFT?

Solution

(a) In this example each index-map equation involves *three* independent variables, so we cannot draw maps in quite the same way as Figure 7.4. Instead we may tabulate values of n and k as follows:

n_1	n_2	n_3	n
0	0	0	0
1	0	0	4
0	1	0	2
1	1	0	6
0	0	1	1
1	0	1	5
0	1	1	3
1	1	1	7

k_1	k_2	k_3	k
0	0	0	0
1	0	0	1
0	1	0	2
1	1	0	3
0	0	1	4
1	0	1	5
0	1	1	6
1	1	1	7

Comparing the two maps, we see that if k is taken through the natural-order sequence 0 to 7, then the corresponding input sequence must be:

$$0, 4, 2, 6, 1, 5, 3, 7$$

(b) The form of the DFT for $N=8$ (compare with equation (7.28)) is:

$$X[k] = \sum_{n=0}^{N-1} x[n]W_N^{kn} = \sum_{n=0}^{7} x[n]W_8^{kn}$$

Substituting for n (compare with equations (7.29) thru (7.31) we obtain:

$$X = \sum_{n_3=0}^{1} \sum_{n_2=0}^{1} \sum_{n_1=0}^{1} xW_8^{k(4n_1+2n_2+n_3)}$$

$$= \sum_{n_3=0}^{1} W_8^{kn_3} \sum_{n_2=0}^{1} W_8^{2kn_2} \sum_{n_1=0}^{1} xW_8^{4kn_1} \qquad (7.42)$$

We could expand this equation to define the 2-point DFTs relevant to three successive processing stages (compare with equations (7.32) thru (7.34)). However, the result is rather cumbersome, so we will infer the form of the signal flow graph directly.

The inner sum of n_1 represents the first stage of processing. Branch transmittances, given by $W_8^{4kn_1}$, take values W_8^0, W_8^4, W_8^8, W_8^{12} ... and so on. Interpreted modulo-8 they all become either W_8^0 or W_8^4, and are shown on the left-hand side of Figure 7.9(a).

The middle sum over n_2 represents the second stage. Branch transmittances, given by $W_8^{2kn_2}$, take values W_8^0, W_8^2, W_8^4, W_8^6,

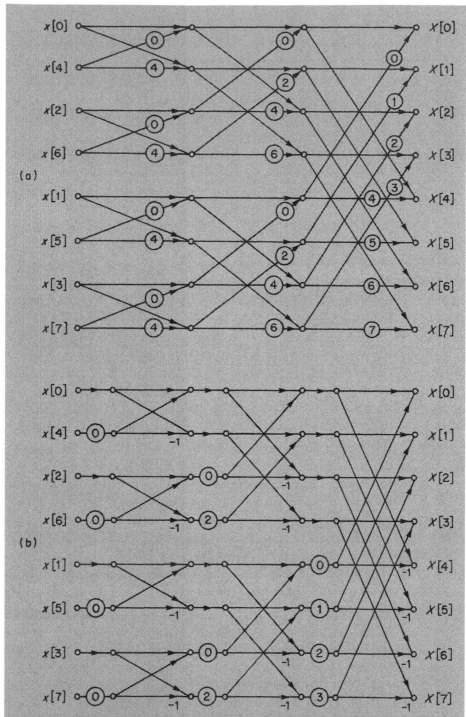

Figure 7.9 8-point, radix-2, in-place, decimation-in-time FFT with shuffled input and natural-order output. (a) Branch transmittances expressed as powers of W_8; (b) recast as basic butterflies and twiddle factors.

W_8^8 ... and so on. Interpreted modulo-8, they are as shown in the figure.

The outer sum over n_3 gives the third stage transmittances which vary between W_8^0 and W_8^7.

It is helpful to compare the overall structure of this graph with that in Figure 7.5(a). There are many similarities. Note how the 'vertical span' of the individual 2-point DFTs doubles each time we enter a new processing stage.

(c) To redraw the signal flow graph in the form of FFT butterflies and twiddle factors, we recall that the basic butterfly incorporates a subtraction (see Figure 7.6(a)). Furthermore $W_8^4 = -1$. Therefore if we have a transmittance equal to (say) W_8^6, we express it as $W_8^4 W_8^2 = (-1) W_8^2$. The '-1' is incorporated in the basic butterfly, leaving W_8^2 as a twiddle factor. Examination of all the transmittances shows that one twiddle factor is sufficient for each 2-point DFT, providing it is taken *outside* the basic butterfly — as in Figure 7.7(b). The resulting scheme is then as shown in part (b) of Figure 7.9.

Since many of the twiddle factors are $W_8^0 = 1$, they could be omitted. However, their inclusion helps underline the structure and symmetry of the graph.

We see that each processing stage involves four twiddle factors, giving twelve in all. However seven of them are unity, and three of them equal $W_8^2 = -j$. It is only the remaining two twiddle factors (W_8^1 and W_8^3) which actually require complex multiplications.

We recommend that you study Figure 7.9 carefully, noting its major features and symmetries, and comparing it with the signal flow graphs of Figure 7.5. Such decimation-in-time schemes are amongst the most widely-used of all FFTs, and come into the class of *Cooley–Tukey* algorithms first described in the mid-1960s.

The butterfly-twiddle factor arrangement of Figure 7.9(b) can also be related to the index-maps for n and k — very much as we did for the 4-point FFT in the previous section. In this case we have:

$$n = 4n_1 + 2n_2 + n_3 \quad \text{and} \quad k = k_1 + 2k_2 + 4k_3 \tag{7.43}$$

Substituting for both n and k, the 8-point DFT may be written as:

$$X = \sum_{n=0}^{7} x W_8^{kn} = \sum_{n_3=0}^{1} \sum_{n_2=0}^{1} \sum_{n_1=0}^{1} x W_8^{(k_1 + 2k_2 + 4k_3)(4n_1 + 2n_2 + n_3)} \tag{7.44}$$

The index of W_8 multiplies out to give:

$$(16k_3n_1 + 8k_2n_1 + 8k_3n_2 + 4k_1n_1 + 4k_2n_2 + 4k_3n_3 + 2k_1n_2 + 2k_2n_3 + k_1n_3) \tag{7.45}$$

Evaluated modulo-8, the first three of these terms always produce $W_8^0 = 1$, and do not affect the computation. We therefore have:

$$X = \sum_{n_3=0}^{1} \sum_{n_2=0}^{1} \sum_{n_1=0}^{1} x W_8^{4k_1n_1} W_8^{4k_2n_2} W_8^{4k_3n_3} W_8^{2k_1n_2} W_8^{2k_2n_3} W_8^{k_1n_3}$$

$$= \sum_{n_3=0}^{1} W_8^{n_3(2k_2+k_1)} W_8^{4k_3n_3} \sum_{n_2=0}^{1} W_8^{2k_1n_2} W_8^{4k_2n_2} \sum_{n_1=0}^{1} x W_8^{4k_1n_1} \qquad (7.46)$$

The three summations correspond to the three stages of processing in the signal flow graph. In each case there are transmittances equal to a power of W_8^4. These always have values ± 1 and are incorporated in the basic FFT butterfly. The extra terms — $W_8^{2k_1n_2}$ in the second stage, $W_8^{n_3(2k_2+k_1)}$ in the third stage — represent the twiddle factors.

We have mentioned previously that in-place FFT algorithms require data shuffling, to ensure that the input and output nodes of each butterfly are horizontally aligned. So far we have always assumed a shuffled *input*, with the output in natural order. However, it will probably not surprise you to know that the input data can be arranged in natural order if we are prepared to accept a shuffled *output*.

Perhaps the easiest way to explain this is to treat a signal flow graph as a *network* with a defined *topology*. No matter how its nodes are rearranged, the same result will be obtained if we use the same connections between nodes, and the same transmittances. It is only the order of data storage and processing which changes. Applying this idea to our 8-point decimation-in-time FFT, we can readily generate the alternative form shown in Figure 7.10(a). First, we enter the input data $x[0]$ to $x[7]$ in natural order down the left-hand side, and the output values in shuffled order down the right-hand side. It is now quite easy to fill in the various branches and transmittances for the first and third processing steps — making sure that exactly the same connections are made between the various nodes as in Figure 7.9(b). Finally, we add the branches and transmittances for the second stage.

It must be emphasized that Figure 7.10(a) is simply a rearrangement of Figure 7.9(b). The same computations are required in both cases, and we still have an in-place, decimation-in-time, algorithm.

It is also quite possible to reorganize the flow graph to give both input and output in natural order. However, the in-place nature of the algorithm must be sacrificed, because the horizontal levels of butterfly input–output nodes cannot be preserved. A scheme of this type is illustrated in part (b) of Figure 7.10, and we suggest you check it as a problem at the end of the chapter.

You may recall that we defined the shuffled input sequence for Figure 7.9 by referring to the appropriate index maps. We assumed the natural-order output sequence:

$$k = 0, 1, 2, 3, 4, 5, 6, 7$$

and by comparing equivalent locations in the two 'boxes', we found the input sequence:

$$n = 0, 4, 2, 6, 1, 5, 3, 7$$

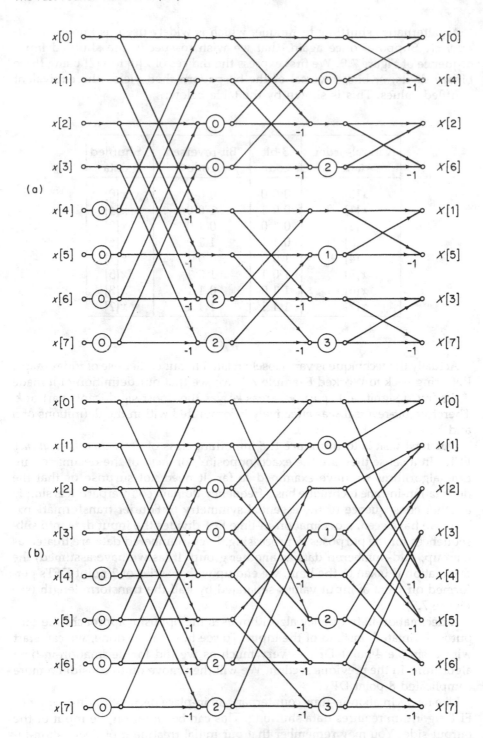

Figure 7.10 Alternative forms for the FFT of Figure 7.9. (a) In-place algorithm with natural-order input and shuffled output; (b) natural-order input and output.

An alternative shuffling technique, which is widely used, is known as *bit-reversal*. Suppose, once again, that we wish to specify the shuffled input sequence of Figure 7.9. We first express the indices of $x[0]$ to $x[7]$ using three binary digits, or bits; reversal of the bit pattern then yields the equivalent shuffled values. This is shown by the table below:

Natural-order data	3-bit code	Bit-reversed code	Shuffled data
$x[0]$	0 0 0	0 0 0	$x[0]$
$x[1]$	0 0 1	1 0 0	$x[4]$
$x[2]$	0 1 0	0 1 0	$x[2]$
$x[3]$	0 1 1	1 1 0	$x[6]$
$x[4]$	1 0 0	0 0 1	$x[1]$
$x[5]$	1 0 1	1 0 1	$x[5]$
$x[6]$	1 1 0	0 1 1	$x[3]$
$x[7]$	1 1 1	1 1 1	$x[7]$

Actually the technique is very closely related to our earlier use of index maps. Referring back to Worked Example 7.1, we see that our definition of n made n_1 its 'most significant' part, whereas k_3 was the 'most significant' part of k. Therefore bit-reversal was effectively incorporated within the definitions of n and k.

Our next task is to introduce the important class of *decimation-in-frequency* FFTs. In a sense they are the exact 'opposite', or *dual*, of the decimation-in-time algorithms we have examined so far. It need not surprise us that the decimation-in-time technique has a frequency-domain counterpart; it is simply another consequence of the essential symmetry of Fourier transformation.

As we have seen, a decimation-in-time FFT divides the input data into sub-sequences for the purposes of processing. Even-numbered data are treated as a group, odd-numbered data as another group. If (as we have assumed) the decimation is taken to the limit, we end up with a series of 2-point DFTs performed on pairs of input values separated by half the transform length (see Figure 7.9).

A decimation-in-frequency algorithm uses the opposite approach: the output is decimated, instead of the input. To see how this is done, we can start with a simple 4-point DFT — very much as we did for decimation-in-time algorithms in the previous section. We will then move on to consider a more complicated 8-point DFT.

Like its decimation-in-time counterpart, an in-place decimation-in-frequency FFT algorithm requires data shuffling. This can be either on the input or the output side. You may remember that our initial treatment of decimation-in-time algorithms assumed *input* data shuffling. To emphasize the dual nature of the decimation-in-frequency approach we will therefore start with an FFT which uses *output* shuffling.

A 4-point decimation-in-frequency FFT may be defined in terms of the following index equations for n and k (compare with equation (7.27)):

$$n = n_1 + 2n_2 \quad \text{and} \quad k = 2k_1 + k_2 \tag{7.47}$$

Substituting for n in the basic 4-point DFT equation, we have:

$$X[k] = \sum_{n=0}^{3} x[n] \, W_4^{kn}$$

$$= \sum_{n_2=0}^{1} \sum_{n_1=0}^{1} x[n_1, n_2] \, W_4^{k(n_1 + 2n_2)} \tag{7.48}$$

We now write $X[k]$ as X, $x[n_1, n_2]$ as x, and rearrange, giving:

$$X = \sum_{n_2=0}^{1} W_4^{2kn_2} \sum_{n_1=0}^{1} x W_4^{kn_1}$$

$$= 1 \sum_{n_1=0}^{1} x W_4^{kn_1} + W_4^{2k} \sum_{n_1=0}^{1} x W_4^{kn_1} \tag{7.49}$$

Each of these sums has the general form of a 2-point DFT. You may like to compare with equation (7.32). Both equations represent the same overall computation, but they are arranged rather differently.

The index maps corresponding to equation (7.47) are shown in Figure 7.11. We see that if $n_2=0$ then $n=0,1$ when $n_1=0,1$; and if $n_2=1$ then $n=2,3$ when $n_1=0,1$. Hence equation (7.49) is equivalent to:

$$X = \{x[0] + x[1] \, W_4^k\} + W_4^{2k} \{x[2] + x[3] \, W_4^k\} \tag{7.50}$$

Interpreting the indices of W_4 modulo-4, and remembering that $W_4^0 = 1$, equation (7.50) may be represented by the signal flow graph of Figure 7.12.

We have assumed a natural-order input and shuffled output. The output is decimated, since $X[0]$ and $X[2]$ are paired, and so are $X[1]$ and $X[3]$. The graph looks like a 'back-to-front' version of the decimation-in-time FFT of Figure 7.5(a).

$n_1 =$	0	1
$n_2 = 0$	0	1
1	2	3

$k_1 =$	0	1
$k_2 = 0$	0	2
1	1	3

Figure 7.11 Index maps for a 4-point decimation-in-frequency FFT.

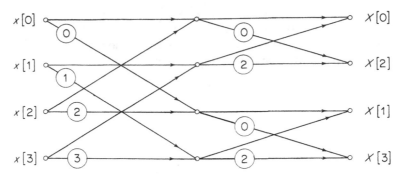

Figure 7.12 4-point, in-place, decimation-in-frequency FFT with natural-order input and shuffled output.

To rearrange Figure 7.12 in the form of FFT butterflies and twiddle factors, we need to combine pairs of branch transmittances which travel *towards*, rather than *away from*, a given node — in contrast to the decimation-in-time case. Therefore the twiddle factors must now *follow* the butterflies, rather than precede them. This means that the basic 2-point butterfly and twiddle factor combination for a decimation-in-frequency algorithm is as shown in Figure 7.13. The corresponding equations are:

$$X_{m+1}(p) = X_m(p) + X_m(q)$$
$$X_{m+1}(q) = \{X_m(p) - X_m(q)\}T_F \qquad (7.51)$$

The rearrangement of Figure 7.12 is shown in Figure 7.14.

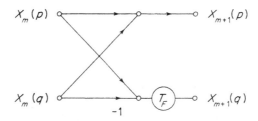

Figure 7.13 General form of butterfly and twiddle factor for a decimation-in-frequency FFT.

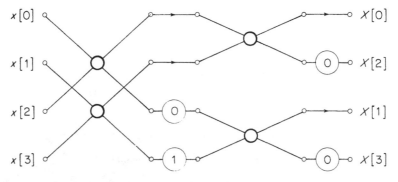

Figure 7.14 Figure 7.12 redrawn in terms of butterflies and twiddle factors.

Now that we have outlined the decimation-in-frequency approach, we can apply it to a rather more complicated example — an 8-point FFT which is the counterpart of the decimation-in-time algorithm explored in Worked Example 7.1.

Example 7.2 The index map equations for an 8-point, radix-2, in-place, decimation-in-frequency FFT are:

$$n = n_1 + 2n_2 + 4n_3 \quad \text{and} \quad k = 4k_1 + 2k_2 + k_3$$

Draw a signal flow graph for the FFT, expressing all branch transmittances as powers of W_8. Draw a second version of the graph in terms of basic (2-point) FFT butterflies and twiddle factors. Assume data shuffling on the output side.

Solution The solution has close parallels with that of Worked Example 7.1. The index maps are similar if n and k are interchanged. Hence the shuffling sequence — applied in this case to the output — is also the same. The FFT decomposition may be expressed as:

$$X = \sum_{n_3=0}^{1} \sum_{n_2=0}^{1} \sum_{n_1=0}^{1} x W_8^{k(n_1 + 2n_2 + 4n_3)}$$

$$= \sum_{n_3=0}^{1} W_8^{4kn_3} \sum_{n_2=0}^{1} W_8^{2kn_2} \sum_{n_1=0}^{1} x W_8^{kn_1}$$

The first processing stage (corresponding to the inner sum over n_1) has branch transmittances equal to powers of W_8^1; in the second stage they are powers of W_8^2; and in the third stage they are powers of W_8^4.

Since this is a decimation-in-frequency algorithm with output data shuffling, its general form must be like Figure 7.12. In particular, the final processing stage must involve basic 2-point transforms on adjacent (shuffled) values of X. If we move back through stage 2 to stage 1, the vertical span of the 2-point transforms will double at each stage.

The above considerations lead to the signal flow graph of Figure 7.15(a), in which all branch transmittances are expressed modulo-8.

To recast the graph in terms of FFT butterflies and twiddle factors, we combine the transmittances of branches travelling towards a given node. For example the node marked N' at the bottom of Figure 7.15(a) has two input branches with transmittances W_8^3 and W_8^7. Since the basic FFT butterfly includes a subtraction, we write W_8^7 as $W_8^4 W_8^3 = (-1) W_8^3$. The W_8^3 term becomes the twiddle factor common to the two input branches.

The overall signal flow graph is redrawn in part (b) of the figure, using the simplified form of butterfly schematic.

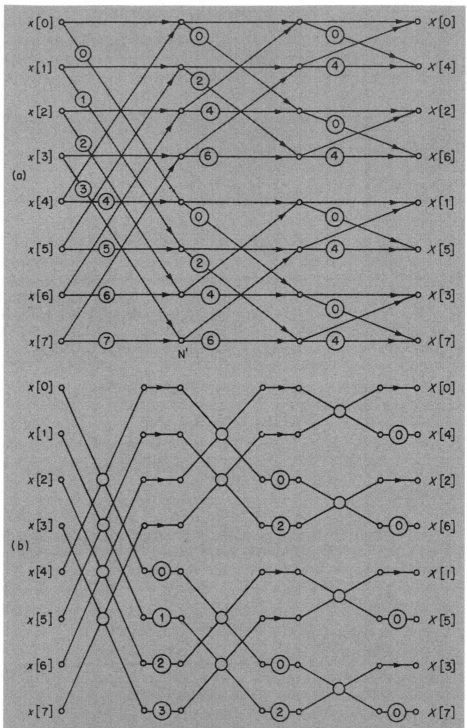

Figure 7.15 8-point decimation-in-frequency FFT. (a) Branch transmittances expressed as powers of W_8; (b) redrawn in terms of butterflies and twiddle factors.

The twiddle factors can also be inferred by substituting for both n and k in the 8-point DFT equation — just as we did for the decimation-in-time FFT of Worked Example 7.1. We suggest this is a problem at the end of the chapter.

There are several possible forms for a given decimation-in-frequency FFT. As long as we preserve the connections between nodes, and use the same branch transmittances, the flow graph can be rearranged in any way we wish. The idea was demonstrated for a decimation-in-time FFT in Figure 7.10, producing two alternative forms. In Figure 7.16 we show one of the alternative forms of the 8-point decimation-in-frequency algorithm — an in-place version with shuffled input and natural order output. There is a problem on this, too, at the end of the chapter.

We have previously mentioned the *duality* of the decimation-in-time and decimation-in-frequency approaches. Specifically, if we compare the signal flow graphs of Figures 7.9 and 7.15, or the FFT butterflies of Figures 7.8 and 7.13, we notice a rather striking feature: each figure can be obtained from the other by simply reversing the signal flow direction and interchanging inputs and outputs. This has important implications for computing the inverse Discrete Fourier Transform (IDFT) — a problem we have not so far discussed.

The IDFT (see equation (7.2)) is given by:

$$x[n] = \frac{1}{N} \sum_{k=0}^{N-1} X[k]\, W_N^{-kn} \tag{7.52}$$

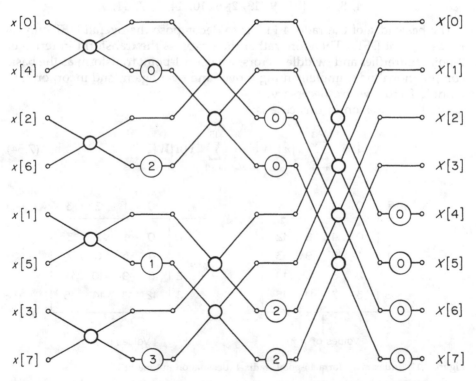

Figure 7.16 An alternative form of Figure 7.15(b) with shuffled input and natural-order output.

Apart from the $1/N$ scaling factor and change of sign in the index of W_N, it is identical to the DFT. It follows that an FFT algorithm can be used to compute the IDFT, if we divide the result by N and use powers of W_N^{-1} rather than W_N. This means that a given algorithm can be used to process either time-domain data, or frequency-domain data, with only very minor modifications. For example, the 8-point decimation-in-time FFT algorithm of Figure 7.9 is readily adapted to give an 8-point decimation-in-frequency *inverse* FFT.

We have concentrated so far on the widely-used type of radix-2 FFT, in which an N-point DFT is decomposed right down to 2-point transforms. However, it is quite possible to specify radix-4 (or higher radix) algorithms, which may be attractive for implementing FFTs efficiently on special-purpose DSP hardware. We therefore end this section by giving a brief introduction to the radix-4 approach from the index-mapping point of view.

Let us take as an example a 16-point DFT, decomposed into a series of 4-point DFTs using an in-place, decimation-in-time, FFT algorithm with data shuffling on the input side. In this case $N = 16$ and $N_1 = N_2 = 4$. The index map equations for n and k are:

$$n = 4n_1 + n_2 \quad \text{and} \quad k = k_1 + 4k_2 \tag{7.53}$$

with n_1, n_2, k_1 and k_2 all taken over the range 0 to 3. The index maps are shown in Figure 7.17. For a natural-order output, the equivalent locations in the 'n-box' show that the input must be shuffled to give the sequence:

$$0, 4, 8, 12, 1, 5, 9, 13, 2, 6, 10, 14, 3, 7, 11, 15$$

The basic idea of the radix-4 FFT is to decompose the overall DFT only as far as 4-point DFTs. Therefore rather than express the transform in terms of 2-point butterflies and twiddle factors, we use a 4-point transform as the basic computational unit, implementing it over and over again, and incorporating twiddle factors where necessary.

The DFT equation in this case is:

$$X[k] = \sum_{n=0}^{N-1} x[n]\, W_N^{kn} = \sum_{n=0}^{15} x[n] W_{16}^{kn} \tag{7.54}$$

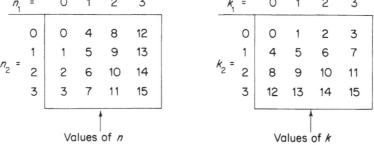

Figure 7.17 Index maps for a 16-point, radix-4, decimation-in-time FFT.

Separate terms representing the 4-point transforms and twiddle factors can be identified by substituting for both k and n using equation (7.53). Writing $X[k]$ as X, and $x[n]$ as x, we obtain:

$$X = \sum_{n_2=0}^{3} \sum_{n_1=0}^{3} x W_{16}^{(k_1+4k_2)(4n_1+n_2)}$$

$$= \sum_{n_2=0}^{3} \sum_{n_1=0}^{3} x W_{16}^{4k_1n_1} \, W_{16}^{16k_2n_1} \, W_{16}^{4k_2n_2} \, W_{16}^{k_1n_2} \qquad (7.55)$$

Now $W_{16}^{16k_2n_1} = W_{16}^{0} = 1$ for all values of k_2 and n_1, and does not affect the computation. This is the decoupling effect noted earlier, which stems from an appropriate choice of parameters in equation (7.53). Also, we may write W_{16}^{4} as W_4. Rearranging, we obtain:

$$X = \sum_{n_2=0}^{3} W_4^{k_2n_2} \, W_{16}^{k_1n_2} \sum_{n_1=0}^{3} x W_4^{k_1n_1} \qquad (7.56)$$

The inner sum over n_1 represents four 4-point DFTs. The outer sum over n_2 represents another four 4-point DFTs, together with twiddle factors given by $W_{16}^{k_1n_2}$. In other words we have defined a radix-4 FFT which can be implemented in two processing stages.

The basic computational unit of this type of algorithm — the 4-point DFT — is obviously rather more complicated than the 2-point butterfly we have used previously. However, if you refer back to Figure 7.5(c), you will see that it requires no multiplications — only additions, subtractions, and changes between real and imaginary. If the 4-point DFT is itself decomposed into 2-point butterflies and twiddle factors, three of the four twiddle factors turn out to be unity (see Figure 7.14). This helps explain why radix-4 decompositions are often highly efficient. Indeed, the argument may be taken at least one step further. An 8-point DFT can be broken down so that it requires only two complex multiplications (the twiddle factors W_8^1 and W_8^3 in Figures 7.9(b) and 7.15(b)). Radix-8 FFTs may also be attractive in some applications.

One danger of illustrating FFT algorithms with 2-point, 4-point, and 8-point signal flow graphs is that we tend to forget the lengthy DFTs often required in practical DSP. For example, we may need to transform signals with many hundreds, or thousands, of sample values. In such cases higher-radix FFTs are a genuine option.

It is also quite possible to use *mixed-radix* algorithms. For example, a lengthy DFT might be decomposed into a mixture of 2-point and 4-point DFTs. Another aspect of FFT design concerns transform lengths which are not integer powers of 2. As we mentioned before, the index-mapping approach is particularly useful in such cases. These and other more advanced FFT topics are covered in several of the books listed in the Bibliography.

Table 7.1 The Discrete Fourier Transform: properties

Property or operation	Signal	DFT
Transformation	$x[n]$	$X[k] = \sum\limits_{n=0}^{N-1} x[n]\, W_N^{kn},$ $0 \leqslant k \leqslant (N-1)$
Inverse transformation	$x[n] = \frac{1}{N}\sum\limits_{k=0}^{N-1} X[k]\, W_N^{-kn},$ $0 \leqslant n \leqslant (N-1)$	$X[k]$
Linearity	$A x_1[n] + B x_2[n]$	$A X_1[k] + B X_2[k]$
Time-shifting	$x[n-n_0]$	$X[k]\, W_N^{kn_0}$
Convolution	$\sum\limits_{m=0}^{N-1} x_1[n]\, x_2[m-n]$	$X_1[k]\, X_2[k]$
Modulation	$x_1[n]\, x_2[n]$	$\frac{1}{N}\sum\limits_{m=0}^{N-1} X_1[m]\, X_2[k-m]$
Real time function	$x[n]$	$X[k] = X^*[-k]$ $\mathcal{R}(X[k]) = \mathcal{R}(X[-k])$ $\mathcal{I}(X[k]) = -\mathcal{I}(X[-k])$

PROBLEMS

SECTION 7.2.1

Q7.1 Suppose you have a program for evaluating a DFT:

$$X[k] = \sum_{n=0}^{N-1} x[n]\, \exp(-j2\pi kn/N), \quad 0 \leq n \leq (N-1)$$

Explain clearly how it can be modified and used to compute the inverse DFT.

Q7.2 Explain:

(a) the distinction between the discrete Fourier Series and the DFT,
(b) the distinction between the discrete-time version of the Fourier Transform and the DFT.

What is the justification for representing an N-sample digital signal by N spectral coefficients?

SECTION 7.2.2

Q7.3 An N-sample signal $x[n]$ has the DFT $X[k]$. Write down expressions for the DFTs of signals:

(a) $3x[n]$;
(b) $x[n-2]$;
(c) $2x[n] + x[n+1]$;
(d) $x[n]x[n-1]$

where all time shifts are assumed periodic.

Q7.4 What statements can you make about the DFT coefficients $X[k]$ of an N-sample signal which is:

(a) real;
(b) real and even;
(c) real and odd;
(d) complex?

Q7.5 Under what minimal set of conditions will an N-point signal produce a DFT which is periodic in $N/2$?

SECTION 7.2.3

Q7.6 Two real signals have the following sample values:

(a) 1, -1;
(b) 1, 2, 1, 3

Estimate the real and imaginary parts of their DFT coefficients $X[k]$.

Q7.7 Suppose you have just computed the 16-point DFT of a real signal, expecting the output to be given in the range $X[0] \ldots X[15]$. However the results are presented as $X[-8] \ldots X[7]$. Relate the coefficients given to those expected.

Q7.8 A signal $x[n] = \cos(18\pi n/40)$ is applied to a 40-point DFT.

(a) Which coefficients have the largest magnitudes?
(b) What are the magnitudes of $X[0]$ and $X[39]$?

Q7.9 Use Program no. 23 in Appendix A1 to compute the DFT of the following signals:

(a) $x[n]$ as listed in the program
(b) $x[n] = 1$, $0 \leq n \leq 63$
(c) $x[n] = \delta[n-1]$, $0 \leq n \leq 63$.

In each case make sure you can explain the form of the screen plot.

Q7.10 Modify Program no. 23 in Appendix A1 to compute DFTs of real signals with (a) 16, (b) 32, (c) 64, and (d) 128 sample values. Compare computation times for the different transform lengths, with and without the table lookup.

SECTION 7.3.1

Q7.11 Tabulate values of W_N^{kn} for a 4-point DFT, with n and k taken over the range 0 to 3.

Q7.12 A real signal $x[n]$ has N sample values, where N is divisible by 2. Let $x_1[n]$ and $x_2[n]$ be the two $N/2$-point sequences defined as:

$$x_1[n] = x[2n], \qquad n = 0, 1, 2 \ldots (N/2 - 1)$$
$$x_2[n] = x[2n+1], \qquad n = 0, 1, 2 \ldots (N/2 - 1)$$

If $x_1[n]$ and $x_2[n]$ transform to $X_1[k]$ and $X_2[k]$ respectively, determine the DFT of $x[n]$ in terms of $X_1[k]$ and $X_2[k]$.

Q7.13 Explain the meaning of the following terms as applied to an FFT algorithm:

(a) radix-2,
(b) decimation-in-time,
(c) in-place,
(d) shuffled-input,
(e) natural-order output,
(f) butterfly,
(g) twiddle factor.

In a 64-point, radix-2, decimation-in-time FFT with shuffled input and natural-order output, what powers of W_{64} would you expect to find associated with the butterflies and twiddle factors in the final processing stage?

Q7.14 Assuming that the time taken to compute a DFT or FFT is proportional to the number of multiplications involved, what speed advantage would you expect a radix-2 FFT to have for the following lengths of signal:

(a) $N = 128$;   (b) $N = 1024$;   (c) $N = 65536$?

SECTION 7.3.2

Q7.15 A 16-point radix-2, in-place, FFT algorithm has the following index map equations:

$$n = 8n_1 + 4n_2 + 2n_3 + n_4 \quad \text{and} \quad k = k_1 + 2k_2 + 4k_3 + 8k_4$$

with all independent variables taken between 0 and 1. Construct index maps for n and k.

The FFT is to be implemented using shuffled input data and natural-order output. Define the shuffled input sequence using (a) the index maps, and (b) the bit-reversal technique.

Q7.16 Use the bit-reversal technique to define the shuffle sequence for a 32-point in-place FFT algorithm.

Q7.17 Check that the FFTs illustrated in Figures 7.9(b) and 7.10(b) of the main text involve the same computations.

Q7.18 Figure Q7.18 shows signal flow graphs for two in-place FFT algorithms. In (a) the input data is shuffled, whereas in (b) the output data is shuffled. What types of algorithm are they, and why?

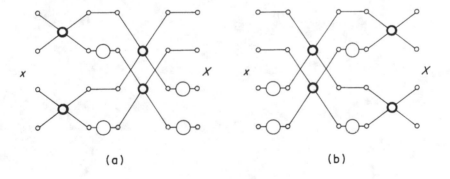

(a)   (b)

Figure Q7.18

Q7.19 Worked Example 7.2 in the main text describes an 8-point decimation-in-frequency FFT. By substituting for n and k in the 8-point DFT equation, identify W_8 terms representing basic (2-point) butterflies and twiddle factors in the various processing stages.

Q7.20   Figure 7.16 in the main text shows an alternative form for the 8-point decimation-in-frequency FFT covered in Worked Example 7.2. Check that its branch connections and transmittances are correct. Draw a signal flow graph for another alternative form with natural-order input *and* output. Why is it not possible to compute this FFT 'in place'?

Q7.21   Explain the essential differences between decimation-in-time and decimation-in-frequency FFT algorithms.

Q7.22   A 64-point, radix-4, in-place, decimation-in-frequency FFT has the following index map equations:

$$n = n_1 + 4n_2 + 16n_3 \quad \text{and} \quad k = 16k_1 + 4k_2 + k_3$$

with all independent variables taken over the range 0 to 3.

Express the FFT as a set of nested summations, and identify W_{64} terms associated with the 4-point DFTs and twiddle factors in the three processing stages.

—— CHAPTER 8 ——
FFT
Processing

8.1 INTRODUCTION

We have already devoted three chapters of this book to frequency-domain aspects of DSP. Chapter 3 introduced many of the basic ideas of digital Fourier analysis and illustrated the spectral properties of digital signals and LTI systems. This work was complemented by coverage of the z-Transform, including z-plane poles and zeros, in Chapter 4. Chapter 7 was mainly concerned with the computational aspects of discrete Fourier transformation, including the design and organization of FFT algorithms.

In this chapter we focus on two of the main application areas of FFT processing. The first – digital spectral analysis – uses the FFT to explore the properties of digital signals and systems. The second – digital filtering by fast convolution – offers a frequency-domain alternative to the time-domain filtering techniques discussed in Chapters 5 and 6.

We noted previously that the development of efficient FFT algorithms from the mid-1960s onwards had a profound effect on DSP. Frequency-domain techniques which had previously been impracticable, due to the heavy computational cost of Fourier transformation, suddenly became quite feasible. The development was accelerated by the ever-increasing performance of digital hardware, making FFT processing valuable in a great variety of applications. FFT software is now widely available on general-purpose computers. FFT techniques are increasingly used in specialist instrumentation — for example, to give fast and accurate spectral analysis. In such fields as radar and sonar, dedicated FFT hardware is now quite common.

Much of the background to digital spectral analysis has already been covered in Chapters 3 and 7. In particular we have seen that:

The DFT of an aperiodic, finite-length, signal involves similar computations to those of the Fourier Series of a periodic version of the same signal.

The DFT coefficients $X[k]$ of an aperiodic signal $x[n]$ may be regarded as samples over one period of the signal's Fourier Transform $X[\Omega]$. The sampling is 'adequate' to define the spectrum completely, according to a frequency-domain version of the Sampling Theorem.

It follows that the spectral properties of digital signals and LTI systems, explored by FFT processing, parallel those described in Chapter 3. You may wish to review this material, and to refer to some specific comments about the relationships between the Fourier Series, Fourier Transform, and DFT/FFT made in Section 7.2.1.

We next make a few general remarks about the use of FFT processing in digital filtering. In Chapters 5 and 6 we described a number of time-domain techniques for designing digital filters. Although we generally started with a frequency response specification, the aim was to devise a time-domain difference equation which gave the required filtering effect. At no stage did we need to find the spectrum of the signal being processed.

An alternative approach, based on the FFT, can be summarized as follows. First, we transform the input signal into the frequency domain. The various spectral coefficients are then modified in magnitude and phase according to the desired filter characteristics. This means that filtering is accomplished by frequency-domain multiplication rather than time-domain convolution. The filtered output signal is finally recovered by inverse transformation. The approach is very flexible and may in principle be used to implement any desired frequency response specification.

Of course, the effectiveness of this method relies on the speed of FFT algorithms. We should be careful not to assume that it is necessarily superior to time-domain filtering. Its flexibility is not always required. Furthermore a time-domain filter with a few recursive coefficients may well be faster than an FFT implementation. In practice the choice between them depends not only on speed, but also on the amount of storage needed, and whether or not true on-line working is required. If we require a filtered output sample to be generated every time a new input sample becomes available, then we must specify a time-domain filter, because effective use of the FFT demands that data is stored and processed in substantial blocks.

Before ending this section, we wish to introduce a general-purpose FFT program which will be used later to demonstrate various aspects of FFT processing. Program no. 24 in Appendix A1 is a radix-2, in-place, decimation-in-time FFT with shuffled input and natural-order output. It includes a data-shuffling routine on the input side. The program can cope with DFTs (and IDFTs) of any length equal to an integer power of 2. We cannot claim that it is optimized for a particular type of computer, nor is a high-level language program as fast as one written in machine code. Nevertheless it allows DFTs and IDFTs up to lengths of (say) 4096 points to be calculated quite speedily on modest personal computers.

In line with our previous practice in this book, we do not intend to describe the program in detail. Rather, we will make some general comments on its organization, to help any of our readers who wish to use it. It may be helpful to say that the program uses three nested loops to implement the actual FFT algorithm. This fairly standard programming technique is described in various books, including those by Stanley, Dougherty and Dougherty, and by Burrus and Parks (see Bibliography). If you are interested in the detailed programming of FFT algorithms, we suggest these as useful references.

The program begins by defining storage arrays for the real and imaginary parts of the input data, the length of the transform required (N), and the parameter $M = \log_2 N$. For demonstration purposes we choose a 512-point transform — although you may, of course, alter this if you wish. A suitable input signal — in this case a 'rectangular pulse' consisting of 32 unit sample values — is loaded, and plotted on the screen (see Figure 8.1(a)). Note that we are assuming a real signal, with all imaginary parts set to zero.

Two control parameters, T and D, determine whether a transform or inverse transform is performed. T sets the sign of the sine terms, and hence of the exponential, in DFT equations (7.1) and (7.2). D provides the $1/N$ scaling factor in equation (7.2). Thus if $T = 1$ and $D = 1$, we compute the transform; if $T = -1$ and $D = N$, the inverse transform. On the first 'pass' through the program, the transform is estimated.

Following input data shuffling, the in-place decimation-in-time FFT is computed. The result is normalized to a suitable peak value, then plotted on the screen. For convenience and ease of interpretation, we have chosen to plot just the magnitude of the transform (rather than magnitude and phase, or real and imaginary parts). Remember, however, that the transform is generally complex, and that its real and imaginary parts remain stored in the relevant arrays.

We have previously emphasized that the process of inverse transformation is almost identical to that of transformation. Our program demonstrates this important fact by carrying out an inverse transform on the spectrum it has just calculated. At the end of the first 'pass', parameters T and D are set to give an inverse transform, and a second 'pass' through the data shuffling and transformation parts of the program takes place. The new result is also plotted on the screen. With a little luck, we should find that the original input signal has been recovered!

The complete screen plot for the input signal we have chosen is reproduced in Figure 8.1. Part (a) shows the rectangular pulse input. Sample values are plotted as adjacent vertical bars, so the 'pulse' appears as a solid block. Part (b) shows the magnitude of the transform, comprising 512 discrete spectral coefficients. Note that, since $x[n]$ is real, the spectrum shows the usual mirror-image pattern. Its half-way point, the 256th spectral coefficient, corresponds to $\Omega = \pi$ in other frequency plots in this book.

In part (c) we see the signal recovered from the spectrum by inverse transformation. The transform-inverse transform process has indeed come full circle.

It is important to remember that parts (b) and (c) of the plot represent *magnitudes*. They are assessed by taking the square root of the sum of squares of the real and imaginary parts. Since we have chosen a signal with only *positive* values for this illustration, parts (a) and (c) of the figure are identical. In general, however, (c) would be a 'rectified' version of (a). Bearing this point in mind, you may like to try other forms of input signal, checking that inverse transformation gets back to the starting point.

It is interesting to assess the FFT program's speed for various transform lengths, and to compare it with a direct DFT implementation. In Section 7.3.1 we noted that the speed advantage of the FFT is expected to be about $N/\log_2 N$, based on the number of multiplications involved. (Of course, many other arithmetic and control operations are needed as well, so the estimate is only approximate.) In Figure 8.2 we show results of some actual speed tests on a

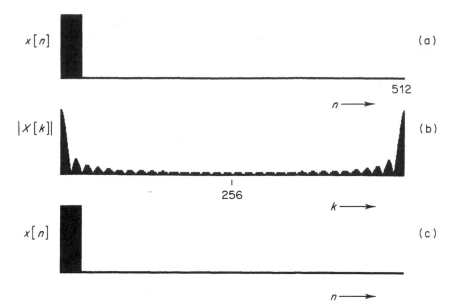

Figure 8.1 (a) A signal $x[n]$, (b) its spectral magnitude $|X[k]|$ found using the FFT, and (c) the signal recovered by inverse transformation *(each abscissa: 512 samples)*.

personal computer. Computer Programs no. 24 (FFT) and no. 23 (DFT), adapted to perform different lengths of transform, were used for these tests. Much less emphasis should be placed on the *absolute* computation times given, than on their *relative* values. This is because the speeds of different computers vary widely, and improve year by year. But the results are broadly in line with the $N/\log_2 N$ factor mentioned above, and confirm a growing advantage for the FFT as transform lengths increase. In fact these tests were carried out several

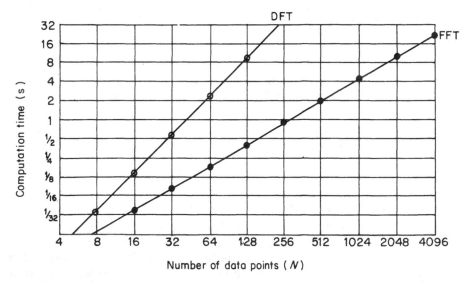

Figure 8.2 A comparison of DFT and FFT computation times on a personal computer.

years ago, and recent personal computers give dramatic reductions in computation time compared with the figure.

8.2 SPECTRAL ANALYSIS

8.2.1 GENERAL CONCEPTS

Digital spectral analysis — the decomposition of a signal into its frequency components using a computer or special-purpose hardware — is a valuable technique in many branches of engineering, applied science, and data processing. An FFT algorithm is the natural choice for this work because of its speed. The basic assumption behind FFT analysis is that a frequency-domain description is likely to reveal important information which is not apparent in the time-domain signal. Note that spectral analysis, unlike digital filtering, is primarily *investigative*. It is not necessarily concerned with *modifying* the signal. Nevertheless the information it yields often leads to important insights or decisions.

In many cases digital spectral analysis is concerned with *naturally-occurring signals*. Examples arise in the analysis of speech, biomedical signals such as the EKG (electrocardiogram), meteorological data, stock market indicators, and so on. Some of these applications involve searching for a wanted signal in the presence of unwanted disturbances or 'noise', on the basis of their different spectral distributions.

A rather different set of applications of spectral analysis is slanted towards investigating *systems*. A system is deliberately disturbed with a suitable input signal — often an impulse or a step function — and its response measured. Spectral analysis then yields information about the frequency-dependent properties of the system. Quite often the input signal has to be repeated many times, and the responses averaged, in order to reduce the effects of noise (see Section 10.5.4). Examples of these techniques arise in such diverse fields as the testing of electronic circuits and filters, the analysis of vibrations in buildings and structures, and in radar, sonar, and seismology.

An FFT algorithm yields a set of spectral coefficients which may be regarded as samples of the underlying continuous spectral function — or as harmonics of a periodic version of the same signal. Although a lengthy transform such as that shown in Figure 8.1(b) may look more or less like a continuous spectrum, it actually consists of a discrete set of coefficients, or harmonics. Just as a digital signal $x[n]$ is sampled in the time domain, so its DFT is sampled in the frequency domain.

We pointed out in Section 3.2.1 that end-on-end repetition of a signal causes no unnatural 'jumps' or discontinuities if all its frequency components have an integral number of periods within the transform length (N). Each component then occupies a definite harmonic frequency, corresponding to a single spectral coefficient. However, the situation is more complicated when $x[n]$ contains frequencies which do not meet this criterion. Repetition causes discontinuities in the time domain, leading to spectral spreading or *leakage* in the frequency domain. We previously demonstrated the effect for a 64-point transform in Figure 3.4.

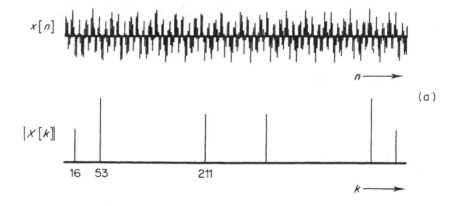

(a)

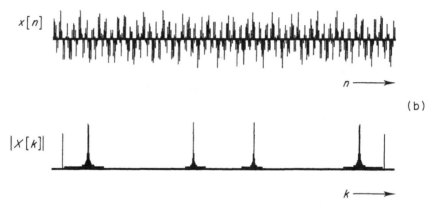

(b)

Figure 8.3 Fourier transformation of (a) a signal containing three exact Fourier harmonics, and (b) a signal containing both harmonic and non-harmonic components *(each abscissa: 512 samples).*

This matter is of such significance for spectral analysis that we illustrate it again in Figure 8.3, using our FFT program. (The plots in the figure were obtained by changing the input signal to the program. Also, the inverse transform included at the bottom of Figure 8.1 has been omitted.) Part (a) of the figure shows the signal:

$$x[n] = 0.1 \sin\left(\frac{2\pi n}{512} 16\right) + 0.2 \sin\left(\frac{2\pi n}{512} 53\right)$$

$$+ 0.15 \cos\left(\frac{2\pi n}{512} 211\right), \quad 1 \le n \le 512 \tag{8.1}$$

Its three frequency components correspond exactly to the 16th, 53rd, and 211th harmonics of the Fourier Series (there is no significance in these *particular* harmonics, which are chosen arbitrarily). The FFT magnitude, shown below, confirms that each component gives rise to a single spectral coefficient with a height proportional to amplitude. Since we are plotting magnitudes, sines and cosines are treated equally. And the FFT is, as usual, symmetrical about its midpoint because $x[n]$ is real.

In part (b) of the figure, we have altered the frequency of two of the components slightly. The signal is now:

$$x[n] = 0.1 \sin\left(\frac{2\pi n}{512} 16\right) + 0.2 \sin\left(\frac{2\pi n}{512} 53.5\right)$$

$$+ \; 0.15 \cos\left(\frac{2\pi n}{512} 211.25\right), \quad 1 \le n \le 512 \tag{8.2}$$

The first component gives a single spectral coefficient, as before. The second now displays a double peak at the 53rd and 54th harmonic frequencies, with spectral leakage to either side. The third component, with a frequency closer to the 211th harmonic than the 212th, gives a relatively large 211th coefficient — and there is again spectral leakage.

There are two rather different, and complementary, ways of explaining spectral leakage. The first, already mentioned, is to think of it as the frequency-domain counterpart of discontinuities in the time-domain. Thus if we wanted to synthesize the signal $x[n]$ in Figure 8.3(b) — including the discontinuities caused by repetition — we would have to superpose all the components indicated in its spectrum.

The other explanation, which gives valuable insight into the nature of discrete Fourier transformation, is to regard the DFT (or FFT) as a type of filtering process. A DFT behaves like a set of elementary bandpass filters which split the signal into its various frequency components. This is illustrated in Figure 8.4(a) for the simple case of an 8-point transform ($N = 8$). The frequency range $0 < \Omega < 2\pi$ is effectively divided into eight overlapping bands, and the 'amount' of input signal falling within each band is estimated.

It is important to notice that the peak response of each filter coincides with

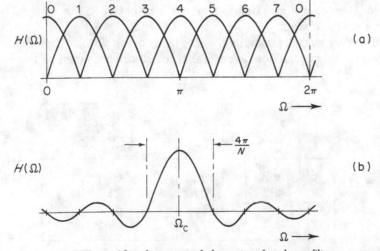

Figure 8.4 An 8-point FFT considered as a set of elementary bandpass filters.

zero response in its neighbors. Thus, for example, a sinusoidal input signal at the center-frequency of filter no. 3 would give a peak output from that filter, but zero output from all the others. Not surprisingly, it turns out that the eight center-frequencies shown in the figure correspond exactly with the harmonics of the 8-point transform.

Part (a) of the figure does not give quite the complete picture, because each elementary filter characteristic has substantial sidelobes to either side of its main lobe. As the transform length increases, each characteristic tends to a sinc function. We show such a function in part (b). The width of its main lobe is $4\pi/N$ radians; the sidelobes are $2\pi/N$ radians wide, with amplitudes decreasing away from the center-frequency Ω_c. Note that the zero-crossings of the sinc function coincide with the center-frequencies of the *other* filters. Thus, once again, a signal component at an exact harmonic frequency only produces an output from one of the filters.

If a component is displaced slightly from the filter's center-frequency, it gives a smaller peak response, plus a whole series of sidelobe responses from adjacent filters. This is the spectral leakage effect. As we show in the following Worked Example, it can quite easily be quantified.

Example 8.1 Predict the spectral leakage effects already illustrated in part (b) of Figure 8.3.

Solution Equation (8.2) defines the signal's two non-harmonic components. One lies midway between the 53rd and 54th harmonics; the other lies a quarter of the way between the 211th and 212th harmonics.

To assess the outputs from the various FFT filters, we first note that all their responses are identical in shape. They are merely displaced from one another by $2\pi/N$ along the frequency axis. Therefore we can base our predictions on the shape of a *single* sinc function, by finding its value at points separated from each other by $2\pi/N$.

Figure 8.5(a) shows the situation for a component lying midway between two harmonics. The responses produced in adjacent filters are given by dots A, B, C, D ..., and by a symmetrical series of dots on the other side of the center-line. The dots lie midway between the zero-crossings of the sinc function.

For convenience let us write the sinc function as $\sin x/x$. Then the values corresponding to the dots in part (a) of the figure are given by the following table:

	x	$\sin x/x$	Magnitude (% of main peak)
A	$\pi/2$	$-2/\pi$	63.7
B	$3\pi/2$	$2/3\pi$	21.2
C	$5\pi/2$	$-2/5\pi$	12.7
D	$7\pi/2$	$2/7\pi$	9.1

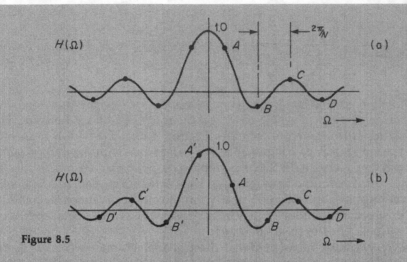

Figure 8.5

The situation for a component lying a quarter of the way between two harmonics is a little more complicated. It is illustrated by Figure 8.5(b). The dots are now offset by a quarter of the distance between zero-crossings, and are not symmetrical about the center-line. The corresponding values are shown below.

	x	$\sin x/x$	Magnitude (% of main peak)
A	$3\pi/4$	$\dfrac{4}{3\pi\sqrt{2}}$	30.0
B	$7\pi/4$	$\dfrac{-4}{7\pi\sqrt{2}}$	12.9
C	$11\pi/4$	$\dfrac{4}{11\pi\sqrt{2}}$	8.2
D	$15\pi/4$	$\dfrac{-4}{15\pi\sqrt{2}}$	6.0
A'	$-\pi/4$	$\dfrac{4}{\pi\sqrt{2}}$	90.0
B'	$-5\pi/4$	$\dfrac{-4}{5\pi\sqrt{2}}$	18.0
C'	$-9\pi/4$	$\dfrac{4}{9\pi\sqrt{2}}$	10.0
D'	$-13\pi/4$	$\dfrac{-4}{13\pi\sqrt{2}}$	6.9

It is rather difficult to compare these predictions with the plot of Figure 8.3(b) because of its small scale and limited vertical resolution. However, we promise you that the two sets of results are in agreement!

One further general point deserves mention. The longer the signal we transform, the greater the value of N, and the more FFT filters there are. The frequency *resolution* therefore increases in proportion. If we wish to resolve closely-spaced frequency components, we must work with a long portion of signal, and use a lengthy FFT. For example a 64-point transform can only be expected to resolve components spaced at least $2\pi/64$ radians apart; a 512-point transform improves the situation nearly tenfold. In applications demanding very high spectral resolution, it may be necessary to use transform lengths up to $N = 4096$ (2^{12}), or even greater.

We have previously mentioned that spectral analysis may provide a useful method for detecting a signal in the presence of noise. The principle can be illustrated quite simply. Suppose we have reason to believe that the 512-point data shown in Figure 8.6(a) contains a periodic square wave buried in noise. Although the human eye is very good at detecting regular patterns, it is difficult to decide whether such a square wave is present. Let us therefore see whether an FFT provides additional evidence.

The transform is shown in part (b). There is a more-or-less even distribution of spectral energy, plus a pronounced peak at the 32nd harmonic, and a lesser peak at the 96th harmonic. Now random noise in which successive time-domain samples are statistically independent ('uncorrelated') has a flat spectrum. It is said to be 'white'. Although its individual spectral lines display chance amplitude variations, their average (or expected) value is constant with frequency. Our FFT is therefore consistent with the view that the time-domain data consists of white noise, plus a signal which is strong in the 32nd and 96th harmonics. Of course, there may be further, smaller, components present which are masked by the noise spectrum.

It is well known in Fourier analysis that a periodic square wave contains spectral components at its fundamental and odd-harmonic frequencies, with diminishing amplitudes. The 3rd-harmonic is one third as great as the fundamental; the 5th-harmonic is one-fifth as great, and so on. (We are here referring to harmonics of the square wave, not the FFT.) We conclude, with reasonable confidence, that a square wave is indeed present in the noise, and that its fundamental corresponds to the FFT's 32nd harmonic. Now the 256th harmonic in a 512-point transform has two samples per period; and the 32nd harmonic has sixteen samples per period. You may like to refer to part (a) of the figure again, to see if you can recognize a square wave with this period.

The reason for the FFT's success in this case is that the signal's spectral energy is well concentrated, whereas the noise is wideband. They cannot be so readily distinguished in the time domain, because they both extend more or less evenly throughout the record. In general we see that such techniques are likely to be valuable whenever signal and noise have substantially different spectral distributions.

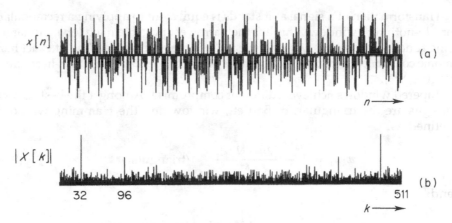

Figure 8.6 Using the FFT to detect a signal in noise.

8.2.2 WINDOWING

In the previous section we saw that a signal component at an exact harmonic frequency of the FFT gives rise to a single, well-defined, spectral line. Unfortunately we cannot expect practical digital signals to cooperate in this way! They normally contain a wide mixture of frequencies, few of them exact harmonics. This means that spectral leakage is generally present, and it may lead to difficulties of interpretation. It is therefore common practice to *taper* the original signal before transformation, reducing any discontinuities at its edges. We do this by multiplying with a suitable *window function*. In Section 3.2.2 we showed that multiplication of two time functions is equivalent to convolving their spectra. It follows that time-domain windowing causes the spectrum of the 'raw' signal $x[n]$ to be convolved with that of the window. By choosing a suitable window, we can arrange that the convolution reduces the undesirable effects of spectral leakage.

You may recall that windows are also of great value in nonrecursive filter design. In Sections 5.3.2–5.3.4 we showed how a variety of windows could be used to truncate an infinite impulse response, so defining a practical filter. By choosing a tapered window we were able to achieve a good compromise between main lobe width and sidelobe levels. Very similar considerations apply in FFT spectral analysis. We can reduce undesirable spectral leakage using a tapered window, at the expense of some broadening around individual spectral lines.

Many different windows have been recommended for this purpose. The choice of a suitable one depends on the nature of the signal or data, and on the type of information to be extracted from its spectrum. The interpretation of results can be a difficult business, even for the experienced, and is often considered more of an art than a science. In general, a good FFT window has a narrow main spectral lobe to prevent 'local' spreading of the spectrum, and low sidelobe levels to reduce 'distant' spectral leakage. Needless to say, the two requirements are in conflict!

Transformation of a signal as it stands is equivalent to applying a rectangular, or 'do-nothing', window. This has the narrowest possible main lobe, but large, sinc-function, sidelobes. As we have seen, it causes no spreading of exact harmonic components; but it produces a lot of spectral leakage with non-harmonics (Figure 8.3).

Tapered windows achieve a different compromise. Amongst the best-known designs are the triangular, or Bartlett, window and the Hamming window, defined as:

$$w[n] = 1 - \frac{|2n - N + 1|}{N} \quad \text{(triangular)} \tag{8.3}$$

and:

$$w[n] = 0.54 + 0.46 \cos\left(\frac{\{2n - N + 1\}\pi}{N}\right) \quad \text{(Hamming)} \tag{8.4}$$

where n is taken from 0 to $(N-1)$ in each case. (We defined these functions slightly differently in equations (5.19) and (5.22), giving windows with $2M + 1$, rather than N, terms.) You may like to refer back to Figures 5.14 and 5.16 which illustrated the corresponding spectra. Note that both types of window offer a wider main lobe, but smaller sidelobes, than the rectangular window. The Hamming function is particularly valuable for its low sidelobe levels, which subtantially reduce distant spectral leakage.

Rather than taper the complete input signal, it is often better to restrict tapering to the edges — for example, just the first and last 15 per cent of sample values. This is commonly done when the signal, and hence the transform, is lengthy. The tapered window function is split into two halves, one being applied to each end of the data. The central portion is left unchanged. In the frequency domain, this approach gives a narrow main lobe, but increased sidelobes, compared with full windowing. As you see, there are many possibilities.

We can illustrate some typical effects of windowing, using a modified version of our FFT program (no. 24 in Appendix A1). It is helpful to take shorter transforms than before (say $N = 128$), giving less condensed screen plots. Also, spectral leakage tends to be more pronounced with shorter transforms, because time-domain discontinuities are a more significant feature of the overall record.

Figure 8.7(a) shows, on the left, a signal of the form:

$$x[n] = \sin\left(\frac{2\pi n}{128} 9\right) + \sin\left(\frac{2\pi n}{128} 24.5\right) + \sin\left(\frac{2\pi n}{128} 51\right)$$

$$+ \sin\left(\frac{2\pi n}{128} 53\right), \quad 1 \leq n \leq 128 \tag{8.5}$$

We see from the equation that there are four spectral components with equal amplitudes: an isolated, exact, harmonic (the 9th); two close-spaced, exact, harmonics (the 51st and 53rd); and a non-harmonic (midway between the 24th and 25th). The magnitude of the FFT is shown on the right. It displays the usual mirror-image pattern, since $x[n]$ is real. The 9th, 51st and 53rd harmonics (and their mirror images) stand out clearly as single spectral lines. However, there is a lot of distant leakage around the non-harmonic component. The main features of this FFT are very similar to the 512-point transform already shown in Figure 8.3(b). Remember that since we have transformed $x[n]$ as it stands, we have effectively used a rectangular ('do nothing') window.

Part (b) of the figure shows the effects of using a triangular window, as defined in equation (8.3). The time function is now tapered down linearly from its center. In the spectrum, the individual spectral lines have broadened, and there are significant sidelobes. This makes it hard to disentangle the 51st and 53rd harmonic terms. However, spectral leakage around the non-harmonic component has been considerably reduced.

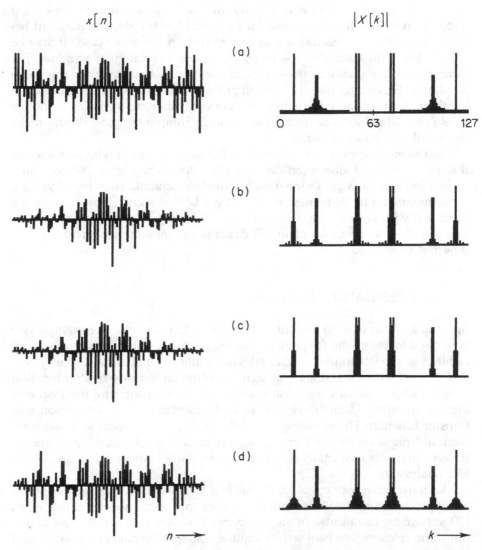

Figure 8.7 The use of windows in FFT analysis: (a) rectangular; (b) triangular; (c) and (d) Hamming *(each abscissa: 128 samples).*

In part (c) we have used a Hamming window (see equation (8.4)). There is again substantial broadening of individual spectral lines, but distant leakage is dramatically reduced. This is because of the low sidelobe levels of the Hamming function.

The effects of tapering just the ends of the data are shown in part (d). In this example, we have tapered the first and last 20 values of $x[n]$ with half-Hamming functions. Note that it is now easier to distinguish the close-spaced 51st and 53rd harmonics, but there is generally more spectral leakage than in part (c).

Incidentally, all the plots in the figure were produced by quite simple changes to Program no. 24, and we suggest you try these as a problem at the end of the chapter.

Perhaps the main conclusion to be drawn is that spectral analysis and windowing are rather complicated! Clearly, the merits of a given window must depend on the application — and, in particular, on the desired trade-off between local spectral resolution and distant leakage. In some cases it may be best to leave the data alone — in effect, to use a rectangular window. For example, if a signal has close spaced components of roughly the same amplitude, the rectangular window will probably offer the best chance of resolving them. Conversely, if the amplitudes are very different, a window with low sidelobes will reduce leakage around the large component, and should make the small one easier to detect.

Windowing offers no magic solutions. The signal we start with is in a sense the 'true' one, and all we can do with a window is modify it. We certainly cannot increase the signal's inherent information content. Therefore if you are recommended a fantastic new window by a DSP designer, you may decide to treat it with healthy skepticism. Finally, it is very important to remember that sensible interpretation of an FFT depends on knowing what form of window has been used.

8.2.3 INVESTIGATING LTI SYSTEMS

Spectral analysis is often used for investigating LTI systems. For example, we may need to assess the frequency-selective properties of an electronic circuit or filter, or the vibration characteristics of a building or structure. The system is disturbed with a suitable input signal — often an impulse or step function — and its time-domain response is recorded. By transforming the response into the frequency-domain we can derive the system's frequency response or transfer function. The growing use of digital instrumentation and measurement techniques makes FFT processing a natural choice for such applications. Indeed, there are now many commercially-available instruments with built-in FFT analyzers.

Like most other aspects of DSP, this is a large field. Fortunately we have already covered most of the underlying concepts: the characterization of an LTI system by its impulse or step response (Section 2.3); the Fourier and z-transform relationships between an impulse response, frequency response, and

transfer function (Sections 3.3.2 and 4.3.3); and, of course, the basic properties of the FFT itself. In this section we will concentrate on some additional
features of FFT analysis as applied to system testing, paying particular attention to transform lengths and spectral resolution. It is worth noting that some
further powerful techniques based on random signals and correlation functions
will be covered in Chapter 10.

Figure 8.8 summarizes the type of investigation considered here. A *wideband
input signal* x[n] disturbs the system under test, and its output or response y[n]
is recorded. (By wideband, we mean that x[n] must contain a significant amount
of all frequencies likely to be transmitted by the system. Only if this condition
is met can we expect to characterize the system *completely*.) For convenience
we have illustrated an all-digital test sequence. If the LTI system under
investigation is analog, appropriate analog-to-digital conversion must be
included.

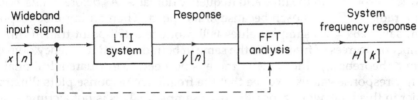

Figure 8.8 Using the FFT to explore system properties.

FFT analysis of the output signal gives the output spectrum. If the spectrum
of the input signal is not accurately known, it too must be determined. Division of the output spectrum by the input spectrum (taking account of phase
as well as magnitude) yields the system's frequency response. Thus:

$$H[k] = \frac{Y[k]}{X[k]}, \quad \text{or} \quad Y[k] = X[k]\,H[k] \tag{8.6}$$

This is equivalent to a periodic convolution of the input signal with the system's
impulse response h[n]:

$$y[n] = x[n] \;\circledast\; h[n] \tag{8.7}$$

The simplest implementation of equations (8.6) and (8.7) involves using an
input impulse. It is spectrally 'white', and simultaneously tests the system's
response to all input frequencies. The output is then the impulse response,
which transforms directly into the frequency response. However impulse-
testing is not always practicable. For one thing, it may be difficult to get enough
energy into a narrow impulse to produce a measurable response — particularly
when noise is present. If we try to overcome the problem by increasing the
strength of the impulse, the system may be driven into nonlinearity. In such
cases a step input may be preferred. If the step response is recorded in digital
form, the corresponding impulse response can be found as its first-order dif-

ference (see equation (2.8)). Alternatively, the unequal spectral distribution of a step signal can be taken into account using equation (8.6).

We have already used a theoretical counterpart of impulse-testing several times in this book. For example we designed various types of nonrecursive filter in Chapter 5, by first specifying the required impulse responses and then computing the equivalent frequency responses. The computations were somewhat faster than a standard DFT, because we made use of the fact that the impulse responses were symmetrical (Programs 13, 15 and 18 in Appendix A1.) Also, we estimated only half of the transform, knowing that the other half must be a mirror image (for example, Figure 5.17). Finally, it is important to remember that we did not enquire how many frequency-domain samples were *necessary* to represent the function properly; we merely computed a convenient number in the range $0 \leq \Omega \leq \pi$, giving a screen plot with good resolution.

Since we are now interested in using the FFT to transform a system's impulse response into its frequency response, we must be clear about the relationship between resolution in the time and frequency domains. As before, let us assume that a radix-2 FFT is chosen because of its speed. Then an impulse response with, say, 128 ($= 2^7$) sample values will produce a 128-point transform. The number of degrees of freedom is the same in both time and frequency domains. From a theoretical point of view, there is no need to evaluate additional frequency response values. We see that the frequency response plots illustrated earlier in this book were generally 'over-sampled'. This is not a crime — after all, it makes the plots easier to visualize — but it is not strictly *necessary*, and it involves extra computation.

Of course we cannot expect the impulse response of a system being investigated to have a number of sample values equal to an exact integer power of 2. So the usual approach is to add zeros to the time-domain data to bring it up to the required length. For example if $h[n]$ has 80 values, we add 48 zeros to make it up to 128. If it has 150 values, we add 106 zeros to make it up to 256 — and so on. The addition of zeros is referred to as *zero-filling*, or *zero-padding*.

Figure 8.9 shows the typical effects of zero-filling, for different lengths of transform. The impulse response $h[n]$ we have chosen represents an LTI system with close-spaced humps or 'resonances' in its frequency response, and is therefore useful for illustrating spectral resolution (such a response might well be obtained form an electronic circuit, or a mechanical structure). Apart from this, it has no special importance. Note that in this case $h[n]$ has about 180 sample values of significant size.

In part (a) of the figure we have zero-filled up to 256 points, and plotted the magnitude of the FFT using a modified version of Program no. 24 in Appendix A1. This is the minimum-length radix-2 transform which can be used to define the frequency response 'completely', according to the Sampling Theorem. We see that the system displays two frequency response peaks, and also responds significantly around zero frequency (DC).

Strictly speaking we could use a 180-point DFT or FFT in this case. It would give a frequency response with slightly lower resolution — but the same 'envelope'. Also, its harmonic frequencies would be different from those in the figure. Their spacing would be $2\pi/180$, rather than $2\pi/256$, radians. From an information theory point of view, the 256-point and 180-point transforms are both adequate to define the system completely.

What happens if we use a different length of radix 2 FFT on the same data — either a longer transform to improve the spectral resolution, or a shorter

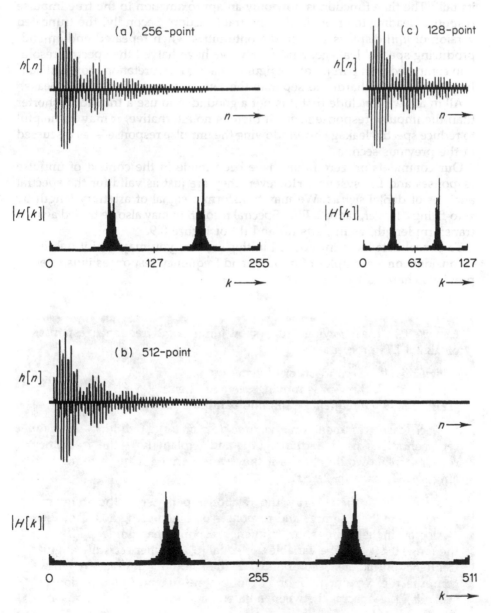

Figure 8.9 Effects of zero-filling on spectral resolution.

one to reduce the amount of computation? In part (b) of the figure we see the effect of using a 512-point transform. In the time-domain, there is a lot more zero-filling; in the frequency-domain, we have 'oversampled', doubling both the number of spectral lines and the resolution. Although this does not improve the basic information content of the frequency response, it may aid interpretation. Of course, we pay for it by having to compute a longer FFT.

Part (c) of the figure illustrates a 128-point transform. There are several interrelated effects. We have had to truncate $h[n]$ somewhat, by 'chopping off its tail'. The time function is now only an approximation to the true impulse response, leading to errors in the spectral function. Secondly, the truncated version of $h[n]$ displays sudden discontinuities when repeated end-on-end, producing spectral leakage. And finally, we have halved the spectral resolution compared with part (a) of the figure, making interpretation more difficult. In particular, it is harder to separate the two close-spaced response peaks.

All in all, we conclude that it is not a good idea to use a transform shorter than the impulse response itself. If there is no alternative, it may be helpful to reduce spectral leakage by windowing the impulse response — as discussed in the previous section.

Our comments on zero-filling have been made in the context of impulse responses and LTI systems. However, they are just as valid for the spectral analysis of digital *signals*. We may transform a signal of arbitrary length by zero-filling a longer, radix-2, FFT. Spectral resolution may also be traded against transform length, as in parts (a) and (b) of Figure 8.9.

To consolidate ideas introduced in this section, you may find it helpful to reconsider some examples of impulse and frequency responses illustrated in previous chapters.

Example 8.2 The following figures show impulse and frequency responses of digital LTI processors:

Figure 4.11   Second-order systems
Figure 5.5   Low-pass moving average filters
Figure 6.19   Frequency-sampling bandpass filter

In each case, comment on the time-domain and frequency domain representations used in the figure, and explain how the frequency response plots would change if they were derived using a minimum-length DFT, or radix-2 FFT.

Solution Let us first clarify the relationship between the frequency responses in these figures and responses obtained using a DFT or FFT.

All plots in the figures show frequency response magnitudes $|H(\Omega)|$ over the range $0 < \Omega < \pi$. The variable Ω is equal to ωT, where ω is the angular frequency in radians/second and T is the sampling interval. When we compute the N-point DFT or FFT of a time function of duration NT seconds, the spacing between adjacent spectral lines (harmonics) is

$1/NT$ Hz, or $2\pi/NT$ radians/second. Hence:

the range $0<\Omega<\pi$ is equivalent to $0<\omega<\pi/T$

and:

the number of spectral lines in this range

$$= \frac{\pi/T}{2\pi/NT} = \frac{N}{2}$$

Therefore the frequency range shown in the figures is equivalent to *half* the length of a DFT or FFT, regardless of the number of values of Ω for which $H(\Omega)$ is *actually* calculated.

Figure 4.11 All the abscissae in this figure have 320 sample values. In effect we have standardized on 320-point DFTs, using zero-filling to make the impulse responses up to the required length. in parts (a), (c) and (d) of the figure there is a lot of zero-filling, so the corresponding frequency responses are substantially over-sampled. In part (b), however, $h[n]$ hardly fits into the 320-point format. We therefore assume that there is slight distortion and spectral leakage in the $|H(\Omega)|$ plot.

If we used a minimum-length DFT in each case, we would obtain 'just-adequate' definition of $|H(\Omega)|$, according to the Sampling Theorem. The transform lengths would be about (a) 70, (b) 320, (c) 20, and (d) 70 points.

Radix-2 FFTs would involve zero-filling. Appropriate transform lengths would be (a) 128, (b) 512, (c) 32, and (d) 128 points.

Figure 5.5 Here, too, 320 values of $|H(\Omega)|$ have been computed, and the plots are greatly over-sampled. Theoretically a 5-term moving-average impulse response can be transformed using a 5-point DFT (or 8-point radix-2 FFT); a 21-term response needs a 21-point DFT (or 32-point radix-2 FFT).

However, it is worth considering what results minimum-length DFTs would yield here. A DFT assumes that the time function being transformed is periodic. The impulse response of a simple low-pass moving-average filter is finite, with all its values equal. (For example, the 5-term filter has five impulse response values equal to 1/5.) Repeated end-on-end, without any zero-filling, it would simply become a sampled DC level. Its transform would contain only one nonzero coefficient, at zero frequency. In other words the minimum-length DFT would reveal nothing of the detail of $H(\Omega)$ shown in the figure, because most of its frequency-domain samples would coincide with nulls (at $\Omega = 2\pi/5, 4\pi/5, \ldots$). We conclude that although such a DFT is *theoretically* adequate, it may be unhelpful for visualizing the detailed shape of a response.

Figure 6.19 The abscissa of the $h[n]$ plot has 160 sample values (separated by gaps). However, it is a finite-impulse-response filter with only 120 nonzero sample values, so a 120-point DFT, or 128-point FFT, would be theoretically adequate. Since we have again computed 320 values of $|H(\Omega)|$, there is substantial oversampling, giving a plot which is 'easy on the eye'.

8.3   DIGITAL FILTERING BY FAST CONVOLUTION

8.3.1   BASIS OF THE METHOD

At the beginning of this chapter we mentioned that digital filtering may, in principle, be carried out in the frequency domain. Rather than *convolve* the time-domain signal with the impulse response of the desired filter, we transform the signal and *multiply* it by the equivalent frequency response. A final inverse transform yields the filtered signal. Although the method may seem tortuous, it often proves faster than time-domain convolution. This is due to the relative simplicity of multiplication, and the speed of the FFT. The approach is particularly valuable for digital filtering long sequences of input data.

Quite often we start with an impulse response rather than a frequency response. It is then necessary to transform the impulse response into the frequency domain as well. The complete filtering process is shown in Figure 8.10, and is known as *fast convolution*. Remember that the required multiplication of $X[k]$ by $H[k]$ is complex, involving phase as well as magnitude.

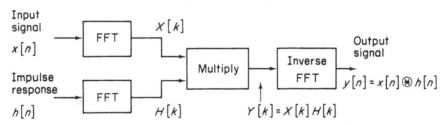

Figure 8.10   Fast convolution.

We have previously emphasized that FFT algorithms assume the functions being transformed are *periodic*, with period N. It follows that the fast convolution shown in the figure produces an output signal $y[n]$ which is the *periodic*, or *circular*, convolution of $x[n]$ and $h[n]$. Circular convolution was first mentioned in Section 3.2.2. We may visualize it as the placing of the N samples of one function around the circumference of a cylinder, and the N samples of the other function in reverse order around a second, concentric, cylinder. One cyclinder is rotated, and coincident sample values are multiplied together and summed. The resulting output signal is also periodic.

In digital filtering we require *linear*, not circular, convolution. That is, we need to convolve an *aperiodic* input signal and impulse response to produce an *aperiodic* output. Fortunately, it turns out that one period of the output signal derived by circular convolution gives the correct result, providing we extend the lengths of the transforms in Figure 8.10 using an appropriate amount of zero-filling.

The problem and its solution are illustrated by Figures 8.11 and 8.12. Figure 8.11 shows two functions $x[n]$ and $h[n]$, and below, their circular convolution $y[n] = x[n] \circledast h[n]$. All are assumed periodic. There is no particular significance in the shape of $x[n]$ and $h[n]$; we have chosen 'rectangular pulses' because they are convenient for illustration. One of them has five finite values per period, the other seven, and the period is nine sampling intervals. $y[n]$ is found by reversing either $x[n]$ or $h[n]$ — which makes no difference in this case — followed by shifting, crossmultiplication, and summation over one complete period. Thus:

$$y[n] = x[n] \circledast h[n] = \sum_{i=0}^{N-1} x[i]\, h[n-i], \qquad 0 \le n \le (N-1) \qquad (8.8)$$

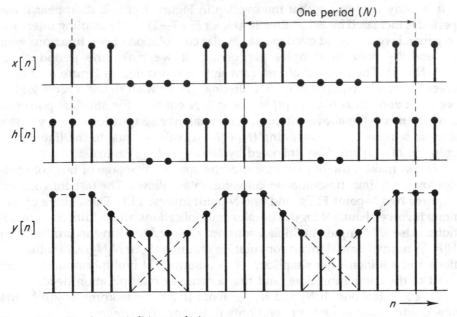

Figure 8.11 Circular (periodic) convolution.

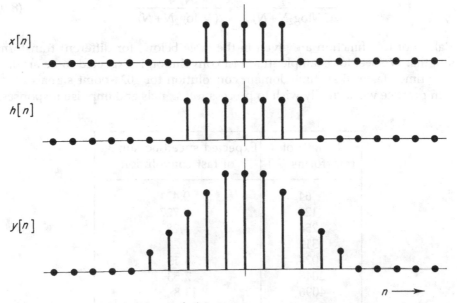

Figure 8.12 Linear (aperiodic) convolution.

We see that in this case $y[n]$ consists of a series of overlapping 'flat-topped triangles', shown dotted in the figure. The overlap occurs because the repetitions of $x[n]$ and $h[n]$ are close together, with few zero values between them. However, if the period is increased by inserting more zeros, the overlap disappears. Each period of $y[n]$ then becomes identical in form to the required *linear* convolution of nonrepetitive versions of $x[n]$ and $h[n]$ — as shown in Figure 8.12.

It is fairly easy to see that the overlap in Figure 8.11 will disappear if the period is increased by zero-filling to at least $(5+7-1) = 11$ sampling intervals. In general we can avoid overlap in a circular convolution of two functions with N_1 and N_2 nonzero samples per period, if we make the period $N \geq (N_1+N_2-1)$. The method of fast convolution illustrated in Figure 8.10 then gives the result required in digital filtering. (If we wish to use a radix-2 FFT, we must zero-fill $x[n]$ and $h[n]$ to a length N equal to the smallest power of 2 which meets the above criterion.) The final inverse transformation yields a sample sequence $y[n]$ containing (N_1+N_2-1) values equal to the linear convolution of $x[n]$ and $h[n]$, followed by $N-(N_1+N_2-1)$ zeros.

We can make a rough assessment of the speed advantage of fast convolution over a normal time-domain convolution as follows. The fast convolution requires two N-point FFTs, and one N-point inverse FFT. These three operations involve about $3N \log_2 N$ complex multiplications and additions/subtractions. Also, N complex multiplications are required to form the product $X[k] H[k]$. A normal time-domain convolution requires about $N_1 N_2$ real multiplications and additions. For simplicity let us assume that multiplications take up most of the computing time, and that a complex multiplication needs twice as long as a real one. If N_1 and N_2 are about half the transform length N, the speed advantage of fast convolution will be approximately:

$$\frac{N_1 N_2}{2(3N\log_2 N + N)} \approx \frac{N^2/4}{2(3N\log_2 N + N)} \tag{8.9}$$

Values of this function are given in the table below, for different transform lengths. We see, for example, that fast convolution is expected to be around four times faster than time-domain convolution for 1024-point signals.

In practice we normally wish to process *real* signals and impulse responses,

Length of transforms (N)	Expected speed advantage of fast convolution
64	0.421
128	0.727
256	1.28
512	2.29
1024	4.13
2048	7.53
4096	13.8

in which case ony half the complete transforms need be calculated. Furthermore, the zero-filling of $x[n]$ and $h[n]$ means that many of the transform multiplications do not, in fact, have to be carried out. Efficient fast convolution algorithms take these factors into account, and may give speed advantages more than twice as great as those listed in the table.

In assessing speed advantage we have assumed that fast convolution is being substituted for *nonrecursive* time-domain filtering. Of course, if we have the option of using a *recursive* digital filter, the advantage is likely to tip back in favor of a time-domain operation.

It is now time to try out the ideas introduced so far in this section. Computer Program no. 25 in Appendix A1 implements the complete fast convolution process summarized by Figure 8.10, for a particular input signal and impulse response, and plots the various time and frequency functions on the screen.

The screen plots are reproduced in Figure 8.13. The program first draws the input signal $x[n]$ and impulse response $h[n]$ — shown in parts (a) and (b). It uses 128-point signals and transforms, although the precise forms of $x[n]$ and

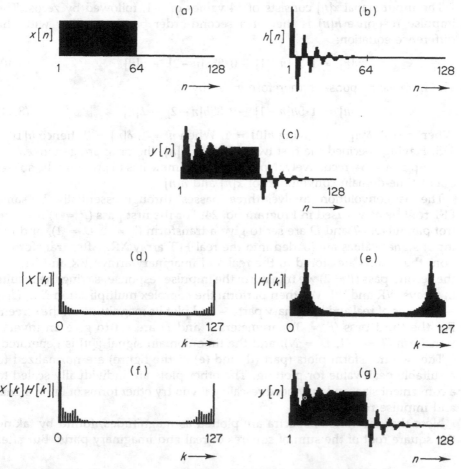

Figure 8.13 Illustration of linear and fast convolution (*each abscissa: 128 samples*).

$h[n]$ are unimportant. We have chosen a 'rectangular pulse' for $x[n]$; and an oscillatory form for $h[n]$, representing a bandpass filter (in fact, the one described in Section 2.3). However, it *is* important that the numbers of nonzero sample values in $x[n]$ and $h[n]$ meet the criterion mentioned earlier, namely $N = 128 \geq (N_1 + N_2 - 1)$. In this example we have specified 64 nonzero values for each function, so that $(N_1 + N_2 - 1) = 127$.

The program next carries out a linear, time-domain, convolution of $x[n]$ and $h[n]$, producing the output signal $y[n]$ shown in part (c) of the figure. We have included it mainly to provide a check on the subsequent fast convolution process.

The screen is then cleared and the fast convolution begins. The FFTs of $x[n]$ and $h[n]$ are calculated in turn, and their magnitudes plotted — see parts (d) and (e) of the figure. Next, the program forms the product $X[k]\,H[k]$, and plots its magnitude in part (f). Finally an inverse FFT is taken and the resulting signal is plotted in part (g). Fortunately, it checks with the output signal derived by the time-domain method!

For those of you who wish to use (or modify) the program, we now make some comments on its organization.

The input signal $x[n]$ consists of 64 values of $+1$, followed by zeros. The impulse response $h[n]$ is that of a second-order bandpass filter with the difference equation:

$$y[n] = 1 \cdot 5y[n-1] - 0 \cdot 85y[n-2] + x[n] \qquad (8.10)$$

The impulse response is therefore given by:

$$h[n] = 1 \cdot 5h[n-1] - 0 \cdot 85h[n-2] + \delta[n] \qquad (8.11)$$

When $n = 0$, $\delta[n] = 1$, giving $h[0] = 1$. When $n = 1$, $\delta[n] = 0$, hence $h[1] = 1 \cdot 5$. Having specified the first two values of $h[n]$, the program generates the rest (up to $n = 64$) recursively. Note how very simple it is to program the subsequent time-domain convolution of $x[n]$ and $h[n]$.

The fast convolution involves three 'passes' through essentially the same FFT routine as we used in Program no. 24. For the first pass ($P = 1$) the control parameters T and D are set to give a transform ($T = 1$, $D = 1$), and the input signal values are loaded into the real FFT array XR. After transformation, the results are stored in the real and imaginary arrays VR and VI. On the second pass ($P = 2$) we transform the impulse response, saving the results in arrays WR and WI. We then perform the complex multiplication $X[k]\,H[k]$ — in terms of real and imaginary parts — and plot its magnitude on the screen. For the third pass ($P = 3$) parameters T and D are set to give an inverse transform ($T = -1$, $D = N$), and the time-domain signal $y[n]$ is generated.

The two transform plots (part (d) and (e) of the figure) are normalized to a suitable peak value for plotting. The other plots are individually scaled to a convenient size, and may need rescaling if you try other forms of input signal and impulse response.

Note that the various spectra are plotted as magnitudes, found by taking the square root of the sum of squares of real and imaginary parts. But after the final inverse transform, we wish to plot negative as well as positive values of the signal $y[n]$. We therefore use just its real part, assuming the imaginary part is zero.

In one respect Program no. 25 is a little disappointing. Although we hope you agree that it demonstrates the principles well, the fast convolution itself is probably slower than the conventional time-domain convolution. There are several good reasons for this. We have used short, 128-point, transforms for which fast convolution is expected to offer little, if any, speed improvement. Furthermore we have not taken advantage of $x[n]$ and $h[n]$ being real, since our FFT routine computes the complete transform. Thirdly, one glance at the program shows that the fast convolution is far more complicated than the time-domain convolution. It involves many control and routing statements, data shuffling, normalization for plotting, and so on. All these add to the computing 'overheads', especially when programmed in a high-level language. A serious fast convolution algorithm, implemented in machine code or by special-purpose hardware, would be much more efficient.

8.3.2 SIGNAL SEGMENTATION

In this section we consider a practical difficulty which often arises when digital filtering by fast convolution. Our discussion so far has assumed that $x[n]$ and $h[n]$ contain roughly the same number of samples, and that they are zero-filled up to the same length of transform ($N = 128$ in Figure 8.13). However, in practice we often need to convolve a long input signal or data sequence with a relatively short impulse response. For example, we might have a signal with several thousand sample values, but an impulse response with less than 100. In such cases it is uneconomic in terms of computing time (and storage) to use the same length of transform for both functions. Furthermore, in real-time applications the use of a very lengthy transform for $x[n]$ may involve an unacceptable processing delay.

The problem may be overcome by segmenting the input signal into sections of manageable length, performing a fast convolution on each section, and combining the outputs. Two approaches are commonly used: the *overlap-add* method; and the *overlap-save* (also known as *select-save*) method. We now make notes on each in turn. Further information can be found in various DSP books and references (for example, references 2 and 3 in the Bibliography).

The overlap-add method is illustrated in Figure 8.14. For simplicity we have assumed the use of very short ($N = 8$) transforms. Also we have chosen an input signal and impulse response with small integer sample values, which makes it easy to check the results of the convolution (we suggest this as a problem for you at the end of the chapter).

The input signal $x[n]$ is first split into non-overlapping segments $x_0[n]$, $x_1[n]$, $x_2[n]$ ..., containing five samples each (of course, individual values may happen to be zero). Since $h[n]$ has just four samples, its convolution with each segment produces an output sequence with $(5+4-1) = 8$ values — the required

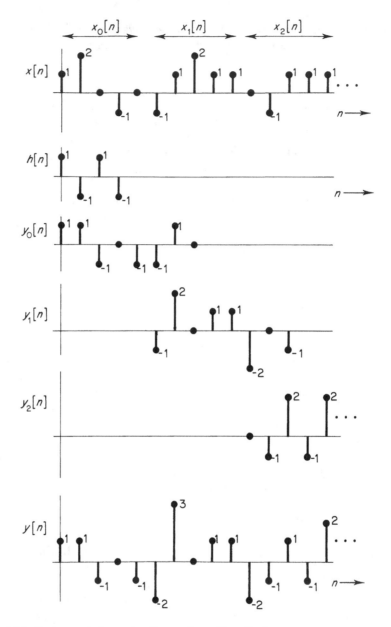

Figure 8.14 Fast convolution using the overlap-add method.

transform length. The various output sequences, or segments, are labeled $y_0[n]$, $y_1[n]$, $y_2[n]$ ... in the figure. Note that they overlap each other by three samples because of the spreading effect of the convolution. The overall output $y[n]$ is found as the addition, or superposition, of the output segments.

We have shown all this as a time-domain operation. But in practice the output segments would be computed by fast convolution, zero-filling the input segments and impulse response up to the required transform length.

In general, the overlap-add method can be summarized as follows. Suppose the impulse response contains N_2 nonzero values, and the input signal is divided up into nonoverlapping segments of length N_1, where $N_1 > N_2$. The ith segment of $x[n]$ is given by:

$$x_i[n] = x[n], \quad iN_1 \le n \le (i+1)N_1 - 1$$
$$= 0, \text{ elsewhere} \tag{8.12}$$

Individual output segments are:

$$y_i[n] = x_i[n] * h[n] = \sum_{k=0}^{N_2-1} h[k] \, x_i[n-k] \tag{8.13}$$

although they are, of course, found by the equivalent frequency-domain multiplication. The overall output is given by:

$$y[n] = x[n] * h[n] = \sum_{i=0}^{N_2-1} h[i]x[n-i] \tag{8.14}$$

As already noted, each output segment $y_i[n]$ is computed by fast convolution, giving a sequence with $(N_1 + N_2 - 1)$ sample values. The overall output is found by superposition, using an overlap of $(N_2 - 1)$ points between adjacent segments.

The alternative *overlap-save*, or *select-save*, technique is a variation on the same theme. In this case the input signal is subdivided into overlapping segments equal to the transform length N. The overlap is $(N_2 - 1)$ points, where as before, N_2 is the length of the impulse response $h[n]$. The latter is zero-filled up to N points. Fast convolution of each input segment with $h[n]$ gives an output segment in which the first $(N_2 - 1)$ values are incorrect, and are discarded. The remaining values are concatenated with other output segments to give the overall output $y[n]$.

We illustrate the process in Figure 8.15, for the same input signal and impulse response used in Figure 8.14. The input segments are eight samples long, and they overlap by $(N_2 - 1) = 3$ sample values. The first three samples of each output segment (shown as hollow 'blobs') are discarded, and the remainder are concatenated to form the output signal $y[n]$. In practice each output segment is computed by the fast convolution technique, and represents one period of a circular convolution.

Note that the first three values of $y[n]$ in the figure are *not* derived from $y_0[n]$. They would be obtained by circular convolution of $h[n]$ with a previous input segment — the one which begins before $n = 0$ and overlaps $x_0[n]$. We cover this point in a problem included at the end of the chapter.

What length of transform (N) should be chosen for fast convolution by the overlap-add or overlap-save techniques? First, we must ensure that N is large enough for circular convolution to give the correct result. This happens if:

$$N \ge (N_1 + N_2 - 1) \quad \text{and} \quad N_1 > N_2 \tag{8.15}$$

If we put $N_1 = (N_2 + 1)$, then $N \ge 2N_2$. Hence the transform length must be at least twice that of the impulse response $h[n]$. Choosing $N = 2N_2$ means that many relatively short convolutions have to be performed. Conversely if N is

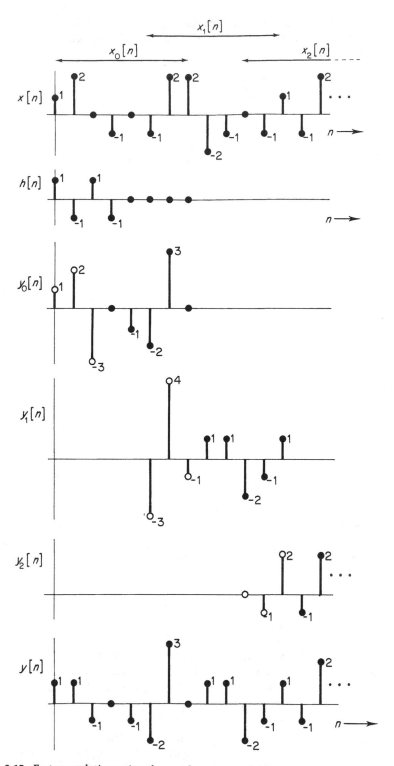

Figure 8.15 Fast convolution using the overlap-save method.

made much larger, we have to perform a small number of lengthy convolutions. It is found that the best compromise depends on the length of the impulse response, and to some extent on the fast convolution algorithm used. Approximate values may be taken from the following table, and further information found in reference 2 in the Bibliography.

Length of impulse response (N_2)	Length of transform (N)
11–17	64
18–29	128
30–52	256
53–94	512
95–171	1024

PROBLEMS

SECTION 8.1

Q8.1 Computer Program no. 24 in Appendix A1 is a general-purpose radix-2 FFT which can readily be modified for different lengths of transform and input signal. Change the program to accommodate the following:

(a) transform length $N = 512$; input signal given by:

$$x[n] = 1, \quad 1 \le n \le 64$$
$$= 0, \quad \text{elsewhere.}$$

(b) transform length $N = 512$; input signal given by:

$$x[n] = 1, \quad n = 1,3,5, \ldots, 63$$
$$= 0\cdot2, \quad n = 2,4,6, \ldots, 64$$
$$= 0, \quad \text{elsewhere.}$$

(c) transform length $N = 256$; input signal given by:

$$x[n] = 0\cdot2 \sin\left(\frac{2\pi n}{256} 9\right) + 0\cdot15 \cos\left(\frac{2\pi n}{256} 100\right), \quad 1 \le n \le 256$$

(d) transform length $N = 256$; input signal given by:

$$x[n] = 0\cdot35 + 0\cdot2 \sin\left(\frac{2\pi n}{256} \frac{100}{3}\right)$$

$$+ 0\cdot15 \cos\left(\frac{2\pi n}{256} \frac{200}{3}\right), \quad 1 \le n \le 256$$

Study the screen plots in each case, making sure you can explain their main features (*Note*: a real input signal should be loaded into array XR; also, remember that the final inverse transform is plotted as a *magnitude*).

Q8.2 Figure 8.2 in the main text shows typical computation times for DFTs and FFTs of various lengths on a personal computer. Repeat the tests on your own computer, by modifying Programs no. 23 and no. 24 in Appendix A1 to accommodate different lengths of transform. Do you obtain similar *relative* speeds to those shown in Figure 8.2?

SECTION 8.2.1

Q8.3 By modifying Program no. 24 in Appendix A1, reproduce the spectral leakage effects shown in Figure 8.3 of the main text.

Q8.4 A 60 Hz sinusoidal signal is sampled at 500 samples per second, and analyzed using a 64-point FFT.

(a) which spectral coefficient will be the largest?
(b) which coefficient will be the next largest, and what is its *relative* size compared with the coefficient in part (a)? Express the answer in decibels (dB).
(c) Find the sampling rate closest to 500 samples/second which will eliminate spectral leakage.

Q8.5 Signals given by:

(a) $x[n] = \sin\left(\dfrac{2\pi n}{256} 43 \cdot 4\right)$

(b) $x[n] = \sin(n)$

are to be analyzed using a 256-point FFT. Predict the three largest spectral coefficients in each case, and estimate their relative magnitudes. Check your results by modifying Computer Program no. 24 in Appendix A1.

Q8.6 (a) A portion of a music signal is sampled at 44 kHz and analyzed using a 2048-point FFT. Approximately what spectral resolution (in Hz) is provided by the transform?

(b) If a 256-point DFT or FFT is considered as a set of elementary bandpass filters, what is the width of each main lobe in the variable Ω (radians)?

(c) The sampling interval in part (b) is 1 second. What is the frequency interval (in Hz) between each filter's center-frequency and the center of its first sidelobe?

Q8.7 Figure 8.6 in the main text illustrates the detection of a square wave in wideband noise using FFT analysis. The transform length is $N = 512$, and the square wave has sixteen samples per period.
    Using Computer Program no. 24 in Appendix A1, and the random noise

generator on your own computer, carry out a similar investigation for different relative amplitudes of signal and noise.

SECTION 8.2.2

Q8.8 Calculate and plot the values of a 16-point Hamming data window. What is the value of the two points immediately to either side of the center-line?

Q8.9 Calculate and plot the values of a 32-point data window with a half-Hamming taper over the first and last five values (that is, the central untapered portion is 22 points long).

Use the values as input data to the FFT program in Appendix A1 (Program no. 24), with a transform length $N = 256$ and zero-filling. Hence plot the window's spectral function. Comment on its merits as an FFT data window, compared with a 32-point rectangular ('do-nothing') window.

Q8.10 Figure 8.7 in the main text illustrates the effects of windowing a signal:

$$x[n] = \sin\left(\frac{2\pi n}{128} 9\right) + \sin\left(\frac{2\pi n}{128} 24 \cdot 5\right) + \sin\left(\frac{2\pi n}{128} 51\right)$$

$$+ \sin\left(\frac{2\pi n}{128} 53\right), \quad 1 \le n \le 128$$

Modify Computer Program no. 24 in Appendix A1 to reproduce the various plots, by multiplying the input signal by the appropriate window function before transformation.

SECTION 8.2.3

Q8.11 A 16-point moving-average low-pass filter has the difference equation:

$$y[n] = \frac{1}{16} \{x[n] + x[n-1] + \dots x[n-15]\}$$

Enter its impulse response as input data to the FFT program in Appendix A1 (Program no. 24). Specify the following lengths of transform:
(a) $N = 16$;   (b) $N = 32$;   (c) $N = 256$.
Use zero-filling where appropriate.

Make sure you understand the effects of transform length on the computed frequency response (magnitude) characteristic. What minimum transform length is theoretically necessary in this case? Does it give a helpful screen plot?

Q8.12 The impulse response of a recursive bandpass filter is shown in Figure 2.4 of the main text. Its difference equation (see equation (2.4)) is:

$$y[n] = 1 \cdot 5y[n-1] - 0 \cdot 85y[n-2] + x[n]$$

Tabulate the first 64 values of its impulse response $h[n]$, *either* by direct measurement (say to the nearest millimeter) from Figure 2.4, *or* by calculation (you may find Program no. 4 in Appendix A1 useful).

Enter the values of $h[n]$ as input data to FFT Program no. 24 using a transform length of (a) $N = 64$, (b) $N = 512$ (with zero-filling). If the sampling rate is 1 kHz, what spectral resolution does each transform provide?

Q8.13 Computer Program no. 10 in Appendix A1 estimates a signal $y[n]$ from a recursive difference equation (see also equation (4.16) and Figure 4.2 in the main text).

Suppose that this signal represents the impulse response $h[n]$ of an LTI processor. By incorporating the relevant program lines from Program no. 10 in Program no. 24, obtain a screen plot of the processor's frequency response. Use a transform length $N = 512$.

The z-transform of the difference equation yields the transfer function of the processor. Its pole-zero configuration is shown in Figure Q8.13. Is the screen plot you have obtained consistent with this set of poles and zeros?

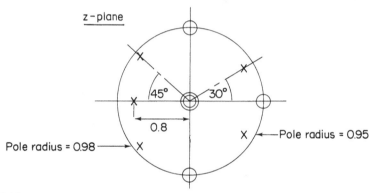

Figure Q8.13

SECTION 8.3

Q8.14 Specify one period of the function obtained by circular convolution of the periodic functions $x[n]$ and $h[n]$ illustrated in Figure Q8.14.

Also perform a linear convolution of $x[n]$ and $h[n]$, assuming them to be *aperiodic*.

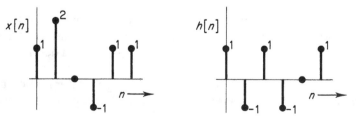

Figure Q8.14

If $x[n]$ and $h[n]$ represent the aperiodic input signal and impulse response of a digital LTI processor, under what conditions will the correct output signal $y[n]$ be obtained using the fast convolution technique?

Q8.15 Use Computer Program no. 25 in Appendix A1 to find the output signal from a digital filtering operation specified as follows:

$$\begin{aligned} \text{input signal } x[n] &= 1, & n &= 1,3,5 \ldots 63 \\ &= -1, & n &= 2,4,6 \ldots 64 \\ &= 0, & &\text{elsewhere.} \\ \text{impulse response } h[n] &= 1, & 1 &\le n \le 16 \\ &= 0, & &\text{elsewhere.} \end{aligned}$$

Note the screen plots carefully, checking that the result given by conventional linear convolution is also given by fast convolution if a sufficiently long FFT is used.

Q8.16 Figure 8.14 in the main text illustrates fast convolution by the overlap-add method. Convince yourself that:

(a) Each output segment may be obtained by convolving the appropriate input segment with the impulse response $h[n]$.
(b) The overall output $y[n]$ is found by superposition.

Q8.17 Figure 8.15 in the main text illustrates fast convolution by the overlap-save method. Show that:

(a) Circular convolution of each 8-point input segment with the 8-point, zero-filled, impulse response yields one of the output segments shown.
(b) Concatenation of the output segments gives the correct output signal $y[n]$, providing the first three values of each segment are discarded.
(c) The first three values of $y[n]$ may be found by circular convolution of $h[n]$ with the input segment beginning before $n = 0$ which overlaps $x_0[n]$.

CHAPTER 9
Random Digital Signals

9.1 INTRODUCTION

In this and the next chapter we turn our attention to the description and processing of random signals. These are in marked contrast to the signals we have already met in this book, which have precisely defined sample values and are generally referred to as *deterministic*. For example, the steps, impulses, and sinusoids described in Section 1.3 are deterministic because they have definite, known values at each sampling instant. By contrast, the individual values of random signals are not defined, nor can we predict their future values with certainty.

You may already be experiencing a little anxiety about where we are heading, for two main reasons. Firstly, random functions and probability are widely regarded as difficult topics, both conceptually and mathematically. Secondly, you may find it hard to understand why random signals need covering in an introductory book on DSP at all, since the emphasis should surely be on designing and using practical processors for signals whose time and frequency domain properties are well-defined? Before getting down to detail, we will therefore spend a few moments trying to lay any such anxiety to rest.

It is certainly true that random phenomena can be difficult to understand and describe. Many books dive straight into mathematics, without attempting any conceptual introduction or discussion. Our aim here is rather different. We will offer a strictly introductory treatment, both of random signals and random DSP, with plenty of illustrations. We will also link our discussion strongly to the work of earlier chapters, showing how many of the central concepts of time and frequency domain analysis can be extended to cover random signals. Our aim is to give you a clear grasp of some basic ideas and techniques which have achieved a central, indeed classical, status in modern signal processing. These will give you the confidence to tackle other, more advanced, topics which you may meet later.

But why should we trouble ourselves with random signals at all? At first sight random functions might seem rather irrelevant. Indeed, a common initial reaction is that signals with ill-defined properties should have little place in any modern scientific theory of signals and systems. In fact the opposite is nearer the truth, for three main reasons:

We often have insufficient knowledge of a signal, or of the process which is producing it, to be able to describe or predict it completely. Alternatively, the effort involved in describing it by a precise analytic function may be too great to be worthwhile. In such cases, it is often extremely useful to evaluate some statistical properties of the signal instead.

It is a central concept of modern signal and communication theory that a signal can only convey useful information if it is, to some extent, random and unpredictable. For example, it would be pointless to transmit a continuous periodic sinusoid through a communication channel, since once its frequency, amplitude and phase had been determined at the receiving end, no further information could be conveyed. It is only by switching such a tone on and off in an unpredictable way — as in a Morse Code message — that the person at the receiving end learns anything new. Uncertainty is essential to information flow.

Unwanted random interference, or *noise*, is present to some extent in all communication, measurement, and recording situations. When picked up along a transmission path (for example, a cable or satellite link), it imposes a fundamental limit on information carrying capacity. Random noise is therefore of great technical and economic importance.

We hope that these three points will have convinced you of the importance of random functions. In the field of DSP, we see that a knowledge of such functions is central to understanding both the information content of digital signals, and the properties of digital noise. The underlying principles are those of probability and statistics, which we will use to define useful *average measures* of random sample sequences.

Note that in principle any random sample sequence could represent either a digital signal, or digital noise. It depends upon one's viewpoint. If the sequence conveys useful information, it is a 'signal'; if it represents unwanted interference, it is 'noise'. Fortunately, the statistical measures we shall use to describe it are essentially similar in the two cases. This means that we can develop some very valuable DSP techniques for enhancing signals in the presence of noise, in other words, for improving the *signal-to-noise ratio*. An introduction to this interesting and vital topic will be given in the next chapter.

The type of situation we shall examine in this and the next chapter can be introduced with a simple computer illustration. You are probably aware that most computer languages include a random number generator, which typically produces random numbers equally distributed in the range 0 to 1. We can use a series of such numbers to represent a random digital sequence, as in part (a) of Figure 9.1. You may think of the sequence either as an information-bearing signal, or unwanted noise. In either case, we may need to evaluate such statistical measures as its *average* (or *mean*, or *dc value*), its *mean-square* (also known as the *average power*), and its *variance* (or *ac power*). We may also wish to examine whether successive sample values are truly independent of one another, or whether the sequence is structured in some way. And we may well need to compute its average spectral properties — for example, if it represents a signal to be transmitted along a cable of limited bandwidth.

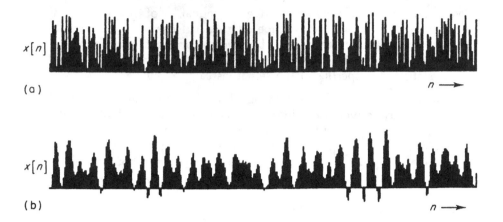

Figure 9.1 Random digital sequences: (a) produced by a random number generator, and (b) after bandpass filtering (*abscissa: 320 samples*).

Other important questions arise if the sample sequence has been modified by linear processing, for example by digital filtering. Some typical effects of filtering (in this example we use a digital bandpass filter) are shown in part (b) of Figure 9.1. You will note that the sequence now has a different range of amplitude values, including some negative ones. In general its mean, mean-square and variance are also modified. Furthermore, it now has a distinctive time-domain structure, with groups of positive and negative values tending to bunch together. This implies that filtering has made the signal more predictable. In the frequency domain, we may expect spectral energy to be concentrated around the center-frequency of the particular bandpass filter we have used. All these effects are potentially important — whether the sequence represents a useful signal or unwanted noise — and we must be able to describe them quantitatively.

A further possibility is that such a sequence represents a wanted signal mixed with unwanted noise. In this case we may wish to process it with a digital filter which gives the best chance of enhancing the signal. This is a particularly important application of DSP which arises in many practical communication, measurement, and recording situations.

Although our treatment will be strictly introductory, we aim in the following pages to give you familiarity with the most widely-used statistical measures of random digital signals and noise. In Chapter 10 we will tackle some valuable techniques for processing them with linear algorithms.

9.2   BASIC MEASURES OF RANDOM SIGNALS

9.2.1   AMPLITUDE DISTRIBUTION

We are often interested in the distribution of amplitude values of a random digital signal or sequence. For example, we may need to know the probability of finding a sample value at or above a certain level, or within a particular range. Such questions assume particular significance in *detection* situations, where the occurrence of a particular amplitude value (often a peak) is used to initiate some action or decision. Also, as we shall see in the next section, the amplitude distribution of a signal may be used to compute such widely-used measures as its mean, mean-square, and variance.

In DSP we normally deal with discrete random variables in which the amplitude values are quantized. In other words, they can only take on a limited number of values. An extreme example is a binary sequence which can only take on two distinct values. Binary sequences are extensively used for coding information in modern computing, communication, and control systems, and are of the greatest practical importance. Note that, even when we are dealing with apparently multivalued signals in a digital computer, the number of possible amplitude values is in fact finite, being limited by the numerical resolution (wordlength) of the machine. In such a situation, the familiar probability density function used to describe a continuous random variable is inappropriate. Instead, we define a *probability mass function* p_{x_n} as:

$$p_{x_n} = probability\{x[n] = x_n\} \tag{9.1}$$

Thus p_{x_n} simply tells us the probability associated with each of the allowed values of the sequence. An associated function, the *probability distribution function*, is

$$P_{x_n} = probability\{x[n] \leq x_n\} = \sum_{-\infty}^{x_n} p_{x_n} \tag{9.2}$$

This takes a staircase form which is a running integral of the probability mass function. Discontinuities in the staircase occur at the allowable amplitude values of the sequence; and the height of each discontinuity represents the probability that the random sequence takes on that particular value.

Before going any further, we should make it clear that such probability functions strictly relate to an infinite set, or *ensemble*, of random variables generated by a given random process. Here we wish to use them to describe the properties of a single sequence — an approach which involves certain assumptions. We will discuss these assumptions, and the ideas underlying them, in Section 9.2.3.

We will now illustrate the above types of probability function with some simple examples.

To start with, let us consider a coin-tossing experiment. We toss the coin many times, generating a sequence of 'heads' and 'tails'. For convenience, let us denote these by $+1$ and -1, respectively, producing a random binary sequence. Assuming the coin is fair and unbiased we expect both 'heads' and 'tails' to occur with probability 0.5. Hence the probability mass function is as illustrated in Figure 9.2. It shows the probabilities of finding the signal at each of its two allowable values. Note that the sum of terms in such a mass function must always equal unity.

The corresponding probability distribution function, which is its running integral, is shown alongside. Such a distribution function is of the cumulative

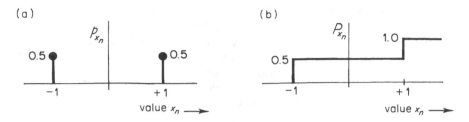

Figure 9.2 (a) Probability mass function, and (b) probability distribution function of a random binary sequence.

type, starting at zero on the left-hand side, and ending at unity on the right. The value corresponding to any given signal level indicates the probability of finding the signal *below* that level.

Example 9.1 A random digital sequence is generated by tossing a six-sided die. Assuming the die is fair, plot the probability mass function and probability distribution function of the resulting sequence.

Solution The sequence has six allowable values, 1 to 6 inclusive. Assuming the die is fair, each occurs with probability 1/6. Hence the probability mass function is as shown in Figure 9.3(a). Note that the sum of all terms is unity.

The probability distribution function is formed as the running integral of the probability mass function. It therefore takes the form of a series of six steps, each of height 1/6, as shown in part (b) of the figure.

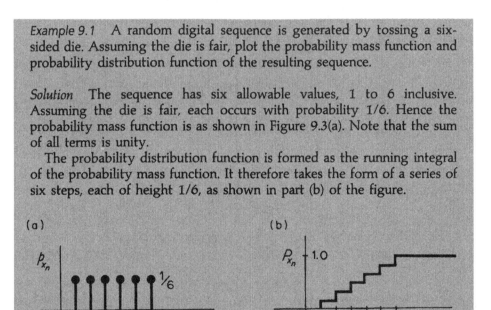

Figure 9.3 Probability functions for a die-tossing experiment.

It is also interesting to consider the form of these two types of probability function for the computer-generated random number sequence already illustrated in Figure 9.1(a). The values of this sequence fall anywhere in the range 0 to 1, with equal probability. At first sight, it may appear that there is an infinite number of possible values; but it is in fact limited by the resolution with which numbers are held in the machine. For example, if the numbers are given to six decimal places (0.000000 to 0.999999), there are one million possible discrete values. Assuming the random number generator is 'fair', they will all have the probability 10^{-6}. We can hardly draw a probability mass function with one million separate probabilities on it, or a staircase distribution function with one million small steps, so we leave these figures to your imagination! (Incidentally, it is worth noting that the finite vertical resolution of the computer screen of Figure 9.1 has reduced the number of plotted values to about 100.)

So far, all the amplitude distributions we have considered are *uniform* or *even*. That is to say, each of the possible signal values has the same probability. We now turn to a very different form of distribution, the *gaussian* distribution, which occurs widely in practical DSP.

It is important to realize that the random signals and noise met in practice do not generally display uniform distributions. In this respect, coin-tossing and die-tossing experiments are untypical. Far more common is the gaussian distribution, named after the German mathematician Johann Gauss, who investigated the random errors arising in astronomical observations. He considered the total error of a given measurement or observation to be the sum of a large number of individual random errors, each of which could be positive or negative in sign. He then showed that the probability density function of the total error takes the form:

$$p(x) = \frac{1}{\sigma\sqrt{2\pi}} \exp\left(\frac{-(x - \bar{x})^2}{2\sigma^2}\right) \tag{9.3}$$

where $\bar{x}$ is the mean value of the error variable, σ^2 is the variance and σ the standard deviation, and the constant $\sigma\sqrt{2\pi}$ is included to make the total area under the curve unity. In deriving this result, Gauss made no attempt to ascertain details of the small contributing errors, and his distribution has found a large number of applications in different fields. It is therefore often called the *normal* distribution.

As far as DSP is concerned, any random sequence which is produced by superposition of many contributing processes is likely to have a gaussian form of amplitude distribution. For example, noise arising in digital communication and computer systems is often of this type. Of course, we must now use a probability mass function in place of the density function of equation (9.3), obtaining a discrete function which follows a gaussian 'envelope'.

Fortunately we can demonstrate these ideas quite easily, because it is straightforward to generate a gaussian digital sequence on a computer using a standard random number generator. It follows from Gauss's result that if we add together many independent, uniformly-distributed, random numbers to form a new random number, and repeat the exercise over and over again to form a sequence, that sequence will be approximately gaussian. The more random numbers we add together, the better the approximation. However a compromise is clearly necessary because we need to limit the amount of computation.

Computer Program no. 26 in Appendix A1 illustrates the above points, and produces estimates for two probability mass functions: one for a sequence whose values are uniformly distributed; the other for a gaussian sequence. It works as follows. First, it uses the computer's random number generator to produce a sequence of 3000 uniformly-distributed integer numbers in the range 0 to 300, and generates a histogram of 'scores' which builds up on the screen. A typical histogram is shown in part (a) of Figure 9.4. Sample values are plotted as

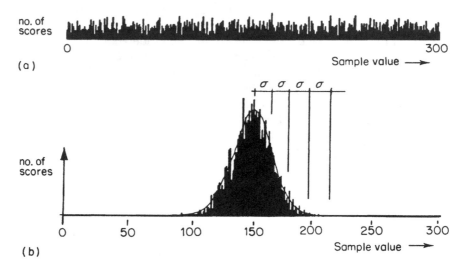

Figure 9.4 Estimates of probability mass functions for (a) a uniform distribution and (b) a gaussian distribution (*abscissa: 301 samples*).

the abscissa, and the number of times each score occurs as the ordinate. The latter is therefore an indication of relative probability, and we see that it is indeed more or less uniform (of course, in a limited 'trial' we can only expect a result whose shape *tends towards* the underlying probability distribution).

The program next generates a gaussian sequence, also with 3000 sample values. Each is formed by adding together 30 uniformly-distributed random numbers in the range 0 to 1 (giving a new random number in the range 0 to 30), and then multiplying by 10 to give a number in the range 0 to 300. This allows comparison with part (a) of the figure. Once again, a histogram of scores is generated, as shown in part (b).

We notice straight away that although the complete range 0 to 300 is theoretically possible, in fact the numbers almost all fall between about 100 and 200. The clustering of values around the mean (in this case 150), with a rapid fall-off towards the 'tails', is typical of a gaussian distribution. In our example it would be extremely rare, but still just possible, to get a value below about 75 or above about 225.

For the sake of completeness, we have shown the theoretical form of the underlying gaussian curve as a full line in the figure, and indicated intervals equal to one standard deviation (σ). We will explain this further in the next section. For the moment we simply ask you to note the general shape of the curve. You may also like to try (and perhaps modify) the program for yourself, and run it several times to see the variability obtained in such a simulation.

Gaussian and uniform amplitude distributions are not the only types met in practice but they are the most common. The gaussian case, in particular, has received a great deal of theoretical attention over the years and its properties are well known. It crops up regularly in DSP theory and practice.

9.2.2   MEAN, MEAN-SQUARE AND VARIANCE

The amplitude distribution of a random digital sequence, expressed in terms of a probability mass function, gives us a complete statistical description of its amplitude characteristics. It tells us the probability that any sample value will be found at a particular amplitude level. However, in many practical situations we do not need all this information; it may well be sufficient to know just the mean value, and the average size of fluctuations about the mean.

The mean of a random, quantized, digital sequence is defined as:

$$m = E\{x[n]\} = \sum_{-\infty}^{\infty} x_n p_{x_n} \tag{9.4}$$

where E denotes mathematical expectation. Thus, the mean is found by multiplying each allowed value of the sequence by the probability with which it occurs, followed by summation. It is the same as the *average value* of the sequence, and in electronic engineering is also widely referred to as the *dc value*.

The mean has several simple and useful properties. If every value of a sequence is multiplied by a constant, then its mean is multiplied by the same constant. Also, if we add together two or more random sequences to form a new sequence, then the mean of their sum equals the sum of their means. Note, however, that the mean of the *product* of two random variables is not, in general, equal to the product of their means. (In situations where this does apply, the variables are said to be *linearly independent* or *uncorrelated*. We will develop the very important topic of correlation in later sections.)

Another useful measure is the mean-square. In electrical terminology this is often referred to as the *average power* (from the notion that the mean-square value of a voltage or current waveform, applied to a 1-ohm resistor, gives the average power dissipation in watts). The mean-square of a digital sequence is defined as follows:

$$\overline{m^2} = E\{x[n]^2\} = \sum_{-\infty}^{\infty} x_n^2 p_{x_n} \tag{9.5}$$

Thus we compute the square of each sequence value, multiply by the relevant probability, and sum over all possible values.

The third important measure to be considered is the variance, which refers to fluctuations about the mean and is given the symbol σ^2. Thus:

$$\sigma^2 = E\{(x[n] - m)^2\} = \sum_{-\infty}^{\infty} (x_n - m)^2 p_{x_n} \tag{9.6}$$

The variance is therefore similar to the mean-square, but with the mean removed. Also, since the mean of a sum equals the sum of the means, it may easily be shown that:

$$\text{variance} = \text{mean-square} - (\text{mean})^2 \tag{9.7}$$

Equation (9.7) has a direct counterpart in electrical terms. Since the mean-square is equivalent to the total average power of a random variable, and the square of the mean represents the power in its dc component, it follows that the variance is a measure of the average power in all its other frequency components. We may therefore think of the variance as a measure of the *fluctuation power* or *ac power*.

The square root of the variance is called the *standard deviation* (σ). It is a measure of the average size of fluctuations about the mean. We have previously marked intervals equal to σ on the gaussian curve in part (b) of Figure 9.4. Note that a gaussian sequence spends most of its time within about 2 standard deviations of the mean; the chance of finding it beyond about 4 standard deviations of the mean is extremely small.

We now illustrate the above ideas for a familiar digital sequence — that produced by tossing a six-sided die.

Example 9.2 Starting with the relevant probability mass function, compute the mean, mean-square and variance of the scores produced when a fair six-sided die is tossed repeatedly. Check that the results agree with the expected relationship between these three measures.

Solution The die-tossing experiment yields a sequence of integer numbers between 1 and 6. The relevant probability mass function is as shown in part (a) of Figure 9.3. Using equations (9.4), (9.5) and (9.6) we obtain:

$$\text{mean} = 1(1/6) + 2(1/6) + 3(1/6) + 4(1/6) + 5(1/6) + 6(1/6)$$
$$= 21/6 = 3.5$$
$$\text{mean-square} = 1^2(1/6) + 2^2(1/6) + \ldots 6^2(1/6) = 15.16667$$
$$\text{variance} = (1-3.5)^2(1/6) + (2-3.5)^2(1/6) + \ldots (6-3.5)^2(1/6)$$
$$= 2.91667$$

The expected relationship between these three measures is given by equation (9.7). We have:

$$\text{mean-square} - (\text{mean})^2 = 15.16667 - 3.5^2 = 15.16667 - 12.25$$
$$= 2.91667$$

which, as expected, equals the variance.

We noted previously that if two or more random processes are superposed (added) to form a new process, then their means are additive. This also applies to their variances (but *not* their mean-square values), provided the individual random processes are statistically independent (uncorrelated). In electrical terms, this is equivalent to saying that the ac or fluctuation power of independent, superposed, random processes is additive.

Example 9.3 A random digital signal is formed by repeatedly tossing two fair dice, and adding their scores to give an integer total between 2 and 12. By considering all possible outcomes when two dice are tossed together, derive the probability mass function of the signal.

Compute the mean, mean-square and variance of the resulting signal. Show that the mean and variance are twice the values obtained when a single die is tossed.

Solution   A combined score of 2 (or 12) is obtained only when both dice score 1 (or 6). The probability is therefore $1/6 \times 1/6 = 1/36$.

A combined score of 3 is obtained *either* when one die scores 1 and the other scores 2, or *vice versa*. The associated probability is therefore $(1/6 \times 1/6) + (1/6 \times 1/6) = 2/36$. The same applies to a combined score of 11.

A combined score of 4 is obtained with any of the following individual scores:

die A	die B
1	3
2	2
3	1

The associated probability is therefore 3/36. This probability also applies to a combined score of 10.

Similar arguments apply to the other possible combined scores, leading to the probability mass function shown in Figure 9.5. As we might expect, a total score of 2 or 12 is much less likely than one towards the center of the range.

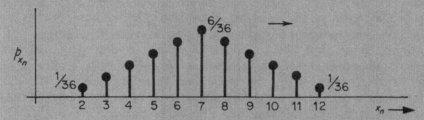

Figure 9.5   Probability mass function for the total score obtained with two dice.

Using equations (9.4), (9.5) and (9.6) we obtain:

$$\text{mean} = 2(1/36) + 3(2/36) + 4(3/36) + 5(4/36) + 6(5/36)$$
$$+ \ 7(6/36) + 8(5/36) + \ldots = 7.0$$

(It is intuitively obvious that the mean is 7.0 because 7 is the most probable score and the distribution is symmetrical about this value.)

$$\text{mean square} = 2^2(1/36) + 3^2(2/36) + 4^2(3/36) + \ldots 12^2(1/36)$$
$$+ \ 54.8333$$

$$\text{variance} = (2-7)^2(1/36) + (3-7)^2(2/36) + (4-7)^2(3/36)$$
$$+ \ldots (12-7)^2(1/36) = 5.8333$$

Comparing with the results derived in Worked Example 9.2, we confirm that, as expected, the mean is twice that obtained when a single die is tossed, and so is the variance. But the mean-square is not.

We have chosen to illustrate this section with die-tossing sequences because of their familiarity and conveniently small number of possible values. However the ideas we have developed apply equally well to a wide range of quantized digital sequences, representing both signals and noise.

9.2.3   ENSEMBLE AVERAGES, TIME AVERAGES, AND FINITE SEQUENCES

In our discussions in this chapter so far, we have implied that the probabilities associated with the various discrete values of a random digital signal or sequence are known in advance, or can be calculated. Furthermore, we have used these probabilities to infer such measures as the mean and variance of the signal. This raises a number of issues which need explanation and justification. In particular, we should clarify the distinction between what are known as *ensemble averages* and *time averages*.

We freely admit that the discussion of such matters can get rather complicated, and many books offer highly mathematical treatments. Here we will attempt a very simplified approach, with the aim of giving you some familiarity with the basic issues and terminology. If you require a more rigorous discussion, you will find it in almost any advanced book on signal processing.

Formally, a random or *stochastic* digital process is one giving rise to an infinite set of random variables, of which a particular sample sequence $x[n]$, $-\infty < n < \infty$, is one realization. The set of all sequences which could be generated by the process is known as an *infinite ensemble*, and it is this to which the probability functions and expected values defined by equations (9.4) to (9.6) strictly refer. Such measures are therefore known as ensemble averages.

However, in practical DSP we generally wish to deal with individual sequences, rather than ensembles. Each sample represents a single value of one of the variables described by the underlying stochastic process. We must therefore make a connection between the properties of an ensemble and those of an individual sequence existing over a period of time.

The way forward is in terms of time averages. Thus we define such measures as the time-averaged mean and variance of an individual sequence:

$$m_x = \langle x[n] \rangle = \lim_{N \to \infty} \frac{1}{2N+1} \sum_{n=-N}^{N} x[n] \tag{9.8}$$

and:

$$\sigma^2 = \langle (x[n] - m_x)^2 \rangle = \lim_{N \to \infty} \frac{1}{2N+1} \sum_{n=-N}^{N} (x[n] - m_x)^2 \tag{9.9}$$

It turns out that, under certain conditions, such time averages may indeed be used in place of ensemble averages.

The first condition is that the limits used in equations (9.8) and (9.9) only exist if the sequence $x[n]$ has a finite mean value and is also *stationary*. A process is said to be *stationary in the strict sense* if all its statistics are independent of time origin. However in the case of linear DSP, we can accept the less stringent condition of *wide sense stationarity*. This requires only that the mean and the correlation functions, which we will meet in later sections, are independent of time origin.

The practical reason for requiring stationarity is straightforward: if we are going to estimate such measures as the mean and variance of a digital sequence from a large number of sample values, it is intuitively clear that the underlying statistics of the sequence must not change as time progresses.

Given wide-sense stationarity, we also require that the digital sequences we are dealing with obey the so-called *ergodic hypothesis*. An ergodic process may indeed be defined as one in which time averages equal ensemble averages. Although simply stated, ergodicity may be very hard to prove. It is generally covered in more advanced texts.

Fortunately, the above conditions are met in a wide range of theoretical and practical problems encountered in linear DSP. In simple situations such as die-tossing, which we have used to illustrate previous sections, the equivalence of ensemble averages and time averages seems intuitively obvious. For example, if we put a very large number of dice in a pot, and shake them all at once, we get an ensemble of individual scores from which we could compute an ensemble average. If we take a single die and toss it very many times, successive scores form a time sequence from which we could compute a time average. Provided all the dies are fair in the first case, and the fairness of the single die does not vary with repeated throws in the second case, common sense suggests that the ensemble-averaged and time-averaged statistics must be the same.

There is one more important issue to be tackled. The infinite limits used in equations (9.8) and (9.9) cannot be realized in practice, because we are always forced (by lack of time, patience, or computer storage!) to deal with sequences of finite duration. All we can do is *estimate* time averages for a *finite* sequence, and assume that our estimates are reasonable approximations to the underlying values.

This raises the question of the quality of estimates. How do we know that an estimate based on a finite set of sample values is reasonably accurate? The usual approach to this question relies on two ideas: the *bias* of an estimate; and its *variance*.

An estimate is said to be unbiased if its expected value is the same as the true, underlying one. Naturally, we wish to make unbiased estimates whenever possible. Then, over a large number of individual trials, we may be sure that the average value of our estimates tends towards the true value we are seeking.

Ideally, we should also like the variance of estimates to be as small as possible. This implies that our estimates will tend to cluster around the true value with the minimum of statistical variability. We note also that an estimation algorithm is referred to as *consistent* if the bias and variance both tend asymptotically to zero as the sequence length on which they are based tends to infinity.

A detailed analysis of the bias and variance of estimates is quite a complicated matter which is outside the scope of this book. But to consolidate the above discussion, you may find it helpful if we illustrate some of the main points with a computer program.

We have previously emphasized the importance of the gaussian distribution, pointing out that gaussian random sequences (and especially gaussian noise)

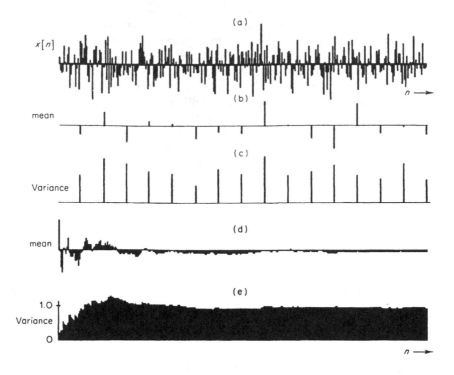

Figure 9.6   Computed estimates of mean and variance for a gaussian random sequence.

often arise in practical DSP. Fortunately, it turns out that unbiased and consistent estimates of gaussian sequences are readily obtained. Our program, listed as Computer Program no. 27 in Appendix A1, starts by generating a gaussian sequence of 320 sample values. These may be thought of as a portion of an infinite sequence with zero mean value and unit variance. (A standard computing technique is used here. We subtract 0.5 from each output of a random number generator to give a value between −0.5 and +0.5. Twelve such values are then added together to produce a new random number; and the process is repeated many times to give an approximately gaussian sequence. It may be shown that the variance of the initial random numbers is 1/12, so that of the final sequence must be unity.) A typical portion of such a zero-mean, unit variance, gaussian sequence is shown in part (a) of Figure 9.6.

Parts (b) and (c) of the figure show a series of estimates for mean and variance, respectively. Each estimate is based on the 20 values of the sequence immediately preceding it. For example, each estimate of the mean is computed by adding together 20 successive sequence values and dividing by 20. It is a well-known theoretical result that estimates formed in this way are unbiased and consistent in the case of a gaussian variable. The expected estimate of the mean is therefore zero, and of the variance unity. However, our individual estimates are quite variable. This is because a small portion of a random sequence (in this case just 20 sample values) cannot be relied upon to be representative of the infinite sequence. We could, of course, reduce the variability by using more sequence values in forming each estimate, but this would require more computation. There is always a trade-off between estimation time and the reliability of the result.

This last point suggests the alternative approach illustrated in parts (d) and (e) of the figure. Rather than divide the sequence up into a series of independent segments and compute estimates for each in turn, we compute a *running mean* and a *running variance*. This is done by continuously updating our estimates of mean and variance as each new sequence value comes in. None of the data is discarded or forgotten. Note that, at the start of the exercise (corresponding to the left-hand side of the plots), we have very few data values to work with, and the estimates are very variable. But as we move towards the right of the figure, they tend more and more convincingly towards the underlying, expected values. Clearly, a longer sequence $x[n]$ would allow us to achieve yet more reliable results.

You may like to try the program, running it a number of times to get a better idea of the chance variations which can occur. A final point to make is that the estimates of variance are only unbiased because we are using the *true* mean (zero) to compute them. If we were to use the *estimated* mean in each case, they would contain bias — although to a diminishing extent as the sequence length increased.

We hope you now have some intuitive grasp of the distinction between ensemble and time averages, and of the way in which they are linked via the notion of stationarity. Further, we hope to have given you some insight into the problems of using a finite portion of an infinite-length sequence to estimate its statistical properties. Such problems are an inherent part of the computer processing of real-life signals.

9.2.4   AUTOCORRELATION

Autocorrelation is one of the most important techniques for exploring the properties of a random digital signal or digital noise. It is essentially a time-domain measure, and we shall see later that it is intimately related to the spectral properties of a sequence in the frequency domain.

At the start of Section 9.2.2 we noted that the amplitude distribution of a random sequence gives a complete statistical description of its amplitude fluctuations. This is true as far as it goes; but it is important to realize that such a distribution is only a *first-order* measure. That is to say, it tells us about the probability of finding an individual sample value at various levels — but nothing about whether successive sample values are *related* to one another.

We can illustrate with two examples. If we toss a fair die, or a coin, we expect successive throws to be independent of one another. Knowledge of previous results does not help us to predict the next one. In such a situation, the amplitude distribution (expressed as a probability mass function or a probability distribution function) does indeed completely define the statistics of the process.

However, a filtered random sequence such as that previously shown in part (b) of Figure 9.1 is rather different. Here, groups of samples with similar values tend to cluster together, so that knowledge of the sequence's recent history would help us predict the next value. In other words the sequence has a time-domain *structure*, and there are many practical situations in which such a structure must be characterized. The most widely used measure for this purpose is the *autocorrelation function*.

You probably recall that first-order measures such as the mean and variance can be formally defined in terms of probability functions and ensemble averages. Then, assuming stationarity and ergodicity, we extend the definitions to encompass time averages taken over a single digital sequence. The same approach is valid with correlation functions. Therefore, rather than go over similar ground again, we will straight away define the autocorrelation function of a sequence, and its close cousin the *autocovariance function*, in terms of time averages:

Autocorrelation:

$$\phi_{xx}[m] = \langle x[n] . x[n+m] \rangle$$

$$= \lim_{N \to \infty} \frac{1}{2N+1} \sum_{n=-N}^{N} x[n] . x[n+m] \qquad (9.10)$$

Autocovariance:

$$\gamma_{xx}[m] = \langle (x[n] - m_x)(x[n+m] - m_x) \rangle$$

$$= \lim_{N \to \infty} \frac{1}{2N+1} \sum_{n=-N}^{N} (x[n] - m_x)(x[n+m] - m_x) \qquad (9.11)$$

Equation (9.10) shows that the autocorrelation function (ACF) is the average product of the sequence $x[n]$ with a time-shifted version of itself. It is therefore a *second-order* measure, and is a function of the imposed time shift. Autocorrelation is a valuable measure of statistical dependence between values of $x[n]$ at different times, and summarizes its time-domain structure.

The autocovariance function defined by equation (9.11) is similar to the ACF, except that the mean is removed. In the case of a random variable having zero mean, the autocorrelation and autocovariance functions are identical. But in the more general case, we may write:

$$\gamma_{xx}[m] = \phi_{xx}[m] - m_x^2 \qquad (9.12)$$

As with the first-order measures discussed in previous sections, when computing ACFs or autocovariances we must settle for estimates based on finite sequence lengths. Our estimates will always possess statistical variability, and will only tend towards the true, underlying, functions as the sequence lengths tends to infinity. Note that equations (9.10) and (9.11) define averages taken over $(2N+1)$ sample products, with N tending to infinity. Practical estimates can only approximate such functions. This does not make them useless; but it does mean that we must interpret them with a little caution.

Example 9.4   A coin is tossed 20 times. 'Heads' is denoted by $+1$, 'tails' by -1. The following sequence of scores is obtained:

1 1 −1 1 1 −1 −1 1 1 1 −1 1 −1 −1 1 −1 1 −1 −1 −1

Use these values to estimate the sequence's autocorrelation function for values of the shift parameter m between −5 and +5. Plot the results and comment.

Solution   We need to form the average product of the sequence with a shifted version of itself, for values of shift between +5 and −5. This is tabulated in Figure 9.7. At the top is shown the sequence itself. For convenience each +1 value is shown simply as a plus sign, and each −1 value as a minus sign.

Immediately below is a shifted version of the sequence for $m = −5$, that is, shifted 5 places to the left. We now cross-multiply it by the original sequence, add together all the products, and divide by the total number of products (in this case, 15). Note that, whenever a plus aligns with a plus, or a minus with a minus, we obtain a contribution of +1 to the total crossproduct. Whenever a plus aligns with a minus, or *vice versa*, the contribution is −1. With $m = 15$, the sum of all contributions turns out to be +1, so the estimate of the ACF at this value of shift is 1/15 = 0.067, as shown on the right-hand side.

The exercise is repeated for other values of m, as shown in the figure. The value $m = 0$ corresponds to the sequence being exactly aligned with itself. In this case all contributions to the cross-product are +1, there are 20 of them, and the ACF is unity.

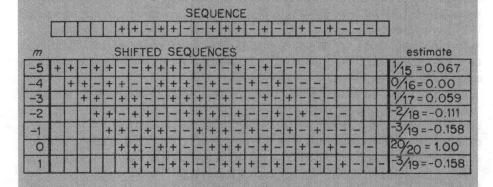

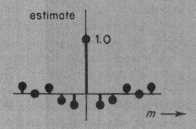

Figure 9.7   Autocorrelation of a coin-tossing sequence.

We continue the tabulation as far as $m = 1$, which gives the same result as $m = -1$. It is intuitively obvious that when we shift a sequence over itself and crossmultiply, the results for positive shifts must be identical to those for negative shifts. In other words, an ACF must be an even function of the shift parameter m. We do not therefore need to continue the tabulation for $m = 2, 3, 4,$ and 5.

The estimated values of the ACF are plotted in the lower part of the figure. The largest value occurs when $m = 0$, that is, when the sequence is perfectly aligned with itself. All the other values are much smaller. We may explain this intuitively as follows. When we cross-multiply the sequence with a shifted version of itself, it is just as likely that a plus value will align with a plus value as with a minus value, so we expect the average product to be zero. In fact it does not turn out as truly zero in this example, but this is simply due to the fact that we are processing a very short sequence. Our estimates cannot be expected to be very reliable. If we tossed the coin a thousand times, we would probably find that the resulting estimates were much closer to zero for all values of m other than $m = 0$. (This assumes, of course, that the coin is 'fair', and that the results of successive tosses are completely independent of one another.)

The above worked example illustrates a number of important points about practical autocorrelation. A key one has already been mentioned: when we estimate an ACF using a limited number of sample values, we end up with an approximation to the true, underlying, function. In fact, forming an ACF from a sequence of just 20 samples would normally be considered quite inadequate. We have done so here simply to illustrate the operations involved in autocorrelation, and to give you some intuitive grasp of its effects. Hundreds, or even thousands, of samples would be more typical. And, of course, hand calculation must be replaced by a computer!

Another point concerns the number of cross-products we form as the shift parameter m increases. Clearly, as the shifted sequence 'slides over' the unshifted one, the number of available products decreases. In the above example, the ACF estimate for $m = 0$ is based on 20 products, whereas that for $m = -5$ (or $m = +5$) is based on only 15. This implies that estimates in the 'tails' of the computed ACF are somewhat less reliable than those in the center. For this reason, it is usual to restrict shift values to no more than about 15–20% of the available sequence length.

As you will appreciate, we have used coin-tossing because of its familiarity and convenience. It provides us with a binary form of signal, and by allocating the values $+1$ and -1 to 'heads' and 'tails', respectively, we obtain a random sequence with zero mean and unit variance. However, we must admit that coin tossing hardly provides true digital signals in the sense we have previously met in this book. When we speak of a digital signal, we usually mean one which has been regularly sampled at a known sampling rate. Sample values are equispaced in time, and this allows us to generate, via Fourier transformation, a whole set of corresponding spectral properties. We can hardly claim to be able to toss a coin with such regularity! What we have generated is simply a numerical sequence, and used it to illustrate some basic ideas. In the DSP context, however, autocorrelation is normally applied to regularly-sampled signals, and from now on we will assume this in our discussion.

A major advantage of autocorrelating a random signal is that in most cases the resulting function tends to die away for large values, both positive and negative, of the time-shift variable m (we see this effect in Figure 9.7). This means that a very lengthy signal can be characterized by a statistical function of reasonable length. Not only is such compactness valuable in its own right but — as we shall see later — it has important implications for estimating the spectral properties of the signal.

Let us now use a computer to illustrate some of the main features of autocorrelation and autocovariance functions. Figure 9.8 shows, on the left-hand side, portions of five different sequences and, on the right, their computed autocorrelation functions. Although we have plotted only 150 values of each sequence, the ACFs were in fact estimated using 1001 values ($N = 500$ in equation (9.10)) to give reasonably reliable estimates. Also, we have computed each ACF over the restricted range $-50 \leq m \leq 50$. This is sufficient to establish its general form without getting involved in too much computation.

Parts (a) and (b) of the figure show two deterministic sequences: a sinusoid, and a cosinusoid plus a dc level. You may find these choices rather surprising, because the whole emphasis of our discussion is on random functions. The reason is that the ACFs of sines and cosines are quite easy to visualize and relate to equation (9.10). They also illustrate some important ideas. When we have discussed them, we can tackle some more complicated examples with confidence.

What happens when we autocorrelate the sinusoid shown in part (a)? As we have already seen, we must multiply a long portion of the sequence by a shifted version of itself, and average the result. The process is then repeated for different values of shift. In this case it is clear that when the shift is zero, positive peaks of the sinusoid align with positive peaks, and negative peaks with negative peaks, giving a large positive product. Hence, we expect the ACF to be large and positive for $m = 0$.

As we start to introduce time shift, the positive and negative peaks begin to lose their alignment. By the time all positive peaks are aligned with negative ones, and *vice versa*, the ACF will have a large negative value. As the shifting process is continued, it is clear that the ACF must trace out a repetitive function, whose period in the time-shift variable m equals the period of the sequence itself in the time variable n.

We conclude that a repetitive sequence $x[n]$ has an ACF which is also repetitive, with the same period. In the special case of a sinusoidal sequence, it turns out that the ACF takes a cosinusoidal form. And since the mean value of the sequence is zero in this case, the ACF is the same as the autocovariance function.

Part (b) of the figure shows a cosine sequence plus a dc level. As before, the ACF is periodic with a peak value at $m = 0$. Note that the finite mean value of the sequence is also represented, the ACF now being offset vertically compared with part (a) of the figure. This illustrates the fact that an ACF contains all the

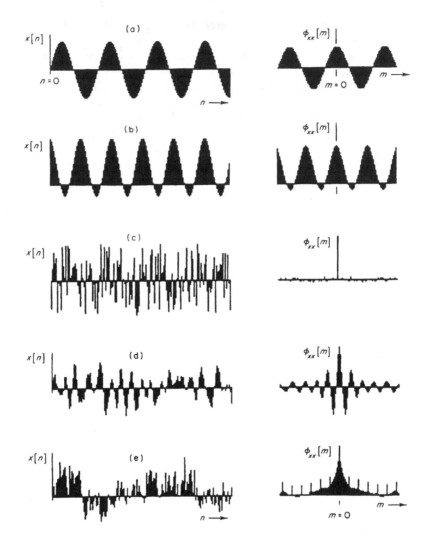

Figure 9.8   Five digital sequences and their estimated autocorrelation functions (*abscissae: 150 and 101 samples*).

frequency components present in the sequence itself. And it may be shown (see, for example, references 29 and 33 in the bibliography) that each component has an amplitude proportional to the square of that in the original sequence — in other words, an amplitude which represents the component's *average power*.

This important point can be developed a little further by noting that the central value of an ACF equals the mean square value of the sequence, and is therefore a measure of its total power. Thus, setting $m = 0$ in equation (9.10) we obtain:

$$\phi_{xx}[0] = \langle x[n]^2 \rangle \qquad (9.13)$$

The central value is always a maximum value. It may be equalled at other values of time shift, as in parts (a) and (b) of Figure 9.8, but it can never be exceeded.

Similarly, by putting $m = 0$ in equation (9.11), we see that the central value of an autocovariance function equals the variance of the corresponding sequence, which is equivalent to its ac power.

Actually we should be a little careful about using the word 'power', because the idea of average power really only applies to signals which continue for ever. In the case of practical, time-limited signals we should talk *either* about power averaged over their actual duration, *or* about their total energy.

Let us now consider ACF estimates for three random functions. Part (c) of Figure 9.8 shows, on the left, a sequence of uniformly-distributed random numbers in the range -1 to $+1$. Successive values may be assumed independent. When such a sequence is autocorrelated, we get a large central peak at $m = 0$, with zero values to either side (actually, our estimate shows small residues to either side — once again, the typical effect of working with a finite-length sequence). Note the essential similarity between this result and the coin-tossing experiment of Worked Example 9.4. Again, we may readily visualize the formation of the ACF. At zero time shift, positive and negative values align with themselves, giving a large positive product. But at any other value of time shift, a given sample is just as likely to align with one of the opposite sign and their cross-product averages out to zero. In other words, there is no correlation between adjacent sample values. The sequence is completely random.

For reasons which will become clear later, a sequence with this type of ACF is often called *white noise*. Note that this term refers to the form of the ACF (with its isolated central spike), not to the amplitude distribution of the underlying sequence. Thus the present sequence, and the coin tossing sequence of Worked Example 9.4, and a zero-mean gaussian sequence, can all be described as 'white' provided successive sample values are truly independent.

Part (d) of the figure shows a similar sequence after bandpass filtering. This has had some major effects. The ACF now spreads considerably to either side of $m = 0$, reflecting correlation between adjacent sample values. In general, the amount of spread depends on the severity of the filtering operation (the filter's bandwidth). In this example there is quite a strong correlation up to shifts of about $m = \pm 10$.

In part (e) we have processed white noise with a low-pass filter, and then added a repetitive impulse train with a period of 8 sampling intervals. This is designed to show some of the typical effects of mixing repetitive and random sequences. You may or may not be able to detect the impulse train in the sequence itself; but autocorrelation brings it out clearly. Note that the filtered noise only contributes to the ACF around $m = 0$, whereas the pulse train, being strictly repetitive, contributes over the complete range of time shift. This suggests a very important practical application of autocorrelation — the detection of a repetitive signal in the presence of unwanted noise.

We hope you now have some insight into the process of autocorrelation, and the forms of ACF likely to be met in practice. You have probably noticed that an ACF is always an even function of time shift. Intuitively this is because the average result of cross-multiplication and summation is independent of the direction of shift. It follows that we can generally save on computation by estimating only 'one half' of the function. There are also important implications for the spectral properties of random sequences which we explore in the next section.

9.2.5   POWER SPECTRUM

As we have already seen many times in this book, every time-domain function has a counterpart in the frequency domain. In the case of an autocorrelation function (ACF), which is a function of the time-shift variable m, the counterpart is called a *power spectral density* or, more simply, a *power spectrum*. It indicates how the sequence's power or energy is distributed in the frequency domain, and is a widely-used measure of random signals and noise. Whether we use a sequence's ACF or its power spectrum in a particular case is largely a matter of convenience. They contain equivalent information in the time domain and the frequency domain, respectively.

You will probably not be surprised to learn that the ACF and power spectrum of a digital sequence are formally related as a Fourier Transform pair (see Section 3.3.1). Thus the power spectrum is defined as:

$$P_{xx}(\Omega) = \sum_{m=-\infty}^{\infty} \phi_{xx}[m] \exp(-j\Omega m) \tag{9.14}$$

An alternative definition, found in some texts, specifies the autocovariance function instead of the ACF. This avoids problems which arise if the underlying sequence has a non-zero mean value (giving rise to a rather awkward zero-frequency impulse in the power spectrum). However, in the following discussion we will assume zero-mean sequences for which the autocorrelation and autocovariance functions are identical.

Example 9.5   Derive expressions for the power spectra of the following types of random sequence, and comment on their form:

(a) a very long coin-tossing sequence in which 'heads' and 'tails' are denoted by the values $+1$ and -1, respectively, as in Worked Example 9.4. Treat the resulting sequence as 'regularly-sampled'.
(b) a random digital sequence whose ACF has value 2 at $m = 0$, and value 1 at both $m = 1$ and $m = -1$. It is zero at all other values of m.

Solution
(a) We estimated the ACF of such a coin-tossing sequence in Worked Example 9.4, and illustrated it at the bottom of Figure 9.7. Remember, however, that the estimate was based on a short sequence of just 20 values. We are now asked to consider the ACF of a very long coin-tossing sequence, which we assume will closely approximate the true, underlying, function. This consists of a unit

sample value at $m = 0$, and zero values elsewhere. Hence in the summation given by equation (9.14), there is just one finite term $\phi_{xx}[0]$ which equals unity. Assuming a regularly-sampled sequence we may therefore write:

$$P_{xx}(\Omega) = \sum_{m=-\infty}^{\infty} \phi_{xx}[m] \exp(-j\Omega m) = \phi_{xx}[0] \exp(0) = 1 \quad (9.15)$$

This result shows that the power spectrum is a constant, in other words its value is not frequency-dependent. This means that the spectral function is 'white' and contains a flat distribution of spectral energy. All frequencies are equally represented.

(b) We are told that the ACF has the values 1,2,1, centered on $m = 0$, with zero values elsewhere. Clearly, this represents a digital sequence in which there is some correlation between adjacent sample values. In this case the infinite summation given by equation (9.14) reduces to just three nonzero terms, and we may write:

$$P_{xx}(\Omega) = \sum_{m=-\infty}^{\infty} \phi_{xx}[m] \exp(-j\Omega m)$$

$$= \phi_{xx}[-1] \exp(j\Omega) + \phi_{xx}[0] \exp(0) + \phi_{xx}[1] \exp(-j\Omega)$$

$$= 1. \exp(j\Omega) + 2 + 1. \exp(-j\Omega) = 2(1 + \cos \Omega) \quad (9.16)$$

This power spectrum is certainly not white. The cosine term signifies a variation of spectral energy with frequency. When $\Omega = 0$, $P_{xx}(\Omega) = 4$, and when $\Omega = \pi$, $P_{xx}(\Omega) = 0$. Hence the random sequence is strongest in low frequencies around $\Omega = 0$. This could occur if, for example, it represents 'white noise' which has passed through a simple low-pass filter.

Just as a Fourier Transform allows us to derive the power spectrum from the ACF, so the inverse transform gives the ACF in terms of the power spectrum:

$$\phi_{xx}[m] = \frac{1}{2\pi} \int_{2\pi} P_{xx}(\Omega) \exp(j\Omega m) \, d\Omega \quad (9.17)$$

Note also that by putting $m = 0$ we obtain:

$$\phi_{xx}[0] = \frac{1}{2\pi} \int_{2\pi} P_{xx}(\Omega) \, d\Omega \quad (9.18)$$

Now $\phi_{xx}[0]$ represents the sequence's total power (see equation (9.13)). We now see that it is also equal to $\frac{1}{2\pi}$ times the area under the power spectrum curve over any 2π interval.

Intuitively, there are two main reasons why the function $P_{xx}(\Omega)$ relates to the spectral power distribution of a digital sequence. Firstly, in forming an ACF we multiply the sequence by a shifted version of itself, giving rise to a second-order, 'amplitude squared' measure with the dimensions of power or energy. And secondly, as we pointed out in the previous section, an ACF is always an even function of time shift. It follows that its spectral equivalent must always be a real function, with no information about the relative phases of the various frequency components present. A measure which relates to power, rather than amplitude and phase, meets this criterion.

We have previously discussed the spectral representation of digital signals in terms of Fourier and z-Transforms in Chapters 3 and 4. One of the conditions for such transforms to exist is that the sequence concerned should have *finite energy*, or it must be possible to multiply by an exponential so that the product has finite energy. This can lead to difficulties with some important classes of sequence, including random noise, which in theory continue for every and therefore possess infinite energy. However, the ACF offers a way forward. As we have seen in Figure 9.8, the ACFs of zero-mean sequences generally decay towards zero to either side of $m = 0$, and may therefore themselves be represented as finite-energy sequences for which Fourier and z-Transforms do exist. The ACF is therefore not only a useful function in its own right; it also provides us with a useful route to the spectral representation of random, infinite-energy, sequences.

However, as we have already noted several times, there may be considerable practical difficulties in computing a sequence's ACF. Basically, the problem is that we can only work with a finite portion of the sequence, and estimate the ACF for a limited set of time-shift, or *lag*, values. If you refer back to parts (c), (d), and (e) of Figure 9.8, you will recall that not only are the computed ACFs subject to random variability, but they are also restricted to a limited range of lag values. In each case we end up with an estimate of the true, underlying function.

What effect does the use of an *estimated* ACF have on a power spectrum obtained using equation (9.14)? We must straight away admit that this is a complicated issue, and all we can do here is introduce some of the problems and the techniques that have been developed to counteract them. If you are interested in the finer points of power spectrum estimation, you will probably find the book by Vaseghi listed in the Bibliography very useful.

Firstly, let us note that if we assume an ACF to be zero outside the estimated range of lag values, we are effectively truncating it with a rectangular window. This tends to produce unwanted sidelobes in the power spectrum as a result of the Gibb's phenomenon (see Section 5.3.2). It is therefore common practice to multiply the estimated ACF by a tapered window function such as a Bartlett (triangular) window before Fourier Transformation. A further advantage of this approach, which is generally referred to as the *Blackman–Tukey Method*, is that tapering has the effect of down-weighting the importance of the 'tails' of the estimated ACF. This is valuable because the tail values are less reliable than those nearer the center, since they are formed from fewer cross-products. However, as always, there is a price to be paid for such windowing: there is a pronounced smoothing effect in the frequency domain, and this reduces spectral resolution.

Some of the above points are illustrated by Figure 9.9, which shows estimated ACF's and power spectra for a bandpass-filtered noise sequence. In

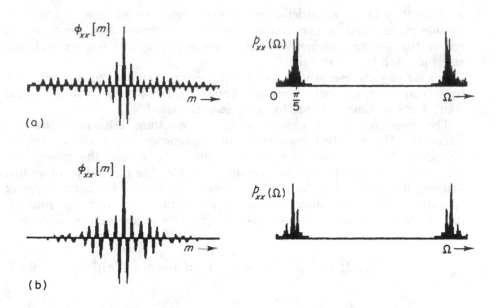

Figure 9.9 Autocorrelation and power spectral estimates for a bandpass-filtered noise sequence (*abscissae: 256 samples*).

part (a) the ACF has been formed by cross-multiplying 1024 values of an infinite sequence, for lag values in the range $-128 \le m \le 128$. We may expect the resulting ACF to be fairly reliable; but, of course, we have had to perform quite a lot of computation (including some $1024 \times 256 = 262,144$ multiplications). Using the FFT listed as Computer Program No. 24 in Appendix A1, we obtain the power spectrum estimate on the right-hand side. Although it appears to contain some chance variations, we may be confident about its overall shape — including the clear peak around $\Omega = \pi/5$ (corresponding to frequency components with about 10 samples per period, as indicated by the form of the ACF). In fact, a smooth curve drawn through the various frequency-domain 'samples' would, in this case, get close to the underlying power spectrum.

However in some cases we do not have so much data to work with, or we cannot afford so much computation. Part (b) of the figure shows another estimated ACF, this time based on a portion of the sequence containing just 128 sample values. Assuming the sequence is zero outside the observed interval, we are effectively multiplying the infinite sequence by a rectangular window. Then, sliding the time-limited sequence over itself to estimate the ACF, we get less and less overlap as the lag increases. As we approach lag values of $m = \pm128$, the ACF values become very unreliable. It is therefore only sensible to taper the result, reducing the importance attached to values in the 'tails'. In this example, we have tapered with a triangular (Bartlett) window.

The power spectrum estimate on the right-hand side was obtained by Fourier Transformation of the windowed ACF, again using a 256-point FFT. It has the same general form as that in part (a), but there is a loss of detail and a tendency for spurious peaks and troughs to occur. Although these results should be taken as illustrative rather than definitive, they indicate the problem of obtaining reliable power spectral estimates via the ACF of limited data sequences. Of course, the problem becomes worse if we are forced to work with even shorter sequences and fewer lag values.

In spite of such reservations, the power spectra of random signals and noise have been widely estimated using the above approach. Because it works via the ACF, it is sometimes referred to as the *indirect method*.

The alternative *direct method* of power spectrum estimation involves calculating the so-called *periodogram* of a sequence — a classic method developed by Sir Arthur Schuster in 1898, long before the power and convenience of digital computers became available. The method involves first deriving the Fourier Transform of a sequence $x[n]$ to give the normal type of spectrum $X(\Omega)$. Then, discarding phase information, we find the squared-magnitude and divide by the sequence length. This yields the periodogram:

$$P'_{xx}(\Omega) = \frac{1}{N}|X(\Omega)|^2 = \frac{1}{N}|\sum_{n=0}^{N-1} x[n]\,\exp(-j\Omega n)|^2 \qquad (9.19)$$

Note that this method assumes we have knowledge of the individual sample values of the random sequence, rather than the time-averaged statistics represented by its ACF. As with power spectral estimates derived from autocorrelation functions, periodograms suffer from random variability due to finite sequence lengths, and a set of parallel techniques have been evolved to enhance their reliability, including windowing and averaging.

Both indirect and direct methods for power spectrum estimation have been widely used in practice, mainly because of the speed and convenience of the FFT. However, they do not always give sufficient spectral resolution. In effect, both methods assume that the random sequence, or its autocorrelation function, are zero outside the range for which data is available. They are referred to as *nonparametric* methods. A very different *parametric* approach assumes that the sequence can be modelled as the output of a system excited with white noise. This means that the output power spectrum is shaped entirely by the frequency response of the model, and that the parameters of the model may therefore be used to characterize the sequence. A series of powerful techniques based on this approach have been developed. They include the so-called *moving-average, autoregressive moving-average*, and *maximum entropy* techniques which are well described in references 7 and 34 listed in the Bibliography.

9.2.6   CROSS-CORRELATION

We have seen how autocorrelation may be used to characterize a sequence's time-domain structure, revealing statistical relationships between successive sample values. Cross-correlation is an essentially similar process; but instead of comparing a sequence with a shifted version of itself, it compares two different sequences. It gives us statistical information about the time-domain relationships between them.

The *cross-correlation function (CCF)* of two sequences $x[n]$ and $y[n]$, and its close cousin the *cross-covariance function*, may be defined in terms of time averages:

Cross-correlation

$$\phi_{xy}[m] = \langle x[n]\, y[n+m] \rangle$$

$$= \lim_{N \to \infty} \frac{1}{2N+1} \sum_{n=-N}^{N} x[n]\, y[n+m] \qquad (9.20)$$

Cross-covariance

$$\gamma_{xy}[m] = \langle (x[n] - m_x)(y[n+m] - m_y) \rangle$$

$$= \lim_{N \to \infty} \frac{1}{2N+1} \sum_{n=-N}^{N} (x[n] - m_x)(y[n+m] - m_y) \qquad (9.21)$$

Both these functions are second-order measures. The CCF provides a statistical comparison of two sequences as a function of the time shift between them. Cross-covariance is similar, except that the mean values of the two sequences are removed.

Just as the form of an ACF reflects the various frequency components in the underlying sequence, so the form of a CCF reflects components *held in common* between $x[n]$ and $y[n]$. However, whereas an ACF, being an even function, retains no phase information, a CCF holds information about the relative phases of shared frequency components. It follows that a CCF is not generally an even function of the time-shift parameter m.

Although cross-correlation is normally applied to lengthy portions of random digital sequences, we can illustrate some of the above features, including the computational steps involved, using two simple time-limited signals.

Example 9.6   Two simple digital signals $x[n]$ and $y[n]$ are shown in Figure 9.10(a). Assuming them to be zero at all other values of n, calculate and plot their cross-correlation function.

Solution   We first note that these signals are strictly time-limited, with just a few non-zero sample values. They are finite-energy signals. Therefore rather than form the long-term averages specified by equation (9.20), which would tend to zero, we simply sum all finite cross-products when a shifted version of $y[n]$ is 'slid below' $x[n]$. This gives a version of CCF relevant for such time-limited signals.

The situation for $m = 3$ is shown in part (b) of the figure. Only one of the finite values in $y[n]$ aligns with one in $x[n]$, giving a summed cross-product equal to unity. Shifting $y[n]$ step by step to the right, there is initially increasing overlap of the two signals, then decreasing overlap, and we obtain the following summations for various values of m:

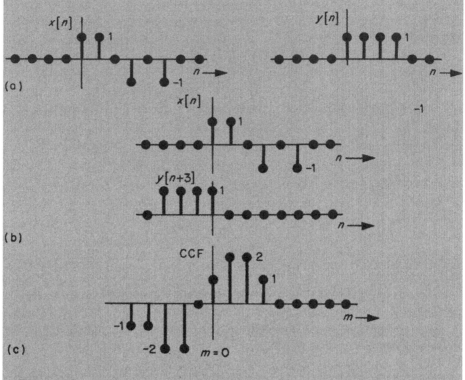

Figure 9.10

$m = 3$	1
$m = 2$	$1 + 1 = 2$
$m = 1$	$1 + 1 + 0 = 2$
$m = 0$	$1 + 1 + 0 - 1 = 1$
$m = -1$	$1 + 0 - 1 + 0 = 0$
$m = -2$	$0 - 1 + 0 - 1 = -2$
$m = -3$	$-1 + 0 - 1 = -2$
$m = -4$	$0 - 1 = -1$
$m = -5$	-1

These results are plotted as the CCF in part (c) of the figure. We see that the function is not even, so it must contain phase as well as amplitude information about shared frequency components in the two signals. It is a little hard to infer much more from its form, although one further point is of interest. We see that the sum of all samples in $y[n]$ is 4, so it must have a substantial dc (zero frequency) component. But the sum of all samples in $x[n]$ is zero, so it has no dc component. Since cross-correlation only reflects frequency components held in common between the two signals, we must expect the CCF to lack a dc component. This is confirmed by noting that the sum of all its sample values is zero.

The above Worked Example is designed as a simple illustration of the basic process of cross-correlation for finite-length signals. However, in practical DSP the technique is generally applied to lengthy random sequences representing signals, noise, or signals mixed with noise. So we now turn to a more realistic situation with the help of a computer.

Part (a) of Figure 9.11 shows portions of two random sequences which are both strong in frequencies having around 12 samples per period. In addition, $y[n]$ contains a substantial amount of high-frequency energy at around two samples per period. We have estimated the CCF shown in part (b) using equation (9.20), but with $N = 500$ and values of m between 0 and 99. We have therefore only obtained a portion of the total function. Nevertheless, it is clear that this estimated CCF reflects only shared frequencies in $x[n]$ and $y[n]$. There is no sign of the high-frequency energy contained in $y[n]$.

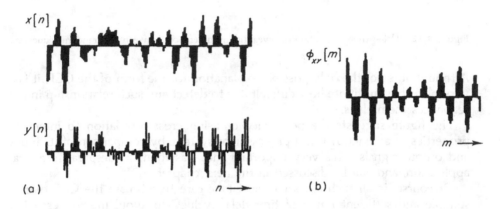

Figure 9.11   Estimating a cross-correlation function.

The fine detail of such a CCF can be very hard to interpret, and when our main interest is in shared frequencies it is generally better to use the equivalent spectral description covered in the next section. However from the practical viewpoint, there is one major type of situation in which a CCF can be particularly revealing — namely when there are *timing differences* between two random sequences. To take a simple case, suppose that $x[n]$ and $y[n]$ are identical white noise sequences which differ only in their time origin. We then expect their CCF to be zero for all values of m, except the one which exactly offsets, or cancels, the timing difference.

You may find the rather more challenging example illustrated in Figure 9.12 both interesting and helpful. Part (a) shows portions of two sequences $x[n]$ and $y[n]$, which we suspect are related in some way. We compute an estimate of their CCF, shown in part (b) of the figure (this was found in exactly the same way as part (b) of Figure 9.11). The result suggests very little correlation between the sequences except for two particular lag values, $m = 27$ and $m = 43$. It implies that $y[n]$ could well have arisen by superposition of two versions of $x[n]$: one delayed by 27 sampling intervals; the other delayed by 43 sampling intervals, and inverted. This is shown by part (c) of the figure.

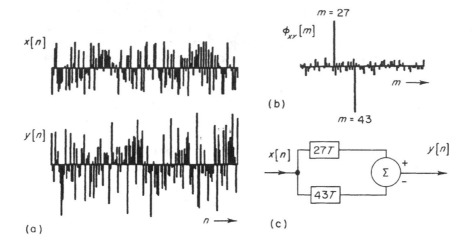

Figure 9.12 Using cross-correlation to reveal timing differences between two random sequences.

Although it is not the only possible explanation for the form of the CCF, it is a very plausible one. Note how difficult it is to detect any such relationship in the sequences themselves.

The figure suggests the possibility of using cross-correlation to infer the properties of an unknown LTI processor or system, by comparison of its input and output signals. This very important idea has found widespread practical application, and will be discussed in the next chapter.

Of course, in all such cases we must take care to estimate the CCF over a suitable and sufficient range of time-delay values, to avoid missing essential features. This is largely a matter of judgement and experience, although we do sometimes have additional *a priori* information which helps us make a sensible choice.

A final point to make about cross-correlation is that it reveals only *linear* relationships between two sequences. Put at its simplest, this means that only if $y[n]$ could have been produced by passing $x[n]$ through an LTI processor (or *vice versa*) may the CCF be used to define their relationship.

9.2.7 CROSS-SPECTRUM

The frequency-domain counterpart of a cross-correlation function (CCF) relating two sequences $x[n]$ and $y[n]$ is known as a *cross-spectral density*, or simply as a *cross-spectrum*. It gives us valuable information about frequencies held in common between $x[n]$ and $y[n]$. Thus if $x[n]$ has a component $A_1 \sin (n\Omega_1 + \theta)$ and $y[n]$ has a component $A_2 \sin (n\Omega_1 + \psi)$, then their cross-spectrum will have magnitude $A_1 A_2 / 2$ and phase $(\theta - \psi)$ at the frequency Ω_1. This shows that a shared component is represented in proportion to the *product* of the individual amplitudes, and with a phase equal to the *difference* between the individual phases. Since a cross-spectrum holds phase information, it is generally a complex function of frequency. In this respect it differs from the power spectrum of an individual sequence, which is always real.

If two sequences $x[n]$ and $y[n]$ have no shared frequencies, or frequency ranges, their cross-spectrum (like their CCF) is zero. They are said to be *linearly independent*, or *orthogonal*. This implies that $y[n]$ could not be produced by passing $x[n]$ through a linear processor, or *vice versa*.

Formally, the cross-spectrum $P_{xy}(\Omega)$ and CCF are related as a Fourier Transform pair. Thus:

$$P_{xy}(\Omega) = \sum_{m=-\infty}^{\infty} \phi_{xy}[m]\ \exp(-j\Omega m) \tag{9.22}$$

and

$$\phi_{xy}[m] = \frac{1}{2\pi} \int_{2\pi} P_{xy}(\Omega)\ \exp(j\Omega m)\ d\Omega \tag{9.23}$$

When $x[n]$ and $y[n]$ both have non-zero mean values, the above definition of $P_{xy}(\Omega)$ results in a rather awkward frequency-domain impulse at $\Omega = 0$. For this reason, the cross-spectrum is sometimes defined as the Fourier Transform of the cross-covariance (see equation (9.21)) rather than the CCF.

In Section 9.2.5 we described the practical difficulty of obtaining a reliable estimate for the power spectrum of a time-limited random sequence. Very similar problems apply to cross-spectrum estimation. Once again it is often helpful to taper the time-domain function (in this case the CCF) with a suitable window function before transformation. A triangular (Bartlett) window is often used.

Apart from its value as a measure of shared frequencies between two sequences, the cross-spectrum relating random input and output signals of an LTI processor can be used to infer the properties of the processor itself. We shall return to this important idea in the next chapter.

PROBLEMS

Q9.1   A random digital sequence takes on three distinct values, 1, 0, and −1, all with equal probability.
(a) Sketch its probability mass and probability distribution functions.
(b) Estimate its mean, mean-square, and variance.

Q9.2   Repeat Q9.1 when the sequence values are 0, 1, and 2, and explain the changes in the mean, mean-square and variance.

Q9.3   A die has been unfairly loaded to give the following probabilities associated with its six possible scores:

score:	1	2	3	4	5	6
probability	0.1	0.15	0.15	0.15	0.15	0.3

(a) Sketch its probability mass and probability distribution functions.
(b) Estimate the mean, mean-square and variance of scores.

Q9.4   Computer Program No. 26 in Appendix A1 estimates probability mass functions for uniformly-distributed and gaussian sequences (see Figure 9.4 in the main text). Run the program several times to observe the chance variations that can occur with such estimates.

Q9.5   Explain clearly:
(a) The distinction between ensemble averages and time averages.
(b) The distinction between autocorrelation and autocovariance functions.

Q9.6   Computer Program No. 27 in Appendix A1 estimates the mean and variance of a gaussian sequence (see Figure 9.6 in the main text).
(a) Run the program several times to observe the chance variations that can occur with such estimates.
(b) Modify the program so that each estimate in parts (b) and (c) of Figure 9.6 is based on the 40 values of the sequence immediately preceeding it. On average, do you observe less variability in the results, compared with those based on 20 sequence values?

Q9.7   Repeat the coin-tossing experiment of Worked Example 9.4 in the main text, generating a different sequence of 20 scores. Then:
(a) Estimate the autocorrelation function, and check that its largest value occurs when $m = 0$.
(b) For the same sequence, denote 'heads' by $+1$ and 'tails' by zero. Estimate the new autocorrelation function, and comment on the differences between it and the ACF found in part (a) above.

Q9.8   Make careful sketches of the autocorrelation functions of the following:
(a) The random digital sequence specified in Problem Q9.1.
(b) The sequence obtained when a fair six-faced die is tossed repeatedly.
(c) The sequence obtained when the unfair die described in Problem Q9.3 is tossed repeatedly.
(d) The sequence illustrated in part (b) of Figure 9.1 in the main text.
In all cases except (d), specify the central peak value of the ACF.

Q9.9   The ACF of a random signal is shown in Figure Q9.9. Find the signal's mean value, mean square value and standard deviation.
   Over approximately what time interval are signal fluctuations significantly correlated?

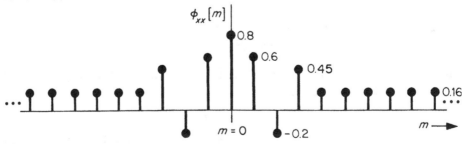

Figure Q9.9

Q9.10 Find expressions for, and sketch, the power spectra corresponding to the ACF's in parts (a) and (b) of Problem Q9.8.

Q9.11 The power spectrum of a random signal over the range $-\pi \leq \Omega \leq \pi$ is given by:

$$P_{xx}(\Omega) = 1, \quad -\pi/4 \leq \Omega \leq \pi/4$$
$$= 0, \quad \text{elsewhere}$$

(a) Find an expression for, and sketch, the signal's ACF.
(b) Find the signal's total average power.

Q9.12 Repeat Problem Q9.11 for a signal having the power spectrum:

$$P_{xx}(\Omega) = 1 + 0.5 \cos \Omega + 0.25 \cos 2\Omega$$

Q9.13 A coin is tossed repeatedly. 'Heads' is counted as $+1$, 'tails' as -1. Assuming the resulting number sequence may be treated as a regularly-sampled random signal, what form do you expect its power spectrum to take, and why?

Q9.14 Discuss the value of cross-correlation for quantifying the relationship between two random sequences. In qualitative terms, what form do you expect the CCF to take in the following cases:
(a) One sequence is a sinusoid, the other is a dc level.
(b) One sequence is a sinusoid, the other is also a sinusoid but of a different frequency.
(c) One sequence is a sinusoid, the other is a cosinusoid at the same frequency.
(d) One sequence is white noise, the other is also white noise but from a different source.
(e) One sequence is white noise, the other is the same white noise but delayed by 5 sampling intervals.
(f) One sequence is white noise, the other is the same white noise but after passing through a low-pass filter.
(g) One sequence is white noise, the other is white noise from a different source after passing through a high-pass filter.
(h) One sequence is white noise after filtering by a low-pass filter. The other sequence is the same white noise after filtering by a high-pass filter. The filter passbands partially overlap.

Q9.15 Long portions of two random digital signals are cross-correlated. The resulting function is found to have the following values:

$$\phi_{xy}[m] = 1, \quad m = 0, 3$$
$$= 2, \quad m = 1, 2$$
$$= 0, \quad \text{elsewhere}$$

Find an expression for the cross-spectrum.

_____ CHAPTER 10 _____

Random DSP

10.1 INTRODUCTION

At the start of the previous chapter we outlined the importance of random
signals and noise in DSP. We then went on to characterize random sequences
— whether representing signals or noise — in terms of such long-term average
measures as the mean, variance, autocorrelation function (ACF) and power
spectrum.

Of course, DSP is concerned primarily with *processing* signals, rather than
merely *describing* them. For example, we may have a digital signal representing
human speech which we need to transmit along an optical fiber. We know the
frequency response of the fiber and the statistical properties of the speech
signal. How are these properties modified as the signal passes along the fiber,
and will it be possible to recover the signal at the receiving end? As another
example, we may have a signal representing electrical activity in the heart
(known as an *electrocardiogram*) which is contaminated by random interference
or *noise*. We know the bandwidth of the signal and the power spectrum of the
noise. How can we specify a digital filter which will be effective in reducing the
noise without significantly degrading the signal? You will probably be able to
think of further practical situations where the effective *processing* of random
sequences is of the greatest importance.

Obvious questions now arise. How are the statistical measures of random
signals and noise which we developed in the previous chapter affected by
processing? If we know such measures at the input to a given filter or processor,
can we predict what they will be a the output? How do we design an optimum
processor for enhancing a signal contaminated by noise?

At first sight, such questions may seem rather daunting. Yet, provided we
restrict ourselves to LTI processors (as we have done so far in this book), it
turns out that the effects of processing can be readily described and predicted.
Furthermore, the concepts and techniques involved fall fair and square within
the framework of time and frequency domain DSP which we have developed
for deterministic signals in earlier chapters. In other words, random DSP can be
treated as an extension of our previous work, rather than as a radical departure
from it. Both conceptually and practically, this is a great advantage.

It is only right to add that the linear processing techniques we shall describe here are a limited set, largely chosen because they illustrate the continuity between deterministic and random signal processing. However many of them have achieved a classical status in DSP theory, as well as being of great practical importance. You may be confident that a clear understanding of this chapter will stand you in good stead for further reading of the DSP literature and for tackling more advanced topics in random signal processing.

10.2 RESPONSE OF LINEAR PROCESSORS

We first examine the general relationships between the input and output of an LTI processor when the input sequence is random. Such a sequence might represent a signal, or noise, or a signal mixed with noise. Some of the mathematics may appear a little heavy at first reading, but it is actually quite straightforward, and we will make a special attempt to illustrate the main results with a computer program. They are the key to understanding the effects of a linear processor, such as a digital filter, on a random digital sequence. As you will see later, we subsequently use them to develop a number of very important practical ideas and techniques.

It is helpful to start in an intuitive way, by reconsidering the idea of digital convolution, but applied this time to a random input signal. We first developed the idea of digital convolution back in Section 2.4, showing how the impulse response $h[n]$ of a processor could be used to assess its response to a deterministic input signal $x[n]$. Figure 2.9 illustrated a graphical interpretation of convolution, in which a time-reversed version of $h[n]$ was laid beneath $x[n]$, followed by cross-multiplication and summation to yield a particular value of the output signal $y[n]$. In principle, the same process is valid for a random input sequence. Figure 10.1 shows a portion of such a sequence and, beneath it, a typical reversed impulse response. Cross-multiplication and summation gives the output sample value relevant to the instant $n = n_1$. By moving the reversed impulse response along step-by-step and repeating the calculation, we could in principle generate the complete output sequence according to the convolution sum (see also equation (2.13)):

$$y[n] = \sum_{k=-\infty}^{\infty} x[n-k]h[k] \qquad (10.1)$$

Of course, the important difference is that the input signal is now random. We do not know its individual sample values, but only its time-averaged statistics. So we cannot calculate the individual output sample values. Nevertheless, there are some valuable insights to be gained from the figure.

The first is that each random output value is formed as a weighted sum of a set of random inputs, the weightings being given by the various impulse response terms. We must therefore expect some correlation between successive outputs (whether or not successive inputs are correlated), because the outputs are not formed independently of one another. Their correlation will increase,

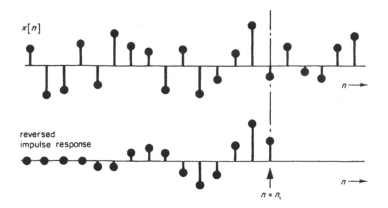

Figure 10.1 Digital convolution with a random input sequence.

the more terms there are in the impulse response. A lengthy impulse response implies a restricted bandwidth in the frequency domain. We therefore see that severe filtering, or shaping, in the frequency domain must tend to produce a highly correlated output sequence in the time domain.

Further insights are possible when the input is *white*, that is, its input samples are completely uncorrelated (and its power spectrum is flat). In this case Figure 10.1 makes clear that each output sample will be formed as a weighted sum of a number of uncorrelated inputs. The effect of a given weight, or impulse response term, is to increase the *amplitude* of that particular input sample by the same factor. Summation to yield each output value is therefore equivalent to adding together a number of weighted, but completely uncorrelated, random inputs. In such circumstances mean values and variances are additive, as noted in Section 9.2.2. We infer that the mean of the random output must be that of the input multiplied by the *sum* of all impulse response terms; and the variance of the output must be that of the input multiplied by the *sum of squares* of all impulse response terms. Already, we are beginning to get some idea of the effects which linear processing can have on the statistical properties of a random input signal.

However, we cannot rely on intuition much further! We must develop a proper analytical approach for describing such effects. Before getting down to

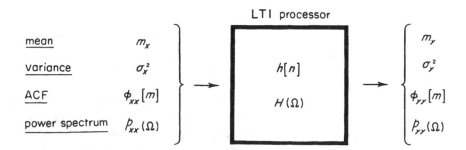

Figure 10.2 Signal statistics across a digital LTI processor.

detail, it is helpful to summarize the general situation we are considering. Figure 10.2 shows a linear processor or filter with impulse response $h[n]$ and frequency response $H(\Omega)$. Input and output signals are described by statistical measures such as mean, variance, AFC and power spectrum. The figure therefore defines the signal statistics across a linear processor.

Equation (10.1) deals with individual sample values, and is not therefore of direct use to us as it stands. However, we can modify it to cope with time-averaged statistics. Thus the mean, or expected, value of the output sequence may be derived as follows:

$$m_y = E\{y[n]\} = \sum_{k=-\infty}^{\infty} E\{x[n-k]\}h[k]$$

$$= m_x \sum_{k=-\infty}^{\infty} h[k] \qquad (10.2)$$

This result uses the fact that the expectation of a sum equals the sum of expectations. It shows, once again and more formally, that the mean of the output is equal to the mean of the input multiplied by the sum of all impulse response terms — a simple and attractive result. Furthermore, in the frequency domain the mean (or dc) value of the output must equal that of the input multiplied by the zero-frequency (or dc) response of the processor. Thus:

$$m_y = m_x H(0) \qquad (10.3)$$

Example 10.1   A random sample sequence with mean value 2.0 forms the input to a simple low-pass filter specified by the recurrence formula:

$$y[n] = x[n] + x[n-1] + x[n-2]$$

Predict the mean value of the output sample sequence and check your result by finding the zero-frequency response of the filter.

Solution   The impulse response $h[n]$ of the filter may be found by replacing $x[n]$ by the unit impulse function $\delta[n]$ and evaluating the output term-by-term. Thus:

$$h[n] = \delta[n] + \delta[n-1] + \delta[n-2]$$

giving

$$h[0] = 1, \quad h[1] = 1, \quad h[2] = 1$$
$$h[n] = 0, \quad \text{elsewhere}$$

Hence, the mean value of the output sample sequence is given by:

$$m_y = m_x \sum_{k=-\infty}^{\infty} h[k] = 2.0(1 + 1 + 1) = 6.0$$

To find the filter's zero-frequency response, we take z-transforms of both sides of the recurrence formula, giving:

$$Y(z) = X(z)(1 + z^{-1} + z^{-2})$$

$$\therefore \quad H(z) = \frac{Y(z)}{X(z)} = 1 + z^{-1} + z^{-2}$$

Hence the filter's frequency response is:

$$H(\Omega) = 1 + \exp(-j\Omega) + \exp(-j2\Omega)$$

The response at zero frequency is:

$$H(0) = 1 + \exp(0) + \exp(0) = 3.0$$

Hence, the mean value of the output sequence is $(3.0 \times 2.0) = 6.0$, as above.

Having quantified the effect of an LTI processor on the mean value of an input signal, you might expect us to tackle the question of variance next. However, it is more satisfactory to deal with input and output autocorrelation functions, returning to the question of variance a little later.

Again using equation (10.1), the output ACF may be written as:

$$\phi_{yy}[m] = E\{y[n]y[n + m]\} \tag{10.4}$$

$$= E\left\{ \sum_{k=-\infty}^{\infty} x[n - k]h[k] \sum_{r=-\infty}^{\infty} x[n + m - r]h[r] \right\}$$

$$= \sum_{k=-\infty}^{\infty} h[k] \sum_{r=-\infty}^{\infty} E\{x[n - k]x[n + m - r]\}h[r] \tag{10.5}$$

The expectation of this last expression equals the value of the input ACF relevant to the time shift $(m + k - r)$. Hence:

$$\phi_{yy}[m] = \sum_{k=-\infty}^{\infty} h[k] \sum_{r=-\infty}^{\infty} \phi_{xx}[m + k - r]h[r] \tag{10.6}$$

It is now helpful to substitute q for $(r - k)$, giving:

$$\phi_{yy}[m] = \sum_{k=-\infty}^{\infty} h[k] \sum_{q=-\infty}^{\infty} \phi_{xx}[m-q]h[k+q]$$

$$= \sum_{q=-\infty}^{\infty} \phi_{xx}[m-q] \sum_{k=-\infty}^{\infty} h[k]h[k+q] \qquad (10.7)$$

If we now write:

$$j[m] = \sum_{k=-\infty}^{\infty} h[k]h[k+m] \qquad (10.8)$$

we obtain:

$$\phi_{yy}[m] = \sum_{q=-\infty}^{\infty} \phi_{xx}[m-q]j[q] \qquad (10.9)$$

Note that $j[m]$ is the autocorrelation sequence of the impulse response $h[n]$. Hence, this last equation gives us an extremely important result which may be stated as follows:

> The ACF of the output sequence of a linear processor may be found by convolving the ACF of its input sequence with a function representing the ACF of the processor's impulse response.

You may recall that, in Chapter 2, convolution was used to find the deterministic output *signal* from a linear processor, given its deterministic input *signal* and the processor's *impulse response*. We now have an equivalent result for random signals, expressed entirely in terms of the autocorrelations of all these functions. It is an elegant and appealing result.

Example 10.2   A random signal has the ACF shown in Figure 10.3(a). It is processed by the low-pass filter specified in previous Worked Example 10.1. Find the ACF of the filtered output signal.

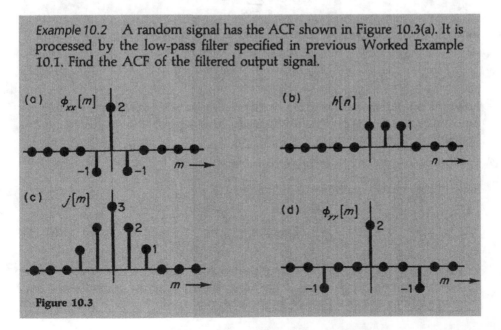

Figure 10.3

> *Solution*   We first need to find the autocorrelation sequence $j[m]$ of the filter's impulse response. As we have already shown, the impulse response consists of three unit sample values and is illustrated in part (b) of the figure. We now slide the function beneath itself, crossmultiplying and summing terms, giving the sequence shown in part (c) of the figure. Finally, we must convolve $\phi_{xx}[m]$ with $j[m]$. Convolution involves sliding a *reversed* version of one function beneath the other function, cross-multiplying and summing terms (since autocorrelation functions are always even functions of time-shift, reversal has no effect in this case and convolution gives the same result as cross-correlation — an important point which we will cover more fully in Section 10.4.3). We obtain the function $\phi_{yy}[m]$ shown in part (d) of the figure, which defines the ACF of the filtered output signal.

We have deliberately adopted an informal approach in the above Worked Example, sliding functions over one another in accordance with the graphical interpretation of correlation and convolution. Our aim is to familiarize you with the import relationships defined by equation (10.9), and to give you confidence in the numerical operations involved. Do not worry if you find the result shown in Figure 10.3(d) hard to visualize or interpret; we will soon use a computer program to illustrate such relationships more realistically.

However, before doing so we should tackle the question of input and output variances. Equation (10.9) provides a key to their relationship. If we put $m = 0$ we obtain the peak central value of the output ACF:

$$\phi_{yy}[0] = \sum_{q=-\infty}^{\infty} \phi_{xx}[-q]j[q] \tag{10.10}$$

Since an ACF is always an even function of time-shift, we may also write:

$$\phi_{yy}[0] = \sum_{m=-\infty}^{\infty} \phi_{xx}[m]j[m] \tag{10.11}$$

Now the output variance is given by the central value of the output covariance function $\gamma_{yy}[m]$ — a point already made towards the end of Section 9.2.4. Equation (9.12) shows that:

$$\sigma_y^2 = \gamma_{yy}[0] = \phi_{yy}[0] - m_y^2 \tag{10.12}$$

Substitution using equation (10.2) and (10.11) yields:

$$\sigma_y^2 = \sum_{m=-\infty}^{\infty} \phi_{xx}[m]j[m] - \left(m_x \sum_{n=-\infty}^{\infty} h[n] \right)^2 \tag{10.13}$$

We have now expressed the output variance in terms of the mean m_x and autocorrelation function $\phi_{xx}[m]$ of the input, and the impulse response $h[n]$ and autocorrelation sequence $j[m]$ of the processor.

Equation (10.13) looks rather complicated. However we quite often deal with random inputs which are white and have zero mean value. In such cases equation (10.13) simplifies greatly and is far easier to visualize. We will develop this topic in the next section.

The other measure we wish to consider is the output power spectrum (see Figure 10.2). This can either be found as the Fourier Transform of the output ACF as given by equation (10.9), or derived directly from the input power spectrum. The basic notion here is that the frequency response $H(\Omega)$ of a processor defines its effect, in amplitude and phase, on any input frequency Ω. Since power is proportional to amplitude-squared and involves no phase information, the equivalent 'power response' of the processor is $|H(\Omega)|^2$. Hence, we may write directly:

$$P_{yy}(\Omega) = |H(\Omega)|^2 \, P_{xx}(\Omega) \qquad (10.14)$$

This rather simple result shows that the output power spectrum is given by the product of the input power spectrum and the power response of the processor. It is certainly easier to visualize than the equivalent time-domain operation specified by equation (10.9), which involves a convolution.

You perhaps find the above results — and especially those involving correlation functions — a little hard to appreciate. So we will now use a computer program to illustrate some of the key input-output relationships in the time domain, and especially those given by equations (10.2), (10.9), (10.12) and (10.13). Graphical outputs from the program are shown in Figure 10.4.

Computer Program No. 28 in Appendix A1 may be summarized as follows:

1. The impulse response values of the desired processor are first entered (up to 10 are allowed). The values listed in the appendix represent a simple bandpass function; but they may, of course, be altered as required.

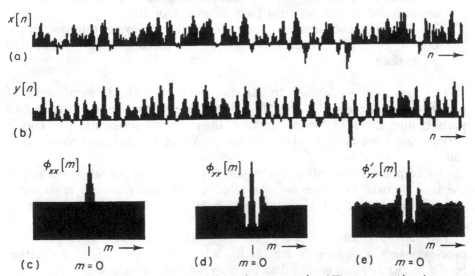

Figure 10.4   (a) Random input signal, and (b) random output of an LTI processor. Also shown are: (c) the input ACF; (d) the output ACF; and (e) a practical estimate of the output ACF based on 1000 samples of $y[n]$ *(abscissae: 320 and 83 samples).*

2. A gaussian noise sequence with zero mean and unit variance is generated. Although we could use this directly as input to the processor, it is more instructive to pass it first through a shaping filter (to give a more interesting input ACF), and add a finite mean. Here we use a simple 4-point shaping filter with low-pass properties, then add 1.0 to each sample value. This gives a sequence $x[n]$ which is suitable for demonstration purposes. 320 values of $x[n]$ are now plotted on the screen, as shown in part (a) of Figure 10.4. Note the low-pass properties of this signal, and its nonzero mean value.

3. The ACF of $x[n]$, denoted by $\phi_{xx}[m]$, is now plotted (see part (c) of the figure). This is a theoretical result based on the known form of the shaping filter's impulse response. Note that it includes a substantial dc pedestal, representing the mean value of $x[n]$.

4. The impulse response of the desired LTI processor, given as data at the start of the program, is now used to calculate the following: its own ACF (represented by $j[m]$ in equation (10.8)); the mean value of the output (see equation (10.2)); and hence the output ACF (see equation 10.9)). The latter is also plotted on the screen — see part (d) of the figure — and represents the convolution of the input ACF with $j[m]$. Note that this is a theoretical, predicted, result based on the known properties of the input signal and the processor.

5. The actual output sample sequence is now found by convolving $x[n]$ and $h[n]$. 320 of its values are plotted in part (b) of the figure. Note that the processor has had a considerable bandpass effect, as expected.

6. A practical estimate of the output ACF is now computed, for comparison with the predicted result shown in part (d) of the figure. The estimate is based on 1000 values of $y[n]$ and is built up on the screen as the computation proceeds (see part (e) of the figure). You may be able to see the function develop as the contributions from successive output samples are included (in fact, we treat them in groups of 10).

7. Finally, the output sample sequence $y[n]$ is used to provide estimates of the output mean and variance. These are compared with the theoretical predictions.

You may find it valuable to run the program with different forms of processor impulse response $h[n]$. Note that, although the shapes of the various parts of the screen plot are significant, their vertical scales are not. This is because we have normalized at various points in the program to ensure a suitable vertical range.

Our computer demonstration concentrates on random signal processing in the time domain. We have not illustrated the frequency-domain relationship given by equation (10.14) because most people find it a great deal easier to visualize.

You may have noticed that we have not so far discussed the effects of linear processing on a sequence's amplitude distribution. Such effects are rather harder to quantify and predict than the ones we have investigated here, but we will have something to say about them towards the end of the following section.

10.3   WHITE NOISE THROUGH A FILTER

In the previous section we saw how various widely-used statistical measures of a random sequence are affected by linear processing. A particular case arises when white noise passes through a digital filter. This situation often arises in DSP theory and practice and merits special attention for the following reasons:

Many of the theoretical results derived in the previous section are greatly simplified when the input sequence is white, with zero mean. This makes them easier to visualize, and gives valuable extra insights into the nature of random signal processing.

It is quite common in practical DSP for white noise to form the input to a digital processor. For example, noise arising along a communications path is often of this type. On reaching the receiving end it passes through the same processing chain as the wanted signal. We must also remember that many practical signals have statistics similar to those of white noise. Such signals, being extremely unpredictable, have a very high information content (a point already made at the beginning of Chapter 9, and mentioned again below).

We first introduced the idea of white noise towards the end of Section 9.2.4. Basically, any random digital sequence with zero mean in which adjacent sample values are completely uncorrelated may be described as 'white'. Its ACF consists of an isolated impulse at $m = 0$ (corresponding to zero time shift). In the frequency domain, its power spectrum is 'flat' — in other words, its spectral energy is evenly distributed throughout the frequency range.

As mentioned earlier, it is the form of the ACF and power spectrum which determines whether or not noise is white. 'Whiteness' does not imply any particular form of amplitude distribution. To reinforce this point we show, in Figure 10.5, portions of three white noise sequences having gaussian, uniform, and binary (two-level) amplitude distributions, respectively. All these sequences

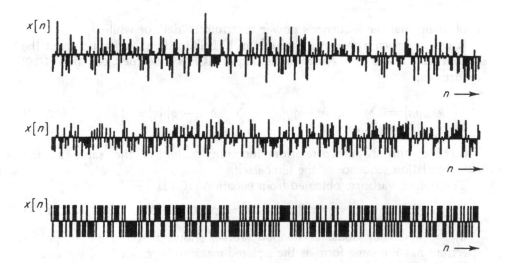

Figure 10.5   Portions of three white-noise sequences (*abscissae: 320 samples*).

have zero mean value, and would autocorrelate to an single impulse at $m = 0$. All have an even distribution of spectral energy.

Although we are concentrating mainly on white noise in this section, it is important to remember that many digital signals — for example, those in digital computers and communication systems — are also more or less white. Indeed, from a communications theory viewpoint a white sequence contains the maximum possible amount of information, since it is completely unpredictable in advance. When such a signal arrives at the receiving end and is decoded, the maximum amount of information is retrieved. This leads us to the apparent paradox that a white sequence, if it represents unwanted noise, is the most chaotic possible; whereas if it represents a signal, it carries the greatest possible amount of information.

Whether such a sequence represents unwanted noise or a wanted signal, it is clearly essential for us to be able to predict the effects of linear processing upon it. If it is noise, we may wish to specify a filter which will reduce it as much as possible; if it is a signal, we may need to enhance or extract it. As before, such measures as the mean, variance, ACF and power spectrum are like to be of most interest.

The various input-output relationships derived in the previous section still hold good; but fortunately they are considerably simplified in the case of a white input sequence. Theoretical treatments of this topic often assume an input with zero mean and unit variance, so we will do the same here. Such a sequence has an ACF consisting of a single, isolated, unit impulse at $m = 0$ (representing its mean-square value, which is the same as the variance since the mean is zero). Thus the statistical properties of the input are:

$$m_x = 0; \quad \sigma_x^2 = 1; \quad \phi_{xx}[m] = \delta[m] \tag{10.15}$$

Furthermore, equation (9.17) gives:

$$P_{xx}(\Omega) = \sum_{m=-\infty}^{\infty} \phi_{xx}[m]\exp(-j\Omega m) = \exp(-j\Omega m)|_{m=0} = 1 \tag{10.16}$$

confirming that the sequence's power spectrum is 'flat', or white.

If we pass such a sequence through an LTI filter, the mean value m_y of the output will also be zero. As far as the output ACF is concerned, equation (10.9) reduces to:

$$\phi_{yy}[m] = \sum_{q=-\infty}^{\infty} \phi_{xx}[m-q]j[q] = \sum_{q=-\infty}^{\infty} \delta[m-q]j[q] = j[m] \tag{10.17}$$

Thus the output autocorrelation function $\phi_{yy}[m]$ is the same as the autocorrelation sequence of the filter itself.

The output variance, obtained from equation (10.12), is:

$$\sigma_y^2 = \phi_{yy}[0] - m_y^2 = \phi_{yy}[0] = j[0] \tag{10.18}$$

In the frequency domain, equation (10.14) shows that the output power spectrum has the same form as the squared-magnitude response of the filter.

We therefore see that both the output ACF and the output power spectrum simply reflect the properties of the LTI processor or filter through which the input sequence has passed. The reason for this is quite straightforward: when the input is white, any output correlation in the time domain, or spectral shaping in the frequency domain, must be entirely due to the processor.

Hence we may summarize the statistics of the output sequence as follows:

$$m_y = 0; \quad \sigma_y^2 = j[0]; \quad \phi_{yy}[m] = j[m]; \quad P_{yy}(\Omega) = |H(\Omega)|^2 \qquad (10.19)$$

Example 10.3 White noise having zero mean and unit variance is processed by a simple high-pass filter with impulse response:

$$h[n] = \delta[n] - \delta[n-1] + \delta[n-2] - \delta[n-3] + \delta[n-4] - \delta[n-5]$$

Specify (a) the mean, (b) the autocorrelation function, and (c) the variance and standard deviation of the output noise from the filter. Also find an expression for its power spectrum.

Solution
(a) Since the input noise has zero mean value and the processor is linear, the mean of the output noise is also zero.
(b) The output ACF is given by the autocorrelation sequence $j[m]$ of the filter. The formula for the impulse response shows that it consists of the following set of sample values, starting at $n = 0$:

1, −1, 1, −1, 1, −1

Sliding this function over itself, cross-multiplying, and summing, we obtain the following finite autocorrelation sequence:

−1, 2, −3, 4, −5, 6, −5, 4, −3, 2, −1

with the peak value 6 occurring at $m = 0$. This function has the same form as the ACF of the output noise.
(c) The variance and standard deviation of the output noise are given by:

$$\sigma_y^2 = j[0] = 6 \quad \text{and} \quad \sigma_y = \sqrt{6}$$

Taking the Fourier Transform of the impulse response, we obtain the filter's frequency response:

$$H(\Omega) = 1 - \exp(-j\Omega) + \exp(-2j\Omega) - \exp(-3j\Omega) + \exp(-4j\Omega) - \exp(-5j\Omega)$$

Hence, the power spectrum of the output noise is given by:

$$P_{yy}(\Omega) = |1 - \exp(-j\Omega) + \exp(-2j\Omega) - \exp(-3j\Omega) + \exp(-4j\Omega) - \exp(-5j\Omega)|^2$$

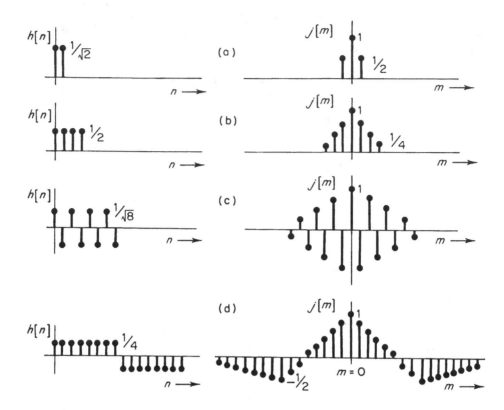

Figure 10.6 Impulse responses and autocorrelation sequences of four digital filters.

You may also find it helpful to have the above ideas illustrated with the help of a computer. Let us consider the typical effects of passing a white noise sequence with zero mean and unit variance into the four different digital filters whose impulse responses are shown on the left-hand side of Figure 10.6. There is no special significance in our choice of these particular filters — except that their impulse responses are simple in form and, as we shall see, neatly illustrate a number of important ideas. The first two filters, labelled (a) and (b), produce simple moving-average (low-pass) effects; the third one, labelled (c), is high-pass; and the fourth is a simple bandpass filter which accentuates input frequencies having around 16 samples per cycle.

The corresponding autocorrelation sequences are shown on the right-hand side of the figure. Each has been formed by sliding the relevant impulse response over itself, cross-multiplying and summing all products.

In each case the central value $j[0]$ of the autocorrelation sequence is unity. This is because we have deliberately chosen the sample heights in $h[n]$ such that their sum of squares is unity. Referring back to equation (10.18), we see that this will produce unit output variance. Thus the average size of output fluctuations will be similar for all four filters, even though their filtering action is quite different.

From equation (10.17), we see that the output ACF in each case is expected to be the same as $j[m]$. Hence, the autocorrelation sequences in Figure 10.6 also define the time-domain statistics of the filtered noise.

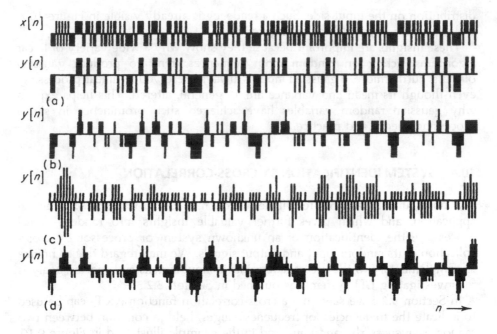

Figure 10.7   White noise processed by the four filters of Figure 10.6 (*abscissae: 320 samples*).

Figure 10.7 shows the typical effects of feeding white noise into each of these filters in turn. A portion of white noise — in this case binary noise with values ±1 — is shown at the top of the figure. It has zero mean and unit variance. Below are shown portions of typical output sequences from the filters. You will notice the simple low-pass, or smoothing, action of filters (a) and (b); the high-pass action of (c); and the bandpass action of (d). All plots have been drawn to the same vertical scale, and their variances are clearly similar as predicted. It is also interesting to refer back to Figure 10.6, noting the visual similarities between the predicted ACF and the actual sequence in each case (please bear in mind that the horizontal scales are very different in the two figures, the sample values in Figure 10.7 being contiguous).

This is a good moment to mention the general effects of linear processing on the amplitude distribution of a random sequence, an issue we have not so far discussed. You may recall that Figure 10.1 illustrated the graphical approach to convolution, in which the impulse response of a processor is laid out backwards beneath the input sequence, followed by cross-multiplication and summation to yield a particular output value. If the input sequence is white noise, it is clear that each output value is formed as a weighted sum of completely uncorrelated input samples. In general, as we increase the number of impulse response terms, producing a more complex filter with reduced bandwidth, each output sample will be formed from a larger number of weighted inputs. As noted in Section 9.2.1, when we add together many independent random variables to produce a new random variable, the result tends to have a gaussian distribution. We may therefore expect that any form of LTI processing of white noise will produce an output which tends towards the gaussian — regardless of the amplitude distribution on the input side. Such a tendency is already visible in Figure 10.7, and especially in part (d) of the figure.

These insights are important because they show why a wide variety of linear algorithms, acting on random input sequences, tend to produce gaussian outputs. Furthermore, a gaussian input is always preserved by linear processing, even though its mean and variance are, in general, altered. This helps explain why gaussian random variables have achieved such prominence in signal processing theory and practice.

10.4   SYSTEM IDENTIFICATION BY CROSS-CORRELATION

We now turn to a DSP technique which has found widespread practical application, and which gives further valuable insights into random signal processing: the identification of an unknown system or processor by cross-collection of its random input and output signals. We may regard this technique as complementary to the use of deterministic input signals and spectral analysis for investigating LTI systems, as outlined in Section 8.2.3.

In Section 9.2.6 we saw how a cross-correlation function (CCF) can be used to indicate the frequencies, or frequency ranges, held in common between two random sequences $x[n]$ and $y[n]$. And in the example illustrated in Figure 9.10, we hinted strongly that if $x[n]$ represents the input to an LTI processor and $y[n]$ represents the output, then the form of their CCF must contain information about the properties of the processor itself.

You may find it surprising that a statistical comparison of *random* input and output signals can yield useful information about a *deterministic* system or processor. The important point, however, is that success depends on working with sufficiently long portions of the signals. Any practical estimate of their CCF will be subject to statistical variability — and only if we reduce this to acceptable levels can we hope to obtain a reliable identification of the system.

Why might we wish to explore an unknown system in this way? There are two main, interrelated reasons:

A system often has a certain amount of random noise present at its input (and therefore at its output) during normal operation, and we may be able to use this to assess the CCF.

On other occasions it may be more convenient to disturb the system with a random input signal, rather than (say) an impulse function or a series of sinusoidal frequencies. Indeed, it is quite often possible to inject a low-level random test signal without affecting normal operation.

Such possibilities have made the technique attractive in a wide range of practical situations, in fields as diverse as electronic circuits, aerospace, biomedicine, and seismology.

It is fairly easy to appreciate that if we are to identify the properties of an unknown system by disturbing it with a random input, then that input must contain a significant amount of all frequencies transmitted by the system. Otherwise identification cannot be complete. In practice, it is usual to employ a white input which has, by definition, a flat spectrum. Then any 'non-whiteness' in the output must be due to the frequency-selective properties of the system itself. Viewed in this way, we may anticipate that a white-noise input will give a particularly simple and attractive relationship between the input-output CCF and the properties of the system under investigation.

We can develop this important idea quantitatively by recalling that the CCF of two sequences equals their average, or expected, cross-product, as a function of the time shift between them. Thus we may write:

$$\phi_{xy}[m] = E\{x[n]y[n+m]\} \tag{10.20}$$

Now the output sequence $y[n]$ may be expressed as the convolution of $x[n]$ with the impulse response $h[n]$ of the system (see equation (10.1)). Therefore:

$$\phi_{xy}[m] = E\left\{ x[n] \sum_{k=-\infty}^{\infty} x[n+m-k]h[k] \right\}$$

$$= \sum_{k=-\infty}^{\infty} \phi_{xx}[m-k]h[k] \tag{10.21}$$

This important result shows that the input-output CCF equals the convolution of the input ACF with the impulse response of the system.

We have already surmised that a white noise input will be particularly suitable for our purposes. Let us therefore assume, for convenience, a zero-mean, white noise input of unit variance. In this case the input ACF consists of a unit impulse at $m = 0$, hence:

$$\phi_{xx}[m] = \delta[m] \tag{10.22}$$

and

$$\phi_{xy}[m] = \sum_{k=-\infty}^{\infty} \delta[m-k]h[k] = h[m] \tag{10.23}$$

We have now derived a very elegant result: given a white noise input, the CCF between input and output becomes identical to the system's impulse response. And, of course, the impulse response of any LTI system or processor completely defines it in the time domain.

You will probably find it helpful if we illustrate this rather surprising result with a practical example. Figure 10.8 shows the technique applied to exploring the properties of a digital bandpass filter. At the top are shown portions of the white-noise input sequence to the filter, and the filter output. Below, on the left, is shown the actual impulse response of the filter (which, we must admit, we knew in advance!); and on the right, a computed estimate of the CCF based on 1000 sample values of $x[n]$ and $y[n]$. In spite of some statistical variability, there is a very good tie-up between the impulse response and the estimated CCF. We could, of course, achieve an even better identification by processing longer portions of the two sequences.

So far we have concentrated on the time domain. There is (as always) a corresponding result in the frequency domain, which is rather easier to visualize.

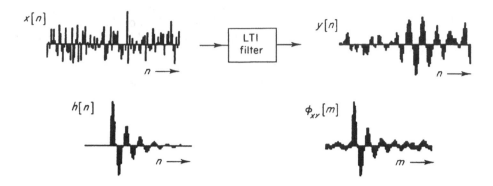

Figure 10.8 Identifying an LTI filter by cross-correlation.

The convolution of input ACF and impulse response specified by equation (10.21) must be equivalent to multiplication of the corresponding spectral functions. These are the input power spectrum and the frequency response of the system, respectively. Therefore we may write:

$$P_{xy}(\Omega) = P_{xx}(\Omega)H(\Omega) \qquad (10.24)$$

In the case of a white-noise input with zero mean and unit variance, the input power spectrum is unity (see equation (10.16)). We now have:

$$P_{xy}(\Omega) = H(\Omega) \qquad (10.25)$$

In this case the cross-spectrum takes the same form as the frequency response of the system.

We noted previously that system identification by cross-correlation has found widespread practical application. Time-domain correlation is particularly appropriate in DSP, because of the ease with which the basic operations required — time shifting, multiplication, and summation — can be performed using digital computers. Because of its practical importance, there is one further aspect of the technique which we feel should be mentioned here: its resistance to certain types of *system nonlinearity*.

At the end of Section 9.2.6 we pointed out that cross-correlation picks out shared frequency components in two sequences, and therefore reveals only linear relationships between them. Remember that an LTI system can only modify the amplitudes and phases of frequency components present in its input signal; it cannot introduce new frequencies. When we use input-output cross-correlation to identify a system, any system nonlinearities will tend to produce frequency components in $y[n]$ which are not present in $x[n]$. Therefore, we must expect the CCF to focus on the linear part of system performance, at the expense of any nonlinear part.

This may or may not be an advantage. If the nonlinearity is spurious or unwanted, we may be pleased to have it suppressed; but if it is an essential aspect of system performance, we clearly need to know about it. The tendency

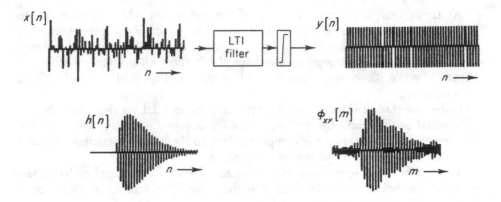

Figure 10.9   Cross-correlation in the presence of non-linearity.

of correlation functions to ignore nonlinearities is therefore valuable in some situations, regrettable in others. But at least we should be aware of it.

Figure 10.9 illustrates the above discussion. The linear processor in this case is a second-order high-pass filter with the impulse response shown in the lower-left part of the figure. Once again, we use gaussian white noise as the filter input, but this time its output is subjected to a severe nonlinearity by being *infinitely-clipped*. In other words each output sample is changed to ±1, depending on whether it is positive or negative. This is equivalent to ignoring all information in the output sequence apart from its zero-crossings. The short portion of the output sequence shown in the figure does indeed suggest that not very much information has been retained! Nevertheless, the computed CCF shown in the lower-right of the figure, again based on 1000 input and output samples, is strikingly like the actual impulse response of the linear part of the system.

We must be careful not to generalize too much from one example, because the degree of resistance to nonlinearity depends very much on the particular case. However, Figure 10.9 does help illustrate a major feature of correlation analysis, and therefore of system identification using random sequences.

10.5   SIGNALS IN NOISE

10.5.1   INTRODUCTION

One of the most important topics in the whole of signal processing concerns the extraction of wanted signals from unwanted noise. This problem arises to some extent in almost all communication, recording, and measurement situations. In communication engineering, the ability of a receiver to detect a weak signal in the presence of noise (either picked up along the transmission path, or in the receiver itself) imposes a fundamental limit on the information capacity of the channel. When recording signals or data, or making sensitive measurements, the reduction of spurious noise to acceptable levels is often a major challenge.

As far as DSP is concerned, many of the techniques which we can use to enhance signal-to-noise ratio have their origins in analog signal processing; but others have been developed from first principles, and offer distinct advantages of their own. In the following sections we give a short introductory account of signals in noise, illustrating a number of key ideas and applications with the aid of a computer.

Before we start, it is worth recalling that in Section 8.2.1 we described the use of spectral analysis and the FFT for detecting a signal in noise. This valuable frequency-domain technique is complementary to the topics discussed here, and you may wish to review it before proceeding.

You will find that terms such as *recovery* and *detection*, when applied to signals in noise, are not entirely standard. It is therefore important to be clear about how we will use them here:

> By *recovery*, we will mean the extraction of a signal from noise, when the signal waveshape is not known in advance and must be preserved as accurately as possible. In such situations, we normally require prior information about the frequency bands occupied by signal and noise.

> The term *detection* will be used when a known signal waveshape of finite duration (such as a pulse) is contaminated by noise, and we wish to find out when (or indeed if) it occurs. This type of problem arises widely in such fields as radar and sonar engineering.

When a signal, contaminated by noise, is to be recovered or detected using an LTI filter or processor, we may fortunately use the Principle of Superposition (see Section 1.5.1). This allows us to compute the processor's effect on signal and noise components separately, and is a great advantage. It assumes, of course, that the noise is *added* to the signal. Other types of noise contamination (such as *multiplicative*) do sometimes arise, but are much harder to analyse and will not be covered here.

It is important to remember that whether a particular digital sequence is regarded as 'signal' or 'noise' depends largely on one's point of view. If it is wanted, it is a signal; if unwanted, it is noise. It follows that most of our previous work in this chapter is relevant to the description of random signals as well as noise.

The topic of signal recovery, as defined above, is extremely wide. Here, we will concentrate on a situation which often arises in practice, and is relatively easy to quantify: a comparatively narrowband signal contaminated by wideband noise. It presents us with the opportunity to discuss some important aspects of time-domain performance which are often ignored in introductory texts, and also illustrates an interesting application of one of the types of digital filter described in Section 6.3.1.

As far as signal detection is concerned, we will introduce you to a widely employed linear technique which is particularly suited to DSP, known as *matched filtering*. And finally, we will consider the enhancement of a repetitive signal by *averaging* over many repetitions — a valuable technique in fields as diverse as pulse radar and biomedicine.

10.5.2   SIGNAL RECOVERY

One of the most common requirements in signal recovery is to separate a comparatively narrowband signal from wideband noise. For example, in many multi-channel communication systems each individual signal is allocated its own narrow frequency slot; but random noise picked up during transmission tends to occupy the complete system bandwidth. One of the main tasks of the receiver, therefore, is to recover narrowband signals from wideband noise.

In such situations we are interested in predicting the improvement in signal-to-noise ratio which can be obtained by linear filtering. As far as the signal is concerned, it is convenient to specify an ideal bandpass filter covering the signal's bandwidth, as indicated in Figure 10.10. If we also assume a linear phase response, the signal will be passed through the filter in undistorted form. Of course, we can only approximate such a filter characteristic in practice — but the ideal case indicates an upper limit to performance.

As far as the noise is concerned, we need to know its spectral distribution (it may or may not be white), and then to assess the reduction in noise level through the filter. This is quite readily done using results derived earlier in this chapter.

Assuming that the noise has zero mean value, then its autocorrelation and autocovariance sequences are identical, and the output noise power, or variance, from the filter is given by an output version of equation (9.18):

$$\phi_{yy}[0] = \frac{1}{2\pi} \int_{2\pi} P_{yy}(\Omega)\, d\Omega \qquad (10.26)$$

Substitution from equation (10.14) gives:

$$\phi_{yy}[0] = \frac{1}{2\pi} \int_{2\pi} |H(\Omega)|^2 P_{xx}(\Omega)\, d\Omega \qquad (10.27)$$

If the filter's squared-magnitude response takes the form shown in the figure, the output noise power becomes:

$$\phi_{yy}[0] = \frac{1}{2\pi} \int_{2\Delta\Omega} P_{xx}(\Omega)\, d\Omega \qquad (10.28)$$

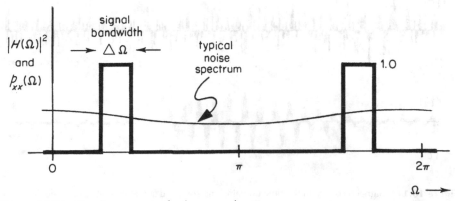

Figure 10.10   Signal recovery in the frequency domain.

In other words, it is found by integrating the output noise power spectrum over the passband of the filter. A further simplification is possible if the input noise to the filter is assumed white, with unit variance. Its input power spectrum is then unity (see equation (10.16)), and we obtain:

$$\phi_{yy}[0] = \frac{1}{2\pi} \int_{2\Delta\Omega} 1 \, d\Omega = \frac{\Delta\Omega}{\pi} \qquad (10.29)$$

We see that the reduction in total noise power (or variance) through the filter is simply given by the ratio of the filter bandwidth to the frequency interval π. Since the signal is assumed to be transmitted unaltered (the filter is ideal), this ratio also gives the improvement in signal-to-noise power ratio. The improvement in the amplitude ratio is its square root.

Although our theoretical treatment may seem straightforward enough, its practical implementation can be rather more awkward. Figure 10.11 simulates some of the problems, as well as the benefits, of signal recovery using an LTI filter. It is essentially a time-domain equivalent of Figure 10.10. The signal we have chosen for illustrative purposes is the switched sinusoid, or 'tone burst', shown in part (a) of the figure. It is basically narrowband, but the switch-on and switch-off cause spectral spreading around the nominal frequency of the sinusoid. In part (b) of the figure we see the signal added to white gaussian noise. It is this sequence we wish to filter to improve the signal-to-noise ratio.

We have used a tenth-order Chebyshev bandpass design (see Section 6.3.1) with its center-frequency at the frequency of the tone burst, a 1 dB passband ripple, and a bandwidth equal to $\pi/36$. This is reasonably close to an ideal filter, so equation (10.29) leads us to expect the average noise power to be reduced about 36 times, and its amplitude (standard deviation) about six times.

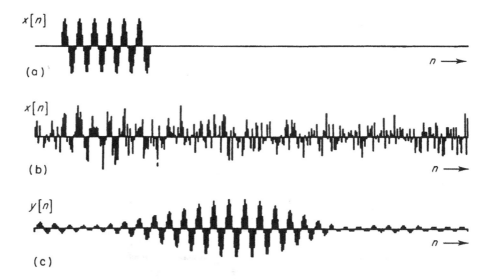

Figure 10.11   Signal recovery in the time domain: a narrowband signal in white noise (*abscissae: 320 samples*).

This is borne out by the form of the filter output shown in part (c) of the figure. Clearly, the noise reduction performance of the filter is very considerable.

However, this is not the whole story. The recovery of the signal is less impressive. Although the signal now stands out clearly from the residual noise, it has a slow onset and decay. Such effects are not predicted by the usual steady-state analysis (including that given above), but they are the real, and inevitable, result of using a linear processor. Basically, the problem is that we have specified a narrowband filter in order to achieve a substantial reduction of the noise; but such a filter has a long-lived transient response, and cannot settle quickly following a sudden change of input. We could, of course, specify a wider filter passband — but it would give an inferior performance as far as the noise is concerned. There is always a compromise to be reached between steady-state noise reduction and transient response to signal fluctuations. Although we have chosen rather a severe example to illustrate the problem, it arises to some extent in many signal-recovery situations.

It is worth making two further points. Firstly, the noise at the filter output is now narrowband (but still gaussian), and occupies the same frequency range as the signal. This is the effect of any such linear processor. Secondly, the signal is considerably delayed by passage through the filter. This is an unavoidable effect, but fortunately not one which causes problems in most practical applications.

We must admit that the above problem is fairly straightforward, since the spectral distributions of signal and noise are so different. This makes the task of separating them with a linear filter relatively easy. Although such situations do quite often arise in practical DSP, on other occasions we are faced with severely overlapping spectra. The way forward is then via the theory of optimal filtering and estimation — a topic beyond the scope of this chapter, but covered in various more advanced texts (for example, the book by Vaseghi listed in the bibliography).

10.5.3   MATCHED FILTER DETECTION

Detection of a signal in noise by use of a *matched filter* is one of the most important topics in linear DSP, from both theoretical and practical points of view. We will give it a fair amount of attention here, because it brings together many of the ideas we have already covered in this book on LTI processors, filters, random sequences and correlation.

In the previous section we considered the problem of recovering a signal of unknown waveshape, using prior information about the frequency bands occupied by signal and noise. A rather different problem arises when we know the signal's waveshape (in effect, we know not only its frequency band, but also the amplitudes and relative phases of its various components within that band), and we need to establish *when*, or indeed *if*, it occurs. A classic instance is pulse radar, let us say for Air Traffic Control. A train of short pulses is sent out by the radar transmitter, and the receiver has to detect echoes of similar shape reflected from distant aircraft. The waveshape is known in advance, and is not itself the

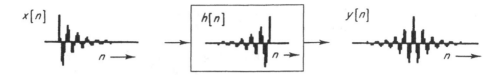

Figure 10.12   A matched filter.

important feature; what matters is detecting the *presence* of echoes, which are often weak and badly contaminated by random noise. In this type of situation the optimum linear filter is the matched filter.

The matched filter is very important in DSP for a further reason: it highlights the close relationships between digital convolution, correlation and filtering. We will explain this aspect first, because it gives us an ideal opportunity to introduce you to the basic properties of a matched filter in the time and frequency domains. We will move on to the detection of signals in noise later in this section.

You have probably noticed that convolution as first described in Section 2.4, and correlation as discussed in Section 9.2.4 and 9.2.6, are very similar operations. In the case of convolution, we compute the output of an LTI processor by laying a reversed version of the processor's impulse response beneath the input signal, then cross-multiplying and summing all terms (as illustrated in Figure 2.9 and again in Figure 10.1). With correlation, we also lay one time function beneath the other, cross-multiply and sum; the only difference is that one of the functions is not time-reversed.

This similarity is neatly illustrated by the operation of a matched filter. Consider the time-limited input signal $x[n]$ shown in figure 10.12, processed by an LTI filter whose impulse response $h[n]$ is a *reversed* version of $x[n]$. This filter is, *by definition*, the matched filter for this particular shape of input signal. We may, as always, find the output signal by convolving $x[n]$ and $h[n]$. We lay a reversed version of $h[n]$ below $x[n]$, cross-multiply and sum all terms to give a particular value of the output $y[n]$. In this case, we see that such an operation is exactly equivalent to forming the autocorrelation function (ACF) of $x[n]$. The resulting form of $y[n]$ is indeed typical of an ACF — symmetrical in form, with a peak central value. We conclude that a matched filter may be thought of as a type of *automatic correlator*, producing an output similar in form to the ACF of the signal to which it is matched. This idea neatly summarises the close relationships between convolution, correlation, and filtering.

Of course, the output from a matched filter is a *signal* and is a function of the time parameter n, rather than the time-shift parameter m which we use for correlation functions. Furthermore, its peak central value occurs at the instant when the complete input signal has just entered the filter, rather than at $n = 0$. But the *form* of the output is identical to the ACF of the input. It is also worth noting that the type of ACF we are discussing here is that relevant to a time-limited signal having finite energy: we compute the summed cross-product, but do not take a long-term average (as we do with an infinite-energy random sequence).

Example 10.4   A time-limited digital signal $x[n]$ has just four nonzero sample values, starting at $n = 0$:

$$3 \quad 2 \quad -1 \quad 1$$

(a) Write down the difference equation of the causal, minimum-delay, matched filter for this signal.
(b) Calculate the output sample values from the filter when the signal is delivered to its input. At what instant does the peak output value occur?

Solution
(a) The impulse response of the matched filter is a time-reversed version of the input signal and must therefore have successive sample values:

$$1 \quad -1 \quad 2 \quad 3$$

For the filter to be causal, the first of these cannot occur before $n = 0$. For the filter to be minimum-delay, its impulse response must start no later than $n = 0$. Hence the value 1 must occur at $n = 0$, and we may write the impulse response as:

$$h[n] = \delta[n] - \delta[n - 1] + 2\delta[n - 2] + 3\delta[n - 3]$$

Rewriting for a general input $x[n]$ and general output $y[n]$, we obtain the following difference equation for the matched filter:

$$y[n] = x[n] - x[n - 1] + 2x[n - 2] + 3x[n - 3]$$

(b) The output from the matched filter has the same form as the ACF of the input signal. The ACF may be found by 'sliding' two versions of the signal over one another, cross-multiplying and summing all products at each step. Thus for time-shift value $m = 3$ we have:

$$3 \quad 2 \quad -1 \quad 1$$
$$\qquad\qquad 3 \quad 2 \quad -1 \quad 1$$

giving a cross-product of 3. When $m = 2$, we obtain a total cross-product of $(1 \times 2) + (-1 \times 3) = -1$, and so on. The complete ACF is the sequence:

$$3 \quad -1 \quad 3 \quad 15 \quad 3 \quad -1 \quad 3$$

The output signal $y[n]$ from the matched filter has the same form as the ACF. Since both input signal and impulse response start at $n = 0$, the first output sample must occur at $n = 0$. The peak value of 15 therefore occurs at $n = 3$. This corresponds to the instant when the complete input signal has just entered the filter.

So far we have concentrated on the time-domain properties of a matched filter, and indeed these are particularly illuminating. However, it is also helpful to consider the frequency domain. Not surprisingly, it turns out that the frequency response of a matched filter bears a simple relationship to the spectrum of the signal to which it is matched. We will now develop this point.

Since the impulse response and frequency response of any LTI filter are related as a Fourier Transform pair, it follows that the frequency response in this case is:

$$H(\Omega) = \sum_{n=-\infty}^{\infty} x[-n]\exp(-j\Omega n)$$

$$= \sum_{k=-\infty}^{\infty} x[k]\exp(-j(-\Omega)k) = X(-\Omega) \tag{10.30}$$

Now the spectrum $X(\Omega)$ of any real sequence can be expressed in the form:

$$X(\Omega) = A(\Omega) + jB(\Omega) \tag{10.31}$$

where $A(\Omega)$ is an even function representing cosine components, and $B(\Omega)$ is an odd function representing sines. Hence:

$$H(\Omega) = X(-\Omega) = A(-\Omega) + jB(-\Omega) = A(\Omega) - jB(\Omega) = X^*(\Omega) \tag{10.32}$$

where the asterisk denotes the complex conjugate. We see that a matched filter's frequency response is given by the complex conjugate of the signal spectrum to which it is matched. Note also that the output signal spectrum is:

$$Y(\Omega) = X(\Omega)H(\Omega) = X(\Omega)X^*(\Omega) = |X(\Omega)|^2 \tag{10.33}$$

This is what we would expect. For we have already noted in Section 9.2.5 that the ACF of a signal contains the same frequencies as the signal itself, but with squared magnitudes and without relative phases. To put this in a slightly different way, we may say that when an input signal enters its matched filter, the filter causes each component to be squared in magnitude, and adjusted in phase so that all phases are perfectly aligned at the output.

We now move on to the second main task of this section — to explain why a matched filter is so often used for detecting the presence of a known, time-limited signal in the presence of noise.

Since the signal waveshape is assumed to be known in advance (if it were not, we could not specify its matched filter!), we do not need a filter which transmits it in undistorted form. A better criterion is that the filter should give the greatest possible instantaneous output whenever the signal waveform occurs. The detection task is then reduced to searching the filtered signal-plus-noise waveform for high peaks. This is easily accomplished by an electronic circuit or a computer program.

It may be shown theoretically that, if the noise has a constant power spectral density over the frequency range occupied by the signal, then the improvement in signal-to-noise (S:N) ratio produced by a matched filter is the best possible (or, more correctly, it is the best possible with any LTI filter: and in the

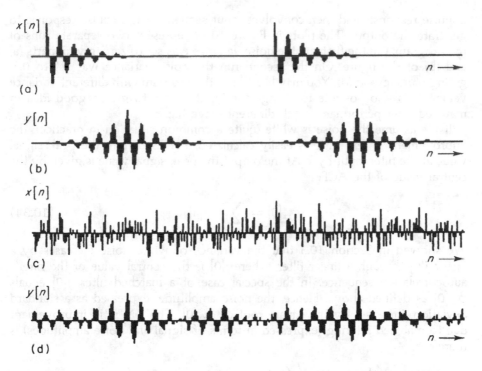

Figure 10.13   Matched filter detection of a repetitive signal in noise (*abscissae: 320 samples*).

restricted case of gaussian noise, it is the best possible with any filter, linear or nonlinear).

We will illustrate the above ideas for you with an example based on Computer Program No. 29 in Appendix A1. Part (a) of Figure 10.13 shows two versions of a finite-length signal, separated from one another along the time axis (as in Figure 10.12, there is no special significance in this choice of signal). When the sequence is fed into the appropriate matched filter, we get an output containing two versions of the ACF, as shown in part (b). Note that, as usual, the peak central value of each ACF coincides with the instant when the complete input signal has just entered the filter.

Now suppose that the same input sequence is badly contaminated by white gaussian noise, as in part (c) of the figure. The signal occurrences are now very difficult to detect, because the noise occasionally rises to levels comparable with the peak signal values. The advantage of processing with the matched filter is confirmed by part (d). Two signal peaks, corresponding to the two central ACF values, now stand out well and would be easy to detect automatically.

Of course, even a matched filter is not perfect, because it inevitably transmits noise lying within its passband. This is mixed with the output signal and may prevent successful detection. You will notice that the residual noise, after filtering, has a similar time-domain structure to the required ACF waveforms. But this is the best we can do with a linear filter.

Computer Program No. 29 generates the input signal, adds a selected amount of gaussian noise, reverses the input signal to produce the required form of impulse response, and then convolves input sequence and impulse response to generate the output. The plots in Figure 10.13 represent two separate runs of the program. On the first run the noise variance was set to zero, giving parts (a) and (b) of the figure. On the second run the noise variance was set to 0.5, giving parts (c) and (d). You may like to run the program with different variance values. It is also possible to change the input signal, to give a good idea of matched filter performance with different waveshapes.

If we assume the noise is white (quite a common situation in practice), the improvement in S:N ratio is easily calculated. Let us denote the peak signal value at the filter input by $\hat{x}$. At the output the peak signal value is given by the central value of the ACF:

$$\phi_{xx}[0] = \sum_{n=-\infty}^{\infty} (x[n])^2 \qquad (10.34)$$

We showed in Section 10.3 that the variance of white noise increases by a factor $j[0]$ through a linear filter, where $j[0]$ is the central value of the filter's autocorrelation sequence. In the special case of a matched filter, $j[0]$ equals $\phi_{xx}[0]$ as defined above. Hence, the noise amplitude, expressed as a standard deviation, increases by the square root of $\phi_{xx}[0]$. The S:N ratio improvement due to matched filtering (expressed in terms of signal and noise amplitudes) is therefore:

$$\left(\frac{\text{output peak signal}}{\text{input peak signal}}\right) \bigg/ \left(\frac{\text{output noise}}{\text{input noise}}\right) = \frac{\phi_{xx}[0]}{\hat{x}} \bigg/ \phi_{xx}[0]^{1/2}$$

$$= \frac{\left\{\sum_{n=-\infty}^{\infty} (x[n])^2\right\}^{1/2}}{\hat{x}} \qquad (10.35)$$

Note that this improvement depends only on the sum of squares of all signal values (in other words, on the signal's total energy) and on its peak value prior to filtering, but not on its detailed waveshape. Computer Program No. 29, described above, uses a signal with a peak value of unity and a sum of squares equal to 5.887. The improvement in S:N ratio, illustrated by Figure 10.13, is therefore:

$$5.887^{1.2} = 2.43, \quad \text{or} \quad 7.7 \text{ dB} \qquad (10.36)$$

Example 10.5   A time-limited digital signal has sample values:

$$1 \quad -1 \quad 1 \quad 1 \quad -1 \quad -1$$

(a) Find the improvement in S:N ratio due to matched filtering when the signal is mixed with white gaussian noise.
(b) Find the improvement in S:N ratio when the filter has an impulse response identical to the signal (not a time-reversed version of it), and comment.

Solution

(a) The sum of squares of all signal values is 6, and the peak signal value is 1. Hence using equation (10.35) we obtain a S:N ratio improvement of $\sqrt{6}/1 = 2.45$, or 7.78 dB.

(b) The output signal may be found by convolving input signal and impulse response. We lay a reversed version of the impulse response below the input signal, cross-multiply and sum. The relevant sequences are:

$$1 \quad -1 \quad 1 \quad 1 \quad -1 \quad -1$$

$$-1 \quad -1 \quad 1 \quad 1 \quad -1 \quad 1$$

Sliding one sequence below the other, we obtain successive output values from the filter:

$$1 \quad -2 \quad 3 \quad 0 \quad -3 \quad 2 \quad 1 \quad -4 \quad -1 \quad 2 \quad 1$$

The peak output value is now 3, rather than 6 obtained with the matched filter, so the improvement in S:N ratio is reduced to $\sqrt{3}/1 = 1.73$, or 4.77 dB.

In this case the signal output from the filter does not take the form of the input ACF, nor is it symmetrical. The filter has the same frequency response *magnitude* as the true matched filter, but not the same *phase* response. This means that it does not optimally realign the phases of the various signal components to produce the largest possible peak output.

It might be claimed that, since the output displays a negative peak value of -4, we would do better by searching the output for *negative* peaks in this case. If this were done the improvement in S:N ratio would be $\sqrt{4}/1 = 2.0$, or 6.02 dB.

You may be wondering whether the output S:N ratio from a matched filter, such as that illustrated by Figure 10.13(d), could be improved by a further matched filtering operation. After all, we know the new form of the 'signal' (it now has the shape of the input ACF), and it is still corrupted by noise. But the answer is that no further improvement is possible, because the signal components have been optimally arranged in phase by the first matched filter to give the largest possible peak, and the signal and noise now have the same spectral distribution. Another linear filter, which could only adjust spectral amplitudes and/or phases, would offer no advantage.

The improvements in S:N ratio we have calculated so far are quite modest, mainly because the signal waveshapes we have chosen are not spread out much in the time domain, and have relatively few sample values (this is particularly true of the above Worked Example). As equation (10.35) has indicated, the advantages of matched filtering improve as the sum of squares of all signal sample values increases in relation to their peak value $\hat{x}$. This raises the interesting question: if we wish to *design* a signal which will give the best performance in a matched filtering situation, what form of signal should we choose?

We clearly need a signal with large total energy (large sum of squares), but as small a peak value as possible. This implies that the signal should be well spread out in the time domain, with plenty of samples of more or less equal magnitude. Furthermore, it will be helpful when it comes to accurate detection if the signal's ACF consists of a large, isolated, central peak with small values to either side. Then not only will the peak stand out clearly from the noise, but there will be little danger of confusing it with one of its neighbors. This will ensure good timing resolution of the detected signal.

These considerations suggest that a white, 'noise-like' signal should do well. Particularly appropriate would be a binary noise-like waveform restricted to the values ± 1. This has an even distribution of energy in the time domain (for the duration of the signal); and an autocorrelation function with an isolated central peak and small values to either side. The best-known and probably most widely used signal of this type is the *pseudo-random binary sequence* (PRBS).

Before we demonstrate the performance of PRBS waveforms in this context, it may be helpful to summarise some practical reasons for their use:

> In pulse radar and sonar, rather than concentrate all the transmitter's energy into narrow pulses, it may be beneficial to 'spread the pulses' using PRBS coding. This reduces the peak power required of the transmitter, while maintaining the total energy of each pulse. Returning echoes, also now spread out, are less susceptible to short-duration bursts of random noise.

> A spread-out signal of low peak amplitude but high total energy is more easy to disguise. This may have advantages in, for example, secure communications.

> Traditional pulse-type signals of large amplitude and short duration may drive a linear processor or filter into nonlinearity as a result of instantaneous overload. The system then fails to operate as intended.

PRBS waveforms, although they appear random, are in fact generated using deterministic algorithms. They are essentially white, with well-defined spectral and autocorrelation properties. The design details are outside the scope of this book, but you will find them discussed in several of the references listed in the bibliography. Basically, the usual type of PRBS has a number of sample values equal to an integer power of 2. Typically, lengths between (say) 32 and 2048 samples are used in practice.

Figure 10.14 is a computer demonstration of the detection of PRBS waveforms in noise, using a matched filter. The particular PRBS we have used has 64 sample values of ± 1, and is shown on the left of part (a) of the figure. To make the detection problem more challenging, we have superposed two further, *overlapping* versions of the PRBS on the right-hand side. It will be interesting to see if the matched filter can disentangle them!

The relevant matched filter, as always, has an impulse response equal to a time-reversed version of the signal. In this case, therefore, its impulse response consists of 64 values ± 1. When the composite signal illustrated in part (a) is fed into this filter, the output is as shown in part (b). The filter's response to the

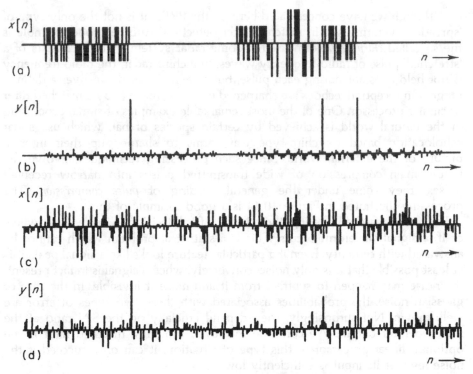

Figure 10.14 Matched filter detection of pseudo-random binary sequences.

single PRBS waveform on the left-hand side has the same form as the ACF of the PRBS. As expected, this displays a high central peak with small values to either side, confirming that the PRBS has spectral properties very like those of white noise. To the right-hand side we see its response to the twin, overlapping PRBS waveforms. Since the filter is linear, this is simply a superposition of two individual ACF waveforms. The peaks are nicely disentangled, each occuring at the precise instant when the relevant PRBS has just entered the filter.

In part (c) we have mixed the composite signal of part (a) with a large amount of gaussian noise (and reduced the figure's vertical scale). it is now impossible to detect occurrences of the PRBS by eye. But after passage through the matched filter, we get the signal-plus-noise waveform shown in part (d) of the figure. The isolated ACF peaks are clearly visible, in just the right places — even though their heights have been somewhat altered by coincident noise. This figure speaks largely for itself.

The sum of squares of input signal values in this case is 64, and their peak value is 1. The improvement in S:N ratio demonstrated by part (d) of the figure is therefore $\sqrt{64/1} = 8$, or 18.1 dB. Use of a longer PRBS would of course give a better result. The improvement is simply proportional to the square root of the number of binary characters in the PRBS.

It is, finally, interesting to note that, in this case, signal and noise have essentially similar forms of energy or power spectrum: both are white. The matched filter is working entirely on the basis of phase or timing information contained in the particular PRBS we have used — an impressive performance.

Although we have concentrated here on the PRBS, it is not the only type of 'spread' waveform used in matched filter detection. Another good example is the so-called *chirp* waveform used in some radar systems, which consists of a 'stretched' pulse of radio-frequency waves. In a chirp radar, the radio frequency is not held constant during each pulse, but sweeps up or down over a defined range. On reception, echoes are sharpened up in the receiver by a matched filter or similar processor. One of the most remarkable examples of signal processing in the natural world is achieved by certain species of bat, which use sonar echolocation based on chirp-type waveforms to sharpen up their internal picture of the external world. Since such processing techniques involve the time-domain compression of wide transmitted pulses into narrow received pulses, they come under the general heading of *pulse compression*. The processing illustrated in Figure 10.14 is a good example of this.

Following matched filtering, there is still a final decision to make about when, or if, a signal waveform occurs. This is a statistical problem which cannot be answered with certainty. Even if a particular feature looks like a signal peak, it is at least possible that it is only noise; conversely, when a signal is in fact present, the noise may happen to subtract from it and make it invisible. In the case of gaussian noise, the probabilities associated with these two types of error are well known. Not surprisingly, they depend crucially on the S:N ratio at the matched filter output. We conclude that even though a matched filter is the optimum linear processor in this type of situation, it can only succeed if the noise level at its input is sufficiently low.

10.5.4   SIGNAL AVERAGING

When recovering or detecting a signal in the presence of noise, we need to take advantage of all its known characteristics. So far our only assumptions have been about bandwidth or waveshape. However, if a deterministic signal of finite duration, obscured by noise, is known to repeat itself at particular instants of time, a further technique known as *signal averaging* may be used to clarify it.

The technique is often used in situations of the *stimulus-response* type, where a series of brief stimuli is applied to a system under investigation. Such situations are quite common in the biological sciences. In the very different field of pulse radar, a train of radio frequency pulses, transmitted at known instants, produces a corresponding series of echoes from a distant target such as an aircraft, and these may be averaged electronically in the receiver or on the radar operator's display to produce an enhanced image. In more conventional electronic engineering, a system such as a linear amplifier or filter may be tested by an input impulse, in order to measure its impulse response. If that response is contaminated by noise, the test may be repeated many times and the results averaged.

Suppose, therefore, that we wish to assess the impulse response of a filter, but our measurements are subject to additive noise. We repeat the input impulse many times, and produce a whole series of responses, as shown in parts (a) and (b) of Figure 10.15 (actually the noise level is so great in the figure that it is

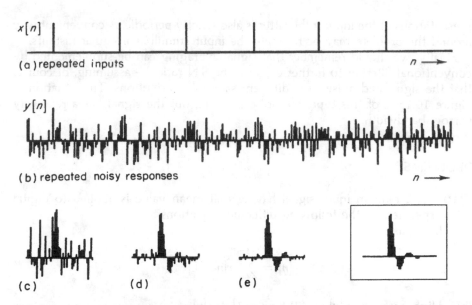

Figure 10.15 Signal averaging (*abscissae: 320 and 50 samples*).

extremely hard to detect individual responses by eye). Provided the noise is largely or wholly uncorrelated between successive responses, we may now add together (or average) successive versions of the response in order to 'rescue' the signal from the background noise. The reason for this is that each response is composed of two parts — a deterministic signal, and unwanted random noise. When we add a number of responses together, the signal portion is truly additive; but noise fluctuations, being uncorrelated, tend to average out to zero.

In the figure, we have shown the typical effect of averaging over (c) 5 responses, (d) 50 responses, and (e) 500 responses. Also shown inset, at the bottom right of the figure, is the pure signal (which, in this example, represents the impulse response of a second-order digital filter). Clearly, the more responses we average, the better the S:N ratio becomes.

If, as in this illustration, the noise is white and therefore completely uncorrelated between successive versions of the response, we can easily calculate the improvement in S:N ratio. As already mentioned in Section 9.2.2, the variance of the sum of uncorrelated random sequences equals the sum of their individual variances. This is also true of a series of uncorrelated portions of the same random sequence. Therefore, if we sum k versions of the noisy response, we get a signal increased in size by a factor k, and a noise variance increased by the same factor. Its standard deviation is increased by $\sqrt{k}$. The S:N ratio, expressed in terms of amplitudes, therefore improves by $\sqrt{k}$. In part (c) of the figure this improvement is $\sqrt{5}$ or 7 dB; in part (d), $\sqrt{50}$ or 17 dB; and in part (e), $\sqrt{500}$ or 27 dB.

If the noise is partly correlated between successive input impulses, the improvement in S:N ratio will generally be less than $\sqrt{k}$. It will also be more difficult to predict. A somewhat different problem arises if the noise contains strictly periodic components, for example at the frequency of the power supply. In such cases, the noise tends to be partly correlated between successive presentations of the input if the latter is also strictly periodic. A convenient way around the problem may be to apply the input stimuli at irregular instants.

Finally, we should remember that signal averaging can be supplemented by conventional filtering to further enhance the S:N ratio — assuming, of course, that the signal and noise have different spectral distributions. The situation in Figure 10.15 is of this type: the noise is white, but the signal has a relatively narrow bandwidth.

PROBLEMS

Q10.1 A random input signal having unit mean value is applied to digital filters specified by the following difference equations:
(a)   Low pass:

$$y[n] = \sum_{k=0}^{5} x[n-k]$$

(b)   High pass: $y[n] = -0.8y[n-1] + x[n]$

(c)   Bandpass: $y[n] = 1.5y[n-1] - 0.85y[n-2] + x[n]$

In each case, find the mean value of the random output.

Q10.2 A digital processor has the transfer function:

$$H(z) = \frac{(z+1)(z^2+1)}{z^2 - 2r \cos \theta z + r^2}$$

If its input signal is a random sequence with unit mean value, express the mean value of the output in terms of r and θ.

Q10.3 A 3-point moving-average filter is used to smooth a random sequence. Figure Q10.3 shows the impulse response of the filter and the autocorrelation function (ACF) of the input sequence.
(a)   What is the variance of the input sequence?
(b)   Sketch the ACF of the output sequence.
(c)   Find the variance of the output sequence.
(d)   Derive an expression for the output power spectrum.

Figure Q10.3

Q10.4 White digital noise with zero mean and unit variance is processed by a digital filter having the following difference equation:

$$y[n] = y[n-1] + x[n] - x[n-8]$$

Specify the ACF of the output noise, and find its variance.

Q10.5 Two different LTI processors are to be identified using random input signals, by the method of input-output cross-correlation. Parts (a) and (b) of Figure Q10.5 show the input ACF and input-output CCF obtained in the two cases.

Find expressions for the frequency responses of the processors.

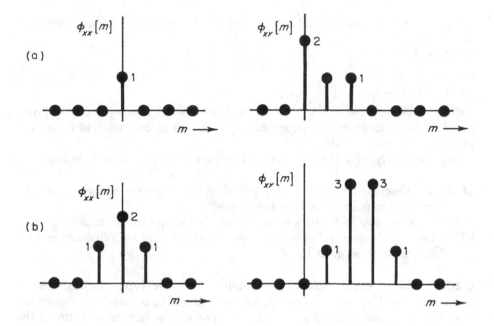

Figure Q10.5

Q10.6 Explain clearly why identification of an unknown LTI processor by the method of input-output cross-correlation is particularly straightforward when the input is white noise.

Q10.7 Discuss the advantages and limitations of system identification by input-out cross-correlation.

Q10.8 A narrowband digital signal is contaminated by white noise. What improvement in signal-to-noise (S:N) ratio may theoretically be achieved using a linear filter with a rectangular response magnitude characteristic, if the signal occupies the following frequency band(s) within the range $0 \leq \Omega \leq \pi$? Express your answers in decibels (dB).

(a) 0 to $\pi/10$

(b) $\pi/50$ to $2\pi/50$

(c) $\pi/12$ to $\pi/10$, and $\pi/24$ to $\pi/20$

Why would these figures not be fully achieved in practice?

Q10.9 Explain how the concept of matched filtering illustrates the links between convolution and correlation, considered as DSP operations.

Q10.10 By modifying Computer Program No. 29 in Appendix A1, investigate matched filter detection for various signal waveshapes and noise levels.

Q10.11 Three time-limited digital signals have the following non-zero sample values:

(a) 4, 3, 2, 1

(b) -2, 4, -2, 1, 1, 2

(c) 1, -1, 1, -1, 1, -1

Each is to be detected in the presence of white noise by using the appropriate matched filter. Estimate the expected improvement in S:N ratio in each case, giving your answer in dB.

Explain clearly why the improvement in part (b) is the same as in part (a).

Q10.12 Discuss the reasons why pseudo-random binary sequences (PRBS) are often used in pulse compression systems.

Would a lengthy PRBS be a suitable input for identification of an unknown LTI system by the method of input-output cross-correlation? What advantages and disadvantages might it have?

Q10.13 In an experiment of the stimulus-response type, the response is obscured by additive white noise. Signal averaging is to be used to enhance the S:N ratio. How many repetitions of the stimulus will be needed to improve the S:N ratio by (a) 20 dB, and (b) 50 dB?

Q10.14 In what circumstances, or under what conditions, are the following true?

(a) A random input sequence, applied to an LTI processor, produces a random output having zero mean value.

(b) The variances of random input and output signals of an LTI processor are the same.

(c) The output ACF from an LTI processor is identical in form (but not necessarily in scale) to the input ACF.

(d) The central, peak values of a random signal's autocorrelation and autocovariance functions are the same.

(e) A noise input to a linear filter gives rise to an output whose ACF is identical to the autocorrelation sequence of the filter itself.

(f) A matched filter fails to give any improvement in signal-to-noise ratio.

Appendix A1
Computer Programs

A1.1 TABLE OF PROGRAMS AND INTRODUCTORY NOTES

The programs are intended to illustrate DSP principles and design methods, rather than programming technique. They are complete and ready to run. Although we do not describe them in any detail, you should have little difficulty in understanding how they are organized if you have followed the relevant section(s) in the main text.

We have tried to keep the programs as simple as possible, using few of the available C or PASCAL functions and procedures. Graphical outputs are all based on a single graphics instruction, which draws a straight line between two points on the screen. We have avoided embellishments such as axis labels (although such labels have been added to the plots reproduced in the main text).

C PROGRAMS

The C programs were developed on an IBM-compatible personal computer using Microsoft Quick C; the program listings refer to this dialect of C. Subsequently they have been tested and compiled using Borland Turbo C++ 3.0. Although the minimum number of graphics commands are used, differences in implementation of the two compilers call for changes to be made to every call to a graphics library function. To facilitate this, a header file, graph.h, is provided with the C program listings. *This is for use with the Borland compilers only*. Inclusion of this file will allow for all differences between the compilers. (The easiest way to do this is to copy the file graph.h to your INCLUDE directory. Alternatively, copy graph.h to your source directory and change the line

#include ⟨graph.h⟩

to

#include "graph.h"

in each source file.) All programs have been designed to display with a resolution of 320 × 200 pixels except for Program 24 (which displays at 640 × 200 pixels), and Programs 7 and 21 (which do not include graphical output).

PASCAL PROGRAMS

The PASCAL programs were originally developed on an IBM-compatible personal computer using Turbo-PASCAL version 4.0. To make them suitable for version 3.0 as well, we used procedures GRAPHMODE (320 × 200 pixels), HIRES (640 × 200 pixels), and DRAW, together with the unit *Graph3*, which is included in a *uses* clause at the start of each program. The programs should also run without problems on subsequent versions of Turbo-PASCAL.

PROGRAM DISK

This disk contains one file — LYNN.EXE. This file is a self-extracting archive that requires Microsoft Windows 3.1x or 95 in order to run (if you do not have Microsoft Windows on your PC please see the NOTE at the end of this section). Simply run the file, either by selecting RUN from the FILE menu (Windows 3.1x) or by selecting RUN from the START menu (Windows 95). The filename is

A:\LYNN.EXE

The installation routine will give you the option to specify where you wish the files to be installed on your hard-disk.

When installed there will be a directory on your hard-disk that contains source programs in C and PASCAL, and compiled programs in C, with extensions .C, .PAS, and .EXE respectively. Each program has a file name of the form DSPnn, where nn refers to the program number between 1 and 29. Thus DSP07.EXE refers to the compiled C-version of Program no. 7, and so on.

The C programs were compiled using Borland Turbo C++ 3.0. To run properly, the files with an extension of .BGI should be copied to the same directory as the program files.

NOTE

If you do not have Microsoft Windows on your PC you can still access the files. You will need to decompress the file LYNN.EXE using an application such as PKUNZIP. The command line is as follows:

pkunzip a:\lynn.exe

TECHNICAL SUPPORT

If you experience any difficulties in using this disk then please contact:

Customer Service Technical Support Group
Tel: (01243) 843312   *International* (+44) 1243 843312
Fax: (01243) 843315   *International* (+44) 1243 843315
email: cs-electronic@wiley.co.uk

A1.2   C PROGRAMS

```c
#include <graph.h>
#include <conio.h>
#include <stdio.h>
#include <math.h>

#define XDIM 320
#define YDIM 320

float    fX[XDIM + 1], fY[YDIM + 1];

void main (void)
    {
    /*
     ********  PROGRAM NO.1    DIGITAL BANDPASS FILTER  ********
     */
    int    iCount;

    _clearscreen  (_GCLEARSCREEN);
    printf ("PROGRAM 1: Digital bandpass filter\n");
    printf ("Press any key to start...\n\n");
    getch();

    /*
     *****  GENERATE INPUT SIGNAL,  AND LOAD INTO ARRAY X  *****
     */
    for (iCount = 1; iCount <= XDIM; iCount++)
        fX[iCount] = sin (3.141593 *
                          (float)iCount *
                          (0.05 + 0.0005 * (float)iCount));

    /*
     ******  ESTIMATE OUTPUT SIGNAL OF RECURSIVE FILTER  *******
     */
    fY[1] = 0.0;
    fY[2] = 0.0;

    for (iCount = 3; iCount <= YDIM; iCount++)
        fY[iCount] = 1.5 *
        fY[iCount - 1] - 0.85 *
        fY[iCount - 2] + fX[iCount];

    /*
     *******  PLOT INPUT AND OUTPUT SIGNALS ON SCREEN  *********
     */

    _setvideomode( _MRES16COLOR );

    for (iCount = 1; iCount <= XDIM; iCount++)
        {
        _moveto (iCount, 50);
        _lineto (iCount, (int)(50.0 - fX[iCount] * 25.0 + 0.5));

        _moveto (iCount, 150);
        _lineto (iCount, (int)(150.0 - fY[iCount] * 4.0 + 0.5));
        }

    getch();

    _setvideomode( _DEFAULTMODE );

    /*
     **********************************************************
     */
    }
```

```c
#include <graph.h>
#include <conio.h>
#include <stdio.h>
#include <math.h>

#define XDIM 100

float   fX[XDIM + 1], fB = 0.0, fW = 0.0;

void main (void)
    {
    /*
     ***   PROGRAM NO.2  SAMPLED EXPONENTIALS AND SINUSOIDS   ****
     */
    int    iCount;

    _clearscreen  (_GCLEARSCREEN);
    printf ("PROGRAM 2: Sampled exponentials and sinusoids\n");

    printf ("Enter value of Beta-Zero :");
    scanf ("%f", &fB);

    printf ("Enter value of Omega :");
    scanf ("%f", &fW);

    for (iCount = 1; iCount <= XDIM; iCount++)
        fX[iCount] = exp (fB * (float)iCount) *
                     cos (fW * (float)iCount);

    /*
     ************** DRAW AXES, THEN SIGNAL SAMPLES *************
     */
    _setvideomode( _MRES16COLOR );

    _moveto (1, 100);    _lineto (300, 100);

    for (iCount = 1; iCount <= XDIM; iCount++)
        {
        _moveto (3 * iCount, 100);
        _lineto (3 * iCount, (int)(100.0 - fX[iCount] * 30.0 + 0.5));
        }

    getch();
    _setvideomode( _DEFAULTMODE );

    /*
     ************************************************************
     */
    }
```

```c
#include <graph.h>
#include <conio.h>
#include <stdio.h>
#include <math.h>

#define XDIM 320

float   fX[XDIM + 1];

void main (void)
    {
    /*
     ***** PROGRAM NO.3   AMBIGUITY IN DIGITAL SINUSOIDS  *****
     */
    int    iCount, iM, iK;

    _clearscreen  (_GCLEARSCREEN);
    /*
     *** GENERATE SIGNAL WITH TEN EQUAL FREQUENCY INCREMENTS ***
     */
    printf ("PROGRAM 3: Ambiguity in digital sinusoids\n");
    printf ("Press any key to start...\n\n");
    getch();

    for (iM = 0; iM < 10; iM++)
        for (iK = 1; iK < 33; iK++)
            {
            iCount = iK + 32 * iM;
            fX[iCount] = cos ((float)iCount * 2.0 *
                        3.141593 * (float)iM / 8.0);
            }

    /*
     ******************** PLOT ****************************
     */
    _setvideomode( _MRES16COLOR );

    for (iCount = 1; iCount <= XDIM; iCount++)
        {
        _moveto (iCount, 50);
        _lineto (iCount, 50 - (int)(fX[iCount] * 25.0  + 0.5));
        }

    getch();
    _setvideomode( _DEFAULTMODE );

    /*
     **********************************************************
     */
    }
```

```
#include <graph.h>
#include <conio.h>
#include <stdio.h>
#include <stdlib.h>
#include <math.h>

#define HDIM 120
#define XDIM 120

float   fH[HDIM + 1], fX[XDIM + 1];

void main (void)
    {
    /*
     ***** PROGRAM NO.4  IMPULSE RESPONSE OF LTI PROCESSOR *****
     */
    int    iCount;
    float  fA1, fA2, fA3, fB0, fB1, fB2, fMax;

    _clearscreen  (_GCLEARSCREEN);
    printf ("PROGRAM 4: Impulse response of LTI processor\n");

    /*
     ****************** LOAD IMPULSE INTO INPUT ARRAY **********
     */
    printf ("Enter 3 recursive multipliers :");
    scanf ("%f, %f, %f", &fA1, &fA2, &fA3);

    printf ("\nEnter 3 nonrecursive multipliers :");
    scanf ("%f, %f, %f", &fB0, &fB1, &fB2);

    fX[10] = 1.0;

    /*
     **** FIND IMPULSE RESPONSE, TAKING N=10 AS TIME ORIGIN ****
     */

    printf ("Press any key to start...\n\n");
    getch();

    for (iCount = 10; iCount <= HDIM; iCount++)
        fH[iCount] = (fA1 * fH[iCount-1] +
                      fA2 * fH[iCount-2] +
                      fA3 * fH[iCount-3] +
                      fB0 * fX[iCount] +
                      fB1 * fX[iCount-1] +
                      fB2 * fX[iCount-2]);

    /*
     **** NORMALISE fH[iCount] TO A MAXIMUM VALUE OF UNITY *****
     */
    fMax = 0.0;
    for (iCount = 1; iCount <= HDIM; iCount++)
        if (fabs (fH[iCount]) > fMax)
            fMax = fabs (fH[iCount]);

    for (iCount = 1; iCount <= HDIM; iCount++)
        fH[iCount] /= fMax;

    _setvideomode( _MRES16COLOR );

    /*
     ********* PLOT HORIZONTAL AXIS AND IMPULSE INPUT **********
     */
    _moveto (0, 50);
    _lineto (300, 50);

    for (iCount = 1; iCount <= 100; iCount++)
        {
        _moveto (iCount * 3, 50);
```

```
        _lineto (iCount * 3, (int)(50.0 - fX[iCount+9] * 25.0 + 0.5));
        }

/*
 ******* PLOT HORIZONTAL AXIS AND IMPULSE RESPONSE *********
 */
_moveto (0, 120);
_lineto (300, 120);

for (iCount = 1; iCount <= 100; iCount++)
    {
    _moveto (iCount * 3, 120);
    _lineto (iCount * 3, (int)(120 - (fH[iCount+9] * 60.0) + 0.5));
    }

getch();
_setvideomode( _DEFAULTMODE );

/*
 ***********************************************************
 */
}
```

```
#include <graph.h>
#include <conio.h>
#include <stdio.h>
#include <stdlib.h>
#include <math.h>

#define XDIM 320
#define YDIM 320
#define HDIM 60

float   fX[XDIM + 1], fY[YDIM + 1], fH[HDIM + 1];

void main (void)
    {
    /*
     ********** PROGRAM NO.5    DIGITAL CONVOLUTION  **********
     */
    int    iCount, iCount2;
    float  fMaxX, fMaxY;

    printf ("PROGRAM 5: Digital convolution\n");
    printf ("Press any key to start...\n");
    getch();

    /*
     ****************** DEFINE INPUT SIGNAL  ******************
     */
    for (iCount = 60; iCount <= XDIM; iCount++)
      fX[iCount] = sin (2.0 * 3.141593 * (float)iCount / 60.0) +
                sin (2.0 * 3.141593 * (float)iCount / 10.0);

    /*
     ***** DEFINE IMPULSE RESPONSE (MAX NO. OF TERMS = 60) *****
     */

    for (iCount = 1; iCount <= 10; iCount++)
        fH[iCount] = 0.1;

    /*
     *********** CONVOLVE INPUT AND IMPULSE RESPONSE ***********
     */
    for (iCount = HDIM + 1; iCount <= YDIM; iCount++)
        for (iCount2 = 1; iCount2 <= HDIM; iCount2++)
            fY[iCount] = fY[iCount] + fH[iCount2] *
                        fX[iCount - iCount2];

    /*
     ** NORMALISE SIGNALS TO A MAXIMUM VALUE OF UNITY FOR PLOT *
     */
    fMaxX = 0.0;
    fMaxY = 0.0;

    for (iCount = 1; iCount <= XDIM; iCount++)
        {
        if (fabs (fX[iCount]) > fMaxX)
            fMaxX = fabs (fX[iCount]);
        if (fabs (fY[iCount]) > fMaxY)
            fMaxY = fabs (fY[iCount]);
        }

    for (iCount = 1; iCount <= XDIM; iCount++)
        {
        fX[iCount] /= fMaxX;
        fY[iCount] /= fMaxY;
        }

    _setvideomode( _MRES16COLOR );

    /*
     ********** PLOT HORIZONTAL AXIS AND INPUT SIGNAL **********
```

```
 */
_moveto (0, 50);
_lineto (XDIM, 50);

for (iCount = 1; iCount <= XDIM; iCount++)
    {
    _moveto (iCount, 50);
    _lineto (iCount, (int)(50 - fX[iCount] * 30.0 + 0.5));
    }

/*
 ********* PLOT HORIZONTAL AXIS AND OUTPUT SIGNAL **********
 */
_moveto (0, 150);
_lineto (YDIM, 150);

for (iCount = 1; iCount <= YDIM; iCount++)
    {
    _moveto (iCount, 150);
    _lineto (iCount, (int)(150 - fY[iCount] * 30.0 + 0.5));
    }

getch();
_setvideomode( _DEFAULTMODE );

/*
 ***********************************************************
 */
}
```

```
#include <graph.h>
#include <conio.h>
#include <stdio.h>
#include <stdlib.h>
#include <math.h>

#define XDIM 320
#define YDIM 320

float   fX[XDIM + 1], fY[YDIM + 1];

void main (void)
    {
    /*
     *****   PROGRAM NO.6      START-UP AND STOP TRANSIENTS  *****
     */
    int    iCount;
    float  fMaxX, fMaxY;

    _clearscreen  (_GCLEARSCREEN);

    printf ("PROGRAM 6: Start-up and stop transients\n");
    printf ("Press any key to start...\n");
    getch();

    /*
     **************** DEFINE INPUT SIGNAL *********************
     */
    for (iCount = 20; iCount <= 200; iCount++)
        fX[iCount] = cos (2.0 * 3.141593 * (float)iCount / 20.0);

    /*
     *********** ESTIMATE OUTPUT FROM BANDSTOP FILTER **********
     */
    for (iCount = 3; iCount <= YDIM; iCount++)
        fY[iCount] = 1.8523 * fY[iCount - 1] - 0.94833 *
                     fY[iCount - 2] + fX[iCount] -
                     1.9021 * fX[iCount - 1] + fX[iCount - 2];

    /*
     ** NORMALISE SIGNALS TO MAXIMUM VALUE OF UNITY FOR PLOT ***
     */
    fMaxX = 0.0;
    fMaxY = 0.0;

    for (iCount = 1; iCount <= XDIM; iCount++)
        {
        if (fabs (fX[iCount]) > fMaxX)
            fMaxX = fabs (fX[iCount]);
        if (fabs (fY[iCount]) > fMaxY)
            fMaxY = fabs (fY[iCount]);
        }

    for (iCount = 1; iCount <= XDIM; iCount++)
        {
        fX[iCount] /= fMaxX;
        fY[iCount] /= fMaxY;
        }

    _setvideomode( _MRES16COLOR );

    /*
     ********** PLOT HORIZONTAL AXIS AND INPUT SIGNAL *********
     */

    _moveto (0, 50);
    _lineto (XDIM, 50);

    for (iCount = 1; iCount <= XDIM; iCount++)
        {
```

```
      _moveto (iCount, 50);
      _lineto (iCount, (int)(50 - fX[iCount] * 35.0 + 0.5));
      }

  /*
   ********  PLOT HORIZONTAL AXIS AND OUTPUT SIGNAL **********
   */

  _moveto (0, 150);
  _lineto (YDIM, 150);

  for (iCount = 1; iCount <= YDIM; iCount++)
      {
      _moveto (iCount, 150);
      _lineto (iCount, (int)(150 - fY[iCount] * 35.0 + 0.5));
      }

  getch();
  _setvideomode( _DEFAULTMODE );

  /*
   ***************************************************************
   */
  }
```

```
#include <graph.h>
#include <conio.h>
#include <stdio.h>
#include <stdlib.h>
#include <math.h>

float   fX[] = {2, 1, -2, 3, -1, -1, 1}, fAR[7], fAI[7];

void main (void)
    {
    /*
     ********  PROGRAM NO.7   DISCRETE-TIME FOURIER  **********
     ********  SERIES FOR SIGNAL WITH 7 SAMPLE VALUES  ********
     */
    int    iCount, iCount2;

    _clearscreen  (_GCLEARSCREEN);
    printf ("PROGRAM 7: Discrete Fourier Series\n");
    printf ("for real signal with 7 sample values\n\n");

    for (iCount2 = 1; iCount2 < 8; iCount2++)
      for (iCount = 1; iCount < 8; iCount++)
        {
        fAR[iCount2-1] = fAR[iCount2-1] + (1.0/7.0) *
              fX[iCount-1] * cos (2.0 * 3.141593 *
              (float)(iCount2-1) * (float)(iCount-1) / 7.0);
        fAI[iCount2-1] = fAI[iCount2-1] - (1.0/7.0) *
              fX[iCount-1] * sin (2.0 * 3.141593 *
              (float)(iCount2-1) * (float)(iCount-1) / 7.0);
        }

    for (iCount2 = 0; iCount2 < 7; iCount2++)
        printf ("%d     % f, % f\n", iCount2,
        fAR[iCount2], fAI[iCount2]);

    /*
     ***************************************************************
     */
    }
```

```
#include <graph.h>
#include <conio.h>
#include <stdio.h>
#include <stdlib.h>
#include <math.h>

#define XDIM      64
#define ARDIM     64
#define AIDIM     64
#define MAGDIM    64
#define PHASEDIM 64

float    fX[XDIM + 1], fAR[ARDIM + 1], fAI[AIDIM + 1],
         fMAG[MAGDIM + 1], fPHASE[PHASEDIM + 1];

void main (void)
    {
    /*
    *****   PROGRAM NO.8   DISCRETE-TIME FOURIER SERIES ******
    *** 64 REAL SAMPLE VALUES, SPECTRAL MAGNITUDE & PHASE  **
    */
    int     iCount, iCount2;
    float   fMax1, fMax2, fMax3, fB;

    _clearscreen  (_GCLEARSCREEN);

    printf ("PROGRAM 8: Discrete Fourier Series\n");
    printf ("64 real sample values, spectral magnitude ");
    printf ("& phase\n\n");

    printf ("Press any key to start...\n");
    getch();

    for (iCount = 1; iCount <= XDIM; iCount++)
        {
        fB = 2.0 * 3.141593 * (float)(iCount - 1);
        fX[iCount] = sin (fB / 64.0) + cos (fB / 16.0) +
                0.6 * cos (fB / 8.0) + 0.5 * sin (fB / 4.0);
        }

    for (iCount = 1; iCount <= XDIM; iCount++)
        for (iCount2 = 1; iCount2 <= XDIM; iCount2++)
            {
            fAR[iCount] = fAR[iCount] + (1.0 / 64.0) *
                    fX[iCount2] * cos (2.0 * 3.141593 *
                    (float)(iCount - 1) * (float)(iCount2 - 1) / 64.0);
            fAI[iCount] = fAI[iCount] - (1.0 / 64.0) *
                    fX[iCount2] * sin (2.0 * 3.141593 *
                    (float)(iCount - 1) * (float)(iCount2 - 1) / 64.0);
            }

    /*
    ******* CONVERT TO SPECTRAL MAGNITUDE AND PHASE ***********
    */
    for (iCount = 1; iCount <= XDIM; iCount++)
        {
        if (fabs (fAI[iCount]) < 0.001)
            fAI[iCount] = 0.0;
        if (fabs (fAR[iCount]) < 0.00001)
            fAR[iCount] = 0.00001;
        fMAG[iCount] = sqrt (pow (fAR[iCount], 2.0) +
                        pow (fAI[iCount], 2.0));
        fPHASE[iCount] = atan (fAI[iCount] / fAR[iCount]);
        }

    /*
    ********** NORMALISE TO MAXIMUM OF UNITY FOR PLOT *********
    */
    fMax1 = fMax2 = fMax3 = 0.001;
```

```
for (iCount = 1; iCount <= XDIM; iCount++)
   {
   if (fabs (fX[iCount]) > fMax1)
       fMax1 = fabs (fX[iCount]);
   if (fabs (fMAG[iCount]) > fMax2)
        fMax2 = fabs (fMAG[iCount]);
   if (fabs (fPHASE[iCount]) > fMax3)
        fMax3 = fabs (fPHASE[iCount]);
   }

for (iCount = 1; iCount <= XDIM; iCount++)
   {
   fX[iCount] /= fMax1;
   fMAG[iCount] /= fMax2;
   fPHASE[iCount] /= fMax3;
   }

/*
 ******** PLOT SIGNAL, SPECTRAL MAGNITUDE AND PHASE ********
 */
_setvideomode( _MRES16COLOR );

for (iCount = 1; iCount <= XDIM; iCount++)
  {
  _moveto (5*(iCount - 1), 50);
  _lineto (5*(iCount - 1), (int)(50 - fX[iCount] * 25.0 + 0.5));
  _moveto (5*(iCount - 1), 120);
  _lineto (5*(iCount - 1), (int)(120 - fMAG[iCount] * 30.0 + 0.5));
  _moveto (5*(iCount - 1), 160);
  _lineto (5*(iCount - 1), (int)(160 - fPHASE[iCount] * 20.0 + 0.5));
  }

getch();
_setvideomode( _DEFAULTMODE );

/*
 ************************************************************
 */
}
```

```c
#include <graph.h>
#include <conio.h>
#include <stdio.h>
#include <stdlib.h>
#include <math.h>

#define XDIM   320
#define YDIM   320

float   fX[XDIM + 1], fY[YDIM + 1];

void main (void)
    {
    /*
     *** PROGRAM NO.9  FREQUENCY RESPONSE OF BANDPASS FILTER ***
     */
    int     iCount;
    float   fW, fA, fB, fMAG, fPHASE;

    _clearscreen   ( _GCLEARSCREEN);
    printf ("PROGRAM 9: Frequency response of ");
    printf ("bandpass filter\n");

    printf ("Press any key to start...\n");
    getch();

    _setvideomode( _MRES16COLOR );

    for (iCount = 1; iCount <= XDIM; iCount++)
        {
        fW = 3.141593 * (iCount - 1) / 320;
        fA = 1.0 - 1.5 * cos (fW) + 0.85 * cos (2.0 * fW);
        fB = 1.5 * sin (fW) - 0.85 * sin (2.0 * fW);
        fMAG = 1.0 / pow ((fA * fA) + (fB * fB), 0.5);
        fPHASE = atan (fB / fA);

        _moveto (iCount, 100);
        _lineto (iCount, (int)(100 - 5.0 * fMAG + 0.5));

        _moveto (iCount, 160);
        _lineto (iCount, (int)(160 - 8.0 * fPHASE + 0.5));
        }

    getch();
    _setvideomode( _DEFAULTMODE );

    /*
     ************************************************************
     */
    }
```

```c
#include <graph.h>
#include <conio.h>
#include <stdio.h>
#include <stdlib.h>
#include <math.h>

#define XDIM   320
#define YDIM   320

float   fX[XDIM + 1], fY[YDIM + 1];

void main (void)
    {
    /*
     **********   PROGRAM NO.10    INVERSE Z-TRANSFORM   **********
     */
    int    iCount;

    _clearscreen  (_GCLEARSCREEN);

    printf ("PROGRAM 10: Inverse z-transform\n");
    printf ("Press any key to start...\n");
    getch();

    fX[5] = 1.0;

    _setvideomode( _MRES16COLOR );

    for (iCount = 5; iCount <= XDIM; iCount++)
        {
        fY[iCount] = -0.54048 * fY[iCount-1] +
                      0.62519 * fY[iCount-2] +
                      0.66354 * fY[iCount-3] -
                      0.60317 * fY[iCount-4] -
                      0.69341 * fY[iCount-5] +
                      fX[iCount] - fX[iCount-1] +
                      fX[iCount-2] - fX[iCount-3];

        _moveto (iCount, 100);
        _lineto (iCount, (int)(100 - 20.0 * fY[iCount] + 0.5));
        }

    getch();
    _setvideomode( _DEFAULTMODE );

    /*
     ************************************************************
     */
    }
```

```
#include <graph.h>
#include <conio.h>
#include <stdio.h>
#include <stdlib.h>
#include <math.h>

#define IDIM     320
#define FDIM     320

float   fI[IDIM + 1], fF[FDIM + 1];

void main (void)
    {
    /*
     *** PROGRAM NO.11  CHARACTERISTICS OF 2ND-ORDER SYSTEMS ***
     */
    int     iCount;
    float   fR, fTHETA, fT, fW, fGAIN, fMax, fA, fB, fC, fD;

    _clearscreen  (_GCLEARSCREEN);

    printf ("PROGRAM 11: ");
    printf ("Characteristics of 2nd-order systems\n");

    /*
     * fI=IMPULSE RESPONSE, fF=FREQUENCY RESPONSE (MAGNITUDE) **
     */
    printf ("Enter RADIUS, ANGLE (degrees) of z-plane poles :");
    scanf ("%f, %f", &fR, &fTHETA);

    fT = fTHETA * 3.141593 / 180.0;

    /*
     ***** LOAD IMPULSE RESPONSE, STARTING AT LOCATION fI[2] ***
     */
    fI[1] = 0.0;
    fI[2] = 1.0;

    for (iCount = 3; iCount <= IDIM; iCount++)
        fI[iCount] = 2.0 * fR * cos (fT) * fI[iCount-1] -
                     fR * fR * fI[iCount-2];

    /*
     ******* NORMALISE TO PEAK VALUE OF UNITY, THEN PLOT *******
     */
    fMax = 0.0;

    for (iCount = 1; iCount <= IDIM; iCount++)
        if (fabs (fI[iCount]) > fMax)
            fMax = fabs (fI[iCount]);

    for (iCount = 1; iCount <= IDIM; iCount++)
        fI[iCount] /= fMax;

    _setvideomode( _MRES16COLOR );

    for (iCount = 1; iCount <= IDIM; iCount++)
        {
        _moveto (iCount, 70);
        _lineto (iCount, (int)(70 - fI[iCount] * 30.0 + 0.5));
        }

    /*
     **************** ESTIMATE PEAK GAIN  *******************
     */
    fA = 1.0 - 2.0 * fR * cos (fT) * cos (fT) +
         fR * fR * cos (2.0 * fT);
    fB = 2.0 * fR * cos (fT) * sin (fT) -
         fR * fR * sin (2.0 * fT);
    fGAIN = 1.0 / pow (fA * fA + fB * fB, 0.5);
```

```
/*
 ****** FIND FREQUENCY RESPONSE (NORMALISED), THEN PLOT ****
 */
for (iCount = 1; iCount <= IDIM; iCount++)
    {
    fW = (float)(iCount - 1) * 3.141593 / (float)IDIM;
    fC = 1.0 - 2.0 * fR * cos (fT) * cos (fW) +
            fR * fR * cos (2.0 * fW);
    fD = 2.0 * fR * cos (fT) * sin (fW) -
            fR * fR * sin (2.0 * fW);
    fF[iCount] = 1.0/(pow(fC * fC + fD * fD, 0.5) * fGAIN);
    }

for (iCount = 1; iCount <= IDIM; iCount++)
    {
    _moveto (iCount, 170);
    _lineto (iCount, (int)(170 - fF[iCount] * 60.0 + 0.5));
    }

getch();
_setvideomode( _DEFAULTMODE );

/*
 ***********************************************************
 */
}




#include <graph.h>
#include <conio.h>
#include <stdio.h>
#include <stdlib.h>
#include <math.h>

void main (void)
    {
    /*
     *** PROGRAM NO.12  F-RESPONSE OF MOVING-AVERAGE FILTER ****
     */
    int    iCount, iCount2, iM;
    float  fH, fJ, fW;

    _clearscreen (_GCLEARSCREEN);
    printf ("PROGRAM 12: F-response of moving-average filter\n");

    printf ("Enter (integer) value of M :");
    scanf ("%d", &iM);

    fJ = 1.0 / (2.0 * (float)iM + 1.0);

    _setvideomode( _MRES16COLOR );

    for (iCount = 1; iCount <= 320; iCount++)
        {
        fW = 3.141593 * (float)(iCount - 1) / 320.0;
        fH = fJ;
        for (iCount2 = 1; iCount2 <= iM; iCount2++)
            fH = fH + 2.0 * fJ * cos ((float)iCount2 * fW);
```

```
        _moveto (iCount, 100);
        _lineto (iCount, (int)(100 - 50.0 * fabs (fH) + 0.5));
        }

getch();
_setvideomode( _DEFAULTMODE );

/*
 ************************************************************
 */
}


#include <graph.h>
#include <conio.h>
#include <stdio.h>
#include <stdlib.h>
#include <math.h>

#define HDIM     200

float    fH[HDIM + 1];

void main (void)
    {
    /*
     ****************** PROGRAM NO.13 ************************
     ** NONRECURSIVE FILTER DESIGN BY FOURIER TRANSFORM METHOD *
     ******* IMPULSE RESPONSE TRUNCATED TO (2M+1) TERMS ********
     */
    int    iCount, iCount2, iM;
    float  fW, fW0, fW1, fA, fB, fH0, fHF;

    _clearscreen (_GCLEARSCREEN);

    printf ("PROGRAM 13: ");
    printf ("Nonrecursive filter design by Fourier\n");
    printf ("Transform method. Impulse response ");
    printf ("truncated to\n(2M+1) terms\n\n");

    printf ("Enter center-frequency (in degrees) :");
    scanf ("%f", &fA);
    printf ("Enter filter bandwidth (in degrees) :");
    scanf ("%f", &fB);
    printf ("Enter (integer) value of M :");
    scanf ("%d", &iM);

    fW0 = fA * 3.141593 / 180.0;
    fW1 = fB * 0.5 * 3.141593 / 180.0;

    _clearscreen (_GCLEARSCREEN);

    printf ("Impulse response values H(0) to H(M) are:\n\n");

    fH0 = fW1 / 3.141593;
    printf ("%10.5f", fH0);

    for (iCount = 1; iCount <= iM; iCount++)
        {
        fH[iCount] = (1.0 / (((float)iCount) * 3.141593)) *
                     sin (((float)iCount) * fW1) *
                     cos (((float)iCount) * fW0);
        printf ("%10.5f", fH[iCount]);
        }

    printf ("\nPress any key to plot frequency response...\n");
    getch();

    _setvideomode (_MRES16COLOR);
```

```c
        for (iCount = 1; iCount <= 320; iCount++)
            {
            fW = 3.141593 * (float)(iCount - 1) / 320.0;
            fHF = fH0;

            for (iCount2 = 1; iCount2 <= iM; iCount2++)
                fHF = fHF + 2.0 * fH[iCount2] *
                cos (((float)iCount2) * fW);

            _moveto (iCount, 100);
            _lineto (iCount, (int)(100 - 70.0 * fabs (fHF) + 0.5));
            }

        getch();
        _setvideomode( _DEFAULTMODE );

        /*
         ************************************************************
         */
        }




#include <graph.h>
#include <conio.h>
#include <stdio.h>
#include <stdlib.h>
#include <math.h>

#define XDIM        320
#define YDIM        320
#define HDIM         21

float   fX[XDIM + 1], fY[YDIM + 1];
float   fH[HDIM] = {-0.007958, -0.025009, -0.017229,  0.021962,
                     0.053052,  0.030746, -0.034458, -0.075026,
                    -0.039789,  0.041192,  0.083333,  0.041192,
                    -0.039789, -0.075026, -0.034458,  0.030746,
                     0.053052,  0.021962, -0.017229, -0.025009,
                    -0.007958};

void main (void)
    {
    /*
     ******************** PROGRAM NO.14 *********************
     ************** TIME-DOMAIN PERFORMANCE OF ***************
     ********* 21-TERM LINEAR-PHASE BANDPASS FILTER *********
     */
    int     iCount, iCount2;

    _clearscreen (_GCLEARSCREEN);

    printf ("PROGRAM 14: Time-domain performance of 21-term\n");
    printf ("linear-phase bandpass filter\n");
    printf ("Press any key to start...\n");
    getch();

    /*
     ****** LOAD INPUT ARRAY WITH TWO SIGNAL COMPONENTS *******
     */
    for (iCount = 108; iCount <= 180; iCount++)
        fX[iCount] = cos ((float)iCount * 5.0 * 3.141593 / 18.0) +
                     cos ((float)iCount * 7.0 * 3.141593 / 18.0);

    /*
     *********************** PLOT ***********************
     */
    _setvideomode (_MRES16COLOR);
```

```
     for (iCount = 1; iCount <= XDIM; iCount++)
        {
        _moveto (iCount, 30);
        _lineto (iCount, (int)(30 - 10.0 * fX[iCount] + 0.5));
        }

/*
 ***** ADD TWO OUT-OF-BAND COMPONENTS TO INPUT SIGNAL ******
 */
     for (iCount = 1; iCount <= XDIM; iCount++)
        fX[iCount] += sin ((float)iCount * 3.141593 / 8.0) +
                      sin ((float)iCount * 3.141593 / 1.8);

/*
 ********************** PLOT ***************************
 */
     for (iCount = 1; iCount <= XDIM; iCount++)
        {
        _moveto (iCount, 90);
        _lineto (iCount, (int)(90 - 10.0 * fX[iCount] + 0.5));
        }

/*
 ** CONVOLVE INPUT SIGNAL AND IMPULSE RESPONSE AND PLOT ****
 */
     for (iCount = 22; iCount <= YDIM; iCount++)
        {
        fY[iCount] = 0;
        for (iCount2 = 0; iCount2 <= 20; iCount2++)
            fY[iCount] += fH[iCount2] * fX[iCount - iCount2];

        _moveto (iCount, 150);
        _lineto (iCount, (int)(150 - 25.0 * fY[iCount] + 0.5));
        }

getch();
_setvideomode (_DEFAULTMODE);

/*
 **************************************************************
 */
}
```

```c
#include <graph.h>
#include <conio.h>
#include <stdio.h>
#include <stdlib.h>
#include <math.h>

void main (void)
    {
    /*
     ********************* PROGRAM NO.15 ********************
     ** SPECTRAL PROPERTIES OF RECTANGULAR WINDOW (LOG PLOT) ***
     */
    int     iCount, iCount2, iM;
    float   fFREQ, fH, fDB;

    _clearscreen  (_GCLEARSCREEN);

    printf ("PROGRAM 15: Spectral properties of rectangular\n");
    printf ("window (log plot)\n\n");

    printf ("Enter (integer) value of M :");
    scanf ("%d", &iM);

    _setvideomode( _MRES16COLOR );

    /*
     ************* DRAW RECTANGULAR GRID FOR PLOT ************
     */
    for (iCount = 0; iCount <= 5; iCount++)
        {
        _moveto (0, 30 + 20 * iCount);
        _lineto (320, 30 + 20 * iCount);
        _moveto (64 * iCount, 30);
        _lineto (64 * iCount, 130);
        }

    _moveto (319, 30);
    _lineto (319, 130);

    /*
     ***** ESTIMATE SPECTRUM, CONVERT TO DECIBELS AND PLOT *****
     */
    for (iCount = 1; iCount <= 320; iCount++)
        {
        fFREQ = 3.141593 * (float)(iCount - 1) / 320.0;
        fH = 1.0;
        for (iCount2 = 1; iCount2 <= iM; iCount2++)
            fH += 2.0 * cos ((float)iCount2 * fFREQ);

        fH /= (1.0 + 2.0 * (float)iM);
        fDB = 20.0 * (float) log (fabs (fH)) * 0.4343;

        if (fDB < -50.0)
            fDB = -50.0;

        _moveto (iCount, 130);
        _lineto (iCount, (int)(30 - 2.0 * fDB + 0.5));
        }

    getch();
    _setvideomode( _DEFAULTMODE );

    /*
     **********************************************************
     */
    }
```

```c
#include <graph.h>
#include <conio.h>
#include <stdio.h>
#include <stdlib.h>
#include <math.h>

#define SEEDIM    200
#define WDIM      320

float fSEE[SEEDIM + 1], fW[WDIM + 1];

void main (void)
    {
    /*
     ********************** PROGRAM NO.16 **********************
     **** SPECTRAL PROPERTIES OF VON HANN & HAMMING WINDOWS ****
     */
    int    iCount, iCount2, iX, iM;
    float  fA, fB, fC, fFREQ, fPEAK, fDB;

    _clearscreen  (_GCLEARSCREEN);

    printf ("PROGRAM 16: Spectral properties of von Hann ");
    printf ("& Hamming windows\n\n");

    printf ("Select window type: 1 = von Hann, 2 = Hamming :");
    scanf ("%d", &iX);

    printf ("\n Enter (integer) value of M :");
    scanf ("%d", &iM);

    /*
     **** LOAD ONE HALF OF WINDOW (EXCLUDING CENTRAL VALUE) ****
     ********************INTO ARRAY fSEE ***********************
     */
    if (iX == 1)
        {fA = 0.5; fB = 0.5; fC = (float)iM + 1.0;}
    else
        {fA = 0.54; fB = 0.46; fC = (float)iM;}

    for (iCount = 1; iCount <= iM; iCount++)
        fSEE[iCount] =
            fA + fB * cos ((float)iCount * 3.141593 / fC);

    _setvideomode( _MRES16COLOR );

    /*
     ************** DRAW RECTANGULAR GRID FOR PLOT ************
     */
    for (iCount = 0; iCount <= 5; iCount++)
        {
        _moveto (0, 30 + 20 * iCount);
        _lineto (320, 30 + 20 * iCount);
        _moveto (64 * iCount, 30);
        _lineto (64 * iCount, 130);
        }

    _moveto (319, 30);
    _lineto (319, 130);

    /*
     ********************** ESTIMATE SPECTRUM ****************
     */
    for (iCount = 1; iCount <= 320; iCount++)
        {
        fFREQ = 3.141593 * (float)(iCount - 1) / 320.0;
        fW[iCount] = 1.0;
        for (iCount2 = 1; iCount2 <= iM; iCount2++)
            fW[iCount] += 2.0 * fSEE[iCount2] *
                          cos ((float)iCount2 * fFREQ);
```

```
    /*
     *** NORMALISE TO PEAK VALUE, CONVERT TO DECIBELS & PLOT ***
     */
        if (iCount == 1)
            fPEAK = fW[iCount];

        fW[iCount] /= fPEAK;
        fDB = 20.0 * (float) log (fabs (fW[iCount])) * 0.4343;

        if (fDB < -50.0)
            fDB = -50.0;

        _moveto (iCount, 130);
        _lineto (iCount, (int)(30 - 2.0 * fDB + 0.5));
        }

    getch();
    _setvideomode( _DEFAULTMODE );

    /*
     ***********************************************************
     */
    }




#include <graph.h>
#include <conio.h>
#include <stdio.h>
#include <stdlib.h>
#include <math.h>

#define SEEDIM    200
#define HDIM      200
#define WDIM      320

float fSEE[SEEDIM + 1], fH[HDIM + 1], fW[WDIM + 1];

void main (void)
    {
    /*
     ***************** PROGRAM NO.17 *************************
     * NONRECURSIVE FILTER DESIGN BY FOURIER TRANSFORM METHOD
     ***** WITH RECTANGULAR, VON HANN, OR HAMMING WINDOW *****
     *** AND DECIBEL PLOT OF FREQUENCY RESPONSE MAGNITUDE ****
     *********** IMPULSE RESPONSE HAS (2M+1) TERMS ***********
     */
    int    iCount, iCount2, iX, iM;
    float  fA, fB, fH0, fW0, fW1, fC, fFREQ, fMax, fDB;

    _clearscreen  (_GCLEARSCREEN);

    printf ("PROGRAM 17: Nonrecursive filter design by ");
    printf ("Fourier\nTransform method with rectangular, ");
    printf ("von Hann or\nHamming window and decibel plot ");
    printf ("of frequency response\nmagnitude. Impulse ");
    printf ("response has (2M+1) terms\n\n");
    printf ("Enter center-frequency (in degrees) :");
    scanf ("%f", &fA);

    printf ("\nEnter filter bandwidth (in degrees) :");
    scanf ("%f", &fB);

    printf ("\nEnter (integer) value of M :");
    scanf ("%d", &iM);
```

```
printf ("\nSelect window:\n");
printf ("0 = Rectangular; 1 = von Hann; 2 = Hamming :");
scanf ("%d", &iX);

fW0 = fA * 3.141593 / 180.0;
fW1 = fB * 0.5 * 3.141593 / 180.0;

/*
 ****************** COMPUTE WINDOW VALUES ******************
 */
switch (iX)
    {
    case 0 :
        fA = 1; fB = 0;
        fC = 1;
        break;

    case 1 :
        fA = 0.5; fB = 0.5;
        fC = (float)iM + 1.0;
        break;

    case 2 :
        fA = 0.54; fB = 0.46;
        fC = (float)iM;
        break;
    }

for (iCount = 1; iCount <= iM; iCount++)
    fSEE[iCount] = fA + fB * cos ((float)iCount *


                        3.141593 / fC);

/*
 *********** COMPUTE IMPULSE RESPONSE VALUES **************
 */
fH0 = fW1 / 3.141593;

for (iCount = 1; iCount <= iM; iCount++)
    fH[iCount] = (1.0 / ((float)iCount *   3.141593)) *
                    sin ((float)iCount * fW1) *
                    cos ((float)iCount * fW0) *
                    fSEE[iCount];

_setvideomode( _MRES16COLOR );

/*
 ************** DRAW RECTANGULAR GRID FOR PLOT ************
 */
for (iCount = 0; iCount <= 5; iCount++)
    {
    _moveto (0, 30 + 20 * iCount);
    _lineto (320, 30 + 20 * iCount);
    _moveto (64 * iCount, 30);
    _lineto (64 * iCount, 130);
    }

_moveto (319, 30);
_lineto (319, 130);

/*
 **************** COMPUTE FREQUENCY RESPONSE *************
 */
for (iCount = 1; iCount <= WDIM; iCount++)
```

```
{
fFREQ = 3.141593 * (float)(iCount - 1) / 320.0;
fW[iCount] = fH0;
for (iCount2 = 1; iCount2 <= iM; iCount2++)
    fW[iCount] += 2.0 * fH[iCount2] *
                    cos ((float)iCount2 * fFREQ);
}
/*
 ***** NORMALISE TO UNITY, CONVERT TO DECIBELS & PLOT ******
 */
fMax = 0;

for (iCount = 1; iCount <= WDIM; iCount++)
    if (fabs (fW[iCount]) > fMax)
        fMax = fabs (fW[iCount]);

for (iCount = 1; iCount <= WDIM; iCount++)
    {
    fDB = 20.0 * log (fabs (fW[iCount]) / fMax) * 0.4343;

    if (fDB < -50.0)
        fDB = -50.0;

    _moveto (iCount, 130);
    _lineto (iCount, (int)(30 - 2.0 * fDB + 0.5));
    }

printf ("Press any key to continue...");
getch();
_setvideomode( _DEFAULTMODE );

printf ("Impulse response values h(0) to h(M)\n");
printf ("(corrected for unity maximum gain) :\n\n");

printf ("%18.5f", (fH0 / fMax));

for (iCount = 1; iCount <= iM; iCount++)
    printf (", %18.5f", (fH[iCount] / fMax));

printf("\n");

/*
 **********************************************************
 */
}
```

```c
#include <graph.h>
#include <conio.h>
#include <stdio.h>
#include <stdlib.h>
#include <math.h>

#define SEEDIM      200
#define HDIM        200
#define WDIM        320

float fSEE[SEEDIM + 1], fH[HDIM + 1], fW[WDIM + 1];

void main (void)
    {
    /*
     ******************* PROGRAM NO.18 *************************
     **** NONRECURSIVE FILTER DESIGN WITH THE KAISER WINDOW ****
     */
    int     iCount, iCount2, iX, iM;
    float   fM, fP, fQ, fR, fS, fT, fA, fA0, fB, fC, fD,
            fALPHA, fBETA, fTEMP, fI,
            fH0, fW0, fW1, fFREQ, fMax, fDB;

    _clearscreen   (_GCLEARSCREEN);
    printf ("PROGRAM 18: Nonrecursive filter design with\n");
    printf ("the Kaiser window\n\n");

    while (1)
        {
        printf ("Enter center-frequency (in degrees) :");
        scanf ("%f", &fP);

        printf ("Enter filter bandwidth (in degrees) :");
        scanf ("%f", &fQ);

        printf ("Enter Kaiser design parameters\n");
        printf ("  Ripple (as a fraction) :");
        scanf ("%f", &fR);

        printf ("\nTransition width (in degrees) :");
        scanf ("%f", &fS);

        fW0 = fP * 3.141593 / 180.0;
        fW1 = fQ * 0.5 * 3.141593 / 180.0;
        fT = fS / 360.0;

        /*
         *************** ESTIMATE KAISER PARAMETERS ************
         */
        fA = -20.0 * log (fR) * 0.4343;
        fM = (fA - 7.95) / (28.72 * fT);
        iM = ((int)(fM+1));

        printf ("\n  No. of impulse response terms = %d\n",
                            2 * iM + 1);
        printf ("\n  Change Parameters, or continue? (0/1) :");
        scanf ("%d", &iX);

        if (iX)
            break;
        }

    if (fA > 49.0)
        fALPHA = 0.1102 * (fA - 8.7);
    else if (fA > 21)
        fALPHA = 0.5842 * pow((fA - 21.0), 0.4) +
                    (0.07886 * (fA - 21.0));
    else
        fALPHA = 0;
```

```c
printf ("\nFor this window, ALPHA = %f\n", fALPHA);
printf ("Press any key to start...");
getch();

/*
 ****************** COMPUTE WINDOW VALUES *****************
 */
for (iCount = 0; iCount <= iM; iCount++)
    {
    fTEMP = 1.0 - pow (((float) iCount / (float) iM), 2.0);
    fBETA = fALPHA * pow (fTEMP, 0.5);
    fI = fB = 1.0;

    for (iCount2 = 1; iCount2 <= 20; iCount2++)
        {
        fC = fB * fBETA * fBETA / 4.0;
        fD = fC / (float)(iCount2 * iCount2);
        fI += fD;
        fB = fD;
        }

    fSEE[iCount] = fI;
    }

fA0 = fSEE[0];
for (iCount = 0; iCount <= iM; iCount++)
    fSEE[iCount] /= fA0;

/*
 *************** COMPUTE IMPULSE RESPONSE VALUES ***********
 */
fH0 = fW1 / 3.141593;

for (iCount = 1; iCount <= iM; iCount++)
    fH[iCount] = (1.0 / ((float)iCount * 3.141593)) *
                 sin ((float)iCount * fW1) *
                 cos ((float)iCount * fW0) *
                 fSEE[iCount];

_setvideomode( _MRES16COLOR );

/*
 ************** DRAW RECTANGULAR GRID FOR PLOT ************
 */
for (iCount = 0; iCount <= 5; iCount++)
    {
    _moveto (0, 30 + 20 * iCount);
    _lineto (320, 30 + 20 * iCount);
    _moveto (64 * iCount, 30);
    _lineto (64 * iCount, 130);
    }

_moveto (319, 30);
_lineto (319, 130);

/*
 *************** COMPUTE FREQUENCY RESPONSE **************
 */
for (iCount = 1; iCount <= WDIM; iCount++)
    {
    fFREQ = 3.141593 * (float)(iCount - 1) / 320.0;
    fW[iCount] = fH0;
    for (iCount2 = 1; iCount2 <= iM; iCount2++)
        fW[iCount] += 2.0 * fH[iCount2] *
                      cos ((float)iCount2 * fFREQ);
    }

/*
 ****** NORMALISE TO UNITY, CONVERT TO DECIBELS & PLOT *****
 */
```

```
fMax = 0;

for (iCount = 1; iCount <= WDIM; iCount++)
    if (fabs (fW[iCount]) > fMax)
        fMax = fabs (fW[iCount]);

for (iCount = 1; iCount <= WDIM; iCount++)
    {
    fDB = 20.0 * log (fabs (fW[iCount]) / fMax) * 0.4343;

    if (fDB < -50.0)
        fDB = -50.0;

    _moveto (iCount, 130);
    _lineto (iCount, (int)(30 - 2.0 * fDB + 0.5));
    }

printf ("Press any key to continue...");
getch();

_setvideomode( _DEFAULTMODE );

printf ("Impulse response values h(0) to h(M)\n");
printf ("(corrected for unity maximum gain) :\n\n");

printf ("%18.5f", (fH0 / fMax));

for (iCount = 0; iCount < iM; iCount++)
    printf (", %18.5f", (fH[iCount] / fMax));

/*
 ************************************************************
 */
}
```

```c
#include <graph.h>
#include <conio.h>
#include <stdio.h>
#include <stdlib.h>
#include <math.h>

#define SEEDIM     200
#define HDIM       200

float fSEE[SEEDIM + 1], fH[HDIM + 1];

void main (void)
    {
    /*
     ********** PROGRAM NO. 19   DIGITAL DIFFERENTIATOR *******
     */
    int     iCount, iCount2, iX, iM;
    float   fA, fB, fC, fW, fHF;

    _clearscreen  (_GCLEARSCREEN);
    printf ("\n\nPROGRAM 19: Digital differentiator\n\n");

    printf ("Enter (integer) value of M :");
    scanf ("%d", &iM);

    printf ("Select window type: 1 Rectangular, 2 Hamming :");
    scanf ("%d", &iX);

    /*
     ******************** COMPUTE WINDOW VALUES ****************
     */
    if (iX == 1)
        {fA = 1.0; fB = 0; fC = 1.0;}
    else
        {fA = 0.54; fB = 0.46; fC = (float)iM;}

    for (iCount = 1; iCount <= iM; iCount++)
        fSEE[iCount] = fA + fB * cos ((float)iCount *
                    3.141593 / fC);

    /*
     **************** COMPUTE IMPULSE RESPONSE ****************
     */
    for (iCount = 1; iCount <= iM; iCount++)
        fH[iCount] = (1.0 / (float)iCount) *
                        pow (-1.0, iCount) * fSEE[iCount];

    /*
     **** COMPUTE AND PLOT FREQUENCY RESPONSE (LINEAR SCALE) ***
     */
    _setvideomode( _MRES16COLOR );

    for (iCount = 1; iCount <= 320; iCount++)
        {
        fW = (float)(iCount - 1) * 3.141593 / 320.0;
        fHF = 0.0;
        for (iCount2 = 1; iCount2 <= iM; iCount2++)
            fHF -= 2.0 * fH[iCount2] *
                    sin ((float) iCount2 * fW);

        _moveto (iCount, 150);
        _lineto (iCount, (int)(150.0 - 20.0 * fHF +0.5));
        }

    printf ("Press any key to continue...");
    getch();

    _setvideomode( _DEFAULTMODE );

    printf ("Remember that h(0) is ZERO!\n\n");
```

```
        printf ("Values h(1) to h(M) are :\n\n");

        for (iCount = 1; iCount <= iM; iCount++)
            printf ("%18.5f, ", fH[iCount]);

    /*
     ***************************************************************
     */
    }




#include <graph.h>
#include <conio.h>
#include <stdio.h>
#include <stdlib.h>
#include <math.h>

#define RPDIM      20
#define RZDIM      20
#define CPDIM      20
#define CZDIM      20
#define HDIM       320

float fH[HDIM + 1], fRP[RPDIM + 1][2], fRZ[RZDIM + 1][2],
        fCP[CPDIM + 1][2], fCZ[CZDIM + 1][2];

void main (void)
    {
    /*
     **** PROGRAM NO. 20  FILTER MAGNITUDE RESPONSE ************
     ****** FROM Z-PLANE POLE AND ZERO LOCATIONS **************
     */
    int    iCount, iCount2, iNRP, iNRZ, iNCP, iNCZ;
    float  fW, fC1, fC2, fS1, fS2, fA, fB, fR, fT, fD, fE,
           fMAX, fDB;

    _clearscreen  (_GCLEARSCREEN);
    printf ("\n\nPROGRAM 20: Filter magnitude response from ");
    printf ("z-plane\npole and zero locations\n\n");

    printf ("Enter no. of SEPARATE real poles :");
    scanf ("%d", &iNRP);

    if (iNRP != 0)
        {
        printf ("\nEnter value, order of each in turn :\n");
        printf ("   Followed by <Return>.\n");

        for (iCount = 1; iCount <= iNRP; iCount++)
            scanf ("%f, %f", &fRP[iCount][0], &fRP[iCount][1]);
        }

    printf ("\nEnter no. of SEPARATE real zeros :");
    scanf ("%d", &iNRZ);

    if (iNRZ != 0)
        {
        printf ("\nEnter value, order of each in turn :\n");
        printf ("   Followed by <Return>.\n");

        for (iCount = 1; iCount <= iNRZ; iCount++)
            scanf ("%f, %f", &fRZ[iCount][0], &fRZ[iCount][1]);
        }
```

```c
printf ("\nEnter no. of complex-conjugate pole pairs :");
scanf ("%d", &iNCP);

if (iNCP != 0)
    {
    printf ("\nEnter radius, angle (in degrees) of ");
    printf ("each in turn :\n");
    printf ("   Followed by <Return>.\n");

    for (iCount = 1; iCount <= iNCP; iCount++)
        scanf ("%f, %f", &fCP[iCount][0], &fCP[iCount][1]);
    }

printf ("\nEnter no. of complex-conjugate zero pairs :");
scanf ("%d", &iNCZ);

if (iNCZ != 0)

{
printf ("\nEnter radius, angle (in degrees) of ");
printf ("each in turn :\n");
printf ("   Followed by <Return>.\n");

for (iCount = 1; iCount <= iNCZ; iCount++)
    scanf ("%f, %f", &fCZ[iCount][0], &fCZ[iCount][1]);
}

/*
 ********** COMPUTE FREQUENCY RESPONSE MAGNITUDE **********
 */
for (iCount = 1; iCount <= HDIM; iCount++)
    {
    fW = 3.141593 * (float)(iCount - 1) / HDIM;
    fC1 = cos (fW);
    fC2 = cos (2.0 * fW);
    fS1 = sin (fW);
    fS2 = sin (2.0 * fW);
    fH[iCount] = 1.0;

    if (iNRP != 0)
        {
        for (iCount2 = 1; iCount2 <= iNRP; iCount2++)
            {
            fA = fRP[iCount2][0];
            fB = fRP[iCount2][1];
            fH[iCount] /= pow ((1.0 - 2.0 * fA * fC1 + fA * fA),
                              (fB / 2.0));
            }
        }

    if (iNRZ != 0)
        {
        for (iCount2 = 1; iCount2 <= iNRZ; iCount2++)
            {
            fA = fRZ[iCount2][0];
            fB = fRZ[iCount2][1];
            fH[iCount] *= pow ((1.0 - 2.0 * fA * fC1 + fA * fA),
                              (fB / 2.0));
            }
        }

    if (iNCP != 0)
        {
        for (iCount2 = 1; iCount2 <= iNCP; iCount2++)
            {
            fR = fCP[iCount2][0];
            fT = fCP[iCount2][1] * 3.141593 / 180.0;
            fD = fC2 - 2.0 * fR * cos (fT) * fC1 + fR * fR;
            fE = fS2 - 2.0 * fR * cos (fT) * fS1;

            fH[iCount] /= pow ((fD * fD + fE * fE), 0.5);
            }
```

```
    if (iNCZ != 0)
        {
        for (iCount2 = 1; iCount2 <= iNCZ; iCount2++)
            {
            fR = fCZ[iCount2][0];
            fT = fCZ[iCount2][1] * 3.141593 / 180.0;
            fD = fC2 - 2.0 * fR * cos (fT) * fC1 + fR * fR;
            fE = fS2 - 2.0 * fR * cos (fT) * fS1;

            fH[iCount] *= pow ((fD * fD + fE * fE), 0.5);
            }
        }
    }

/*
 ************** FIND MAXIMUM VALUE AND PRINT OUT **********
 */
fMAX = 0.0;

for (iCount = 1; iCount <= HDIM; iCount++)
    if (fabs (fH[iCount] > fMAX))
        fMAX = fabs (fH[iCount]);

printf ("\nMaximum gain = %f (%f DB)\n", fMAX,
        20.0 * log (fMAX) * 0.4343);

printf ("Press any key to plot...");
getch ();
_setvideomode( _MRES16COLOR );

/*
 **************** DRAW RECTANGULAR GRID FOR PLOT **********
 */
for (iCount = 0; iCount <= 5; iCount++)
    {
    _moveto (0, 30 + 20 * iCount);
    _lineto (320, 30 + 20 * iCount);
    _moveto (64 * iCount, 30);
    _lineto (64 * iCount, 130);
    }

_moveto (319, 30);
_lineto (319, 130);

/*
 * NORMALISE RESPONSE TO UNITY, CONVERT TO DECIBELS & PLOT *
 */
for (iCount = 1; iCount <= HDIM; iCount++)
    {
    if (fabs (fH[iCount]) < 0.000001)
        fH[iCount] = 0.000001;

    fDB = 20.0 * log (fabs (fH[iCount]) / fMAX) * 0.4343;

    if (fDB < -50.0)
        fDB = -50.0;

    _moveto (iCount, 130);
    _lineto (iCount, (int)(30.0 - 2.0 * fDB + 0.5));
    }

getch();
_setvideomode( _DEFAULTMODE );

/*
 ************************************************************
 */
}
```

```c
#include <graph.h>
#include <conio.h>
#include <stdio.h>
#include <stdlib.h>
#include <math.h>

#define PRDIM     50
#define PIDIM     50

double fPR[PRDIM + 1], fPI[PIDIM + 1];

void main (void)
    {
    /*
     ******** PROGRAM NO. 21    POLES AND ZEROS OF *************
     ****** BUTTERWORTH AND CHEBYSHEV DIGITAL FILTERS **********
     */
    int    iCount, iFF, iFT, iN, iIN, iN1, iN2, iM, iST;
    double fF1, fF2, fF3, fF4, fF5, fW1, fB1, fB2,
           fC1, fC2, fC3, fC4, fC5, fD, fE, fY, fX, fA, fB, fT,
           fR, fI, fSI, fB3, fGR, fAR, fFR, fSR, fFI, fAI, fGI,
           fH1, fH2, fP1I, fP2I, fP1R, fP2R, fTH;

    _clearscreen  (_GCLEARSCREEN);
    printf ("\n\nPROGRAM 21: Poles and zeros of Butterworth and\n");
    printf ("   Chebyshev digital filters\n\n");

    printf ("Choose filter family: 1 = Butterworth, ");
    printf ("2 = Chebyshev :");
    scanf ("%d", &iFF);

    printf ("\nChoose filter type: 1 = Lowpass, 2 = Highpass, ");
    printf ("3 = Bandpass :");
    scanf ("%d", &iFT);

    printf ("\nFilter order :");
    scanf ("%d", &iN);

        switch (iFT)
          {
          case 1 :
          case 2 :
              printf ("\nCut-off frequency (degrees) :");
              scanf ("%lf", &fF1);
              iST = 2;

              if (iFT == 2)
                  fF1 = 180.0 - fF1;

              break;

          case 3 :
              iN /= 2;
              printf ("\nLower, Upper cut-off frequencies ");
              printf ("(degrees) :");
              scanf ("%lf, %lf", &fF2, &fF3);
              fF1 = fF3 - fF2;
              iST = 1;
          }

      if (iFF == 2)
          {
          printf ("\n\nPassband ripple (fraction) :");
          scanf ("%lf", &fD);
          }

    /*
     *********** FIND BUTTERWORTH/CHEBYSHEV PARAMETERS *********
     */
    iIN = iN % 2;
```

```
        iN1 = iN + iIN;
        iN2 = (3 * iN + iIN) / 2 - 1;
        fW1 = 3.141593 * fF1 / 360.0;
        fC1 = sin (fW1) / cos (fW1);
        fB1 = 2.0 * fC1;
        fC2 = fC1 * fC1;
        fB2 = 0.25 * fB1 * fB1;

        if (iFF == 2)
            {
            fE = 1.0 / sqrt (1.0 / ((1.0 - fD) * (1.0 - fD)) - 1.0);
            fX = pow (sqrt (fE * fE + 1.0) + fE, 1.0 / (double)iN);
            fY = 1.0 / fX;
            fA = 0.5 * (fX - fY);
            fB = 0.5 * (fX + fY);
            }

/*
 **** FIND REAL AND IMAGINARY PARTS OF LOW-PASS POLES ******
 */
        for (iCount = iN1; iCount <= iN2; iCount++)
            {
            fT = 3.141593 * (2.0 * (double)iCount + 1.0 -
                (double) iIN) / (2.0 * iN);

            if (iFF == 2)
                {
                fC3 = fA * fC1 * cos (fT);
                fC4 = fB * fC1 * sin (fT);
                fC5 = pow ((1.0 - fC3), 2.0) + (fC4 * fC4);
                fR = 2.0 * (1.0 - fC3) / fC5 - 1.0;
                fI = 2.0 * fC4 / fC5;
                }
            else
                {
                fB3 = 1.0 - fB1 * cos (fT) + fB2;
                fR = (1.0 - fB2) / fB3;
                fI = fB1 * sin (fT) / fB3;
                }

            iM = (iN2 - iCount) * 2 + 1;

            fPR[iM + iIN] = fR;
            fPI[iM + iIN] = fI;
            fPR[iM + iIN + 1] = fR;
            fPI[iM + iIN + 1] = (0.0 - fI);
            }

        if (iIN != 0)
            {
            if (iFF == 1)
                fR = (1.0 - fB2) / (1.0 + fB1 + fB2);
            else
                fR = 2.0 / (1.0 + fA * fC1) - 1.0;

            fPR[1] = fR;
            fPI[1] = 0.0;
            }

        switch (iFT)
            {
            /*
             ********** PRINT OUT Z-PLANE ZERO LOCATIONS **********
             */
            case 1 :
                printf ("\nReal zero, of order %d at z= -1\n", iN);
                break;

            case 2 :
                printf ("\nReal zero, of order %d at z = 1\n", iN);
```

```
              for (iCount = 1; iCount <= iN; iCount++)
                  fPR[iCount] = 0.0 - fPR[iCount];
              break;

         case 3 :
             printf ("\nReal zeros, of order %d at z = 1 and ", iN);
             printf ("z = -1\n");

             /*
              ******* LOW-PASS TO BANDPASS TRANSFORMATION *******
              */
             fF4 = fF2 * 3.141593 / 360.0;
             fF5 = fF3 * 3.141593 / 360.0;
             fA = cos (fF4 + fF5) / cos (fF5 - fF4);
             for (iCount = 1; iCount <= 50; iCount += 2)
                 {
                     fAR = fPR[iCount];
                     fAI = fPI[iCount];
                     if (fabs (fAI) >= 0.0001)
                         {
                         fFR = fA * 0.5 * (1.0 + fAR);
                         fFI = fA * 0.5 * fAI;
                         fGR = fFR * fFR - fFI * fFI - fAR;
                         fGI = 2.0 * fFR * fFI - fAI;
                         fSR = pow (fabs (fGR +
                                 pow ((fGR * fGR + fGI * fGI),
                                 0.5)) / 2.0, 0.5);
                         fSI = fGI / (2.0 * fSR);
                         fP1R = fFR + fSR;
                         fP1I = fFI + fSI;
                         fP2R = fFR - fSR;
                         fP2I = fFI - fSI;
                         }
                     else
                         {
                         fH1 = fA * (1.0 + fAR) / 2.0;
                         fH2 = fH1 * fH1 - fAR;
                         if (fH2 > 0)
                             {
                             fP1R = fH1 + pow (fH2, 0.5);
                             fP2R = fH1 - pow (fH2, 0.5);
                             fP1I = 0;
                             fP2I = 0;
                             }
                         else
                             {
                             fP1R = fH1;
                             fP2R = fH1;
                             fP1I = pow (fabs (fH2), 0.5);
                             fP2I = -1.0 * fP1I;
                             }
                         }

                     fPR[iCount] = fP1R;
                     fPR[iCount + 1] = fP2R;
                     fPI[iCount] = fP1I;
                     fPI[iCount + 1] = fP2I;
                     }
         }
    /*
     ************ PRINT OUT Z-PLANE POLE LOCATIONS ************
     */
    printf ("\nRadii, angles of z-plane poles:\n\n");
    printf ("\n     R        THETA\n\n");

    for (iCount = 1; iCount <= iN; iCount += iST)
        {
        iM = iCount;
        if (iIN != 0)
            if (iCount == 2)
```

```
            iM = iN + 1;

        fR = pow (fPR[iM] * fPR[iM] + fPI[iM] * fPI[iM], 0.5);
        fTH = atan (fabs (fPI[iM]) / fabs (fPR[iM])) *
              (180.0 / 3.141593);
        if (fPR[iM] < 0.0)
            fTH = 180 - fTH;
        printf ("%lf %lf\n", fR, fTH);
        }

   /*
    *****************************************************************
    */
   }




#include <graph.h>
#include <conio.h>
#include <stdio.h>
#include <stdlib.h>
#include <math.h>

#define DIM     320

float fX[DIM + 1], fY[DIM + 1], fP[DIM + 1], fQ[DIM + 1],
      fR[DIM + 1], fS[DIM + 1], fT[DIM + 1], fU[DIM + 1],
      fV[DIM + 1], fW[DIM + 1];

void main (void)
    {
    /*
     *** PROGRAM NO. 22     FREQUENCY-SAMPLING FILTER **********
     */
    int    iCount, iCount2, iM;
    float  fFREQ, fHI, fH, fHR;

    _clearscreen  (_GCLEARSCREEN);
    printf ("PROGRAM 22: Frequency sampling filter\n");

    fX[120] = 1.0;

    /*
     ************** LOAD IMPULSE INTO INPUT ARRAY **************
     */
    _setvideomode( _MRES16COLOR );

    for (iCount = 120; iCount <= 280; iCount++)
        {
        fW[iCount] = fX[iCount] - 0.886867 * fX[iCount-120];
        fP[iCount] = 0.5171211 * fP[iCount-1] - 0.998001 *
                     fP[iCount-2] + 0.5 * fW[iCount];
        fQ[iCount] = 0.415408 * fQ[iCount-1] - 0.998001 *
                     fQ[iCount-2] - fW[iCount];
        fR[iCount] = 0.312556 * fR[iCount-1] - 0.998001 *
                     fR[iCount-2] + fW[iCount];
        fS[iCount] = 0.208848 * fS[iCount-1] - 0.998001 *
                     fS[iCount-2] - fW[iCount];
        fT[iCount] = 0.104567 * fT[iCount-1] - 0.998001 *
                     fT[iCount-2] + fW[iCount];
        fU[iCount] = -0.998001 * fU[iCount-2] - 0.666667 *
                     fW[iCount];
        fV[iCount] = -0.104567 * fV[iCount-1] - 0.998001 *
                     fV[iCount-2] + 0.333333 * fW[iCount];
        fY[iCount] = fP[iCount] + fQ[iCount] + fR[iCount] +
                     fS[iCount] + fT[iCount] + fU[iCount] +
                     fV[iCount];
```

```
    /*
     ************** PLOT IMPULSE RESPONSE VALUE **********
     */
    iM = (iCount - 120) * 2;
    _moveto (iM, 70);
    _lineto (iM, (int)(70 - fY[iCount] * 7.0 + 0.5));
    }

for (iCount = 1; iCount <= 320; iCount++)
    {
    fFREQ = (float) (iCount - 1) * 3.141593 / 320.0;
    fHR = 0.0;
    fHI = 0.0;
    fH = 0.0;

    for (iCount2 = 1; iCount2 <= 120; iCount2++)
        {
        fHR += fY[119 + iCount2] * cos ((float) iCount2 *
            fFREQ);
        fHI += fY[119 + iCount2] * sin ((float) iCount2 *
            fFREQ);
        }

    fH = pow ((fHR * fHR + fHI * fHI), 0.5);

    /*
     ************ PLOT FREQUENCY RESPONSE VALUE **********
     */
    _moveto (iCount, 170);
    _lineto (iCount, (int)(170 - 0.75 * fH + 0.5));
    }

getch();
_setvideomode( _DEFAULTMODE );

/*
 ************************************************************
 */
}
```

```
#include <graph.h>
#include <conio.h>
#include <stdio.h>
#include <stdlib.h>
#include <math.h>

#define DIM     64

float   fX[DIM + 1], fXXR[DIM + 1], fXXI[DIM + 1],
        fMAG[DIM + 1], fPHASE[DIM + 1], fC[DIM + 1], fS[DIM + 1];

/*
*** WHEN COMPARING WITH THE TEXT, NOTE THAT IN THE C PROGRAMS **
****** FLOATING-POINT VARIABLES ARE PREFIXED WITH AN f, *******
************** INTEGER VARIABLES WITH AN i ********************
*/

void main (void)
    {
    /*
     ***** PROGRAM NO. 23  DISCRETE FOURIER TRANSFORM (DFT) ****
     ******* BY DIRECT CALCULATION, AND WITH TABLE LOOKUP ******
     ****** REAL SIGNAL, N=64, SPECTRAL MAGNITUDE & PHASE ******
     */
    int    iCount, iCount2, iP, iI, iL;
    float  fMax1, fMax2, fMax3, fB, fW0;

    _clearscreen   (_GCLEARSCREEN);
    printf ("PROGRAM 23: Discrete Fourier Transform (DFT)\n");
    printf ("by direct calculation, and with table lookup.\n");
    printf ("Real signal, N=64, spectral magnitude & phase.\n");
    printf ("Press any key to start first DFT ...");
    getch();
    /*
     ******************** GENERATE SIGNAL ********************
     */
    for (iCount = 1; iCount <= DIM; iCount++)
        {
        fB = 2.0 * 3.141593 * (iCount - 1);
        fX[iCount] = sin (fB / 64.0) + cos (fB / 16.0) +
                    0.6 * cos (fB / 8.0) + 0.5 * sin (fB / 4.0);
        }

    for (iCount = 1; iCount <= DIM; iCount++)
        for (iCount2 = 1; iCount2 <= DIM; iCount2++)
            {
            fXXR[iCount] = fXXR[iCount] + fX[iCount2] *
                        cos (2.0 * 3.141593 * (float)(iCount - 1) *
                        (float)(iCount2 - 1) / 64.0);
            fXXI[iCount] = fXXI[iCount] - fX[iCount2] *
                        sin (2.0 * 3.141593 * (float)(iCount - 1) *
                        (float)(iCount2 - 1) / 64.0);
            }

    for (iP = 0; iP < 2; iP++)
        {
        _setvideomode( _MRES16COLOR );

        if (iP == 0)
            printf ("\n\nPress any key to start second DFT...");

        if (iP)
            {
            for (iCount = 1; iCount <= DIM; iCount++)
                {
                fXXR[iCount] = 0.0;
                fXXI[iCount] = 0.0;
                }
```

```
/******** FORM TABLE LOOKUP ************************

    fWO = 2.0 * 3.141593 / 64.0;

    for (iCount = 1; iCount <= DIM; iCount++)
        {
        fC[iCount] = cos (fWO * ((float) (iCount - 1)));
        fS[iCount] = sin (fWO * ((float) (iCount - 1)));
        }

    for (iCount = 1; iCount <= DIM; iCount++)
        for (iCount2 = 1; iCount2 <= DIM; iCount2++)
            {
            iI = (iCount -1) * (iCount2 - 1);
            iL = iI % 64;

            fXXR[iCount] = fXXR[iCount] + fX[iCount2] *
                fC[iL + 1];
            fXXI[iCount] = fXXI[iCount] - fX[iCount2] *
                fS[iL + 1];
            }

    }

/*
 ********* CONVERT TO SPECTRAL MAGNITUDE AND PHASE *****
 */
for (iCount = 1; iCount <= DIM; iCount++)
    {
    if (fabs (fXXI[iCount]) < 0.001)
        fXXI[iCount] = 0.0;
    if (fabs (fXXR[iCount]) < 0.00001)
        fXXR[iCount] = 0.00001;
    fMAG[iCount] = sqrt (pow (fXXR[iCount], 2.0) +
                    pow (fXXI[iCount], 2.0));
    fPHASE[iCount] = atan (fXXI[iCount] / fXXR[iCount]);
    }

/*
 ******* NORMALISE TO MAXIMUM OF UNITY FOR PLOT ********
 */
fMax1 = fMax2 = fMax3 = 0.001;

for (iCount = 1; iCount <= DIM; iCount++)
    {
    if (fabs (fX[iCount]) > fMax1)
        fMax1 = fabs (fX[iCount]);
    if (fabs (fMAG[iCount]) > fMax2)
        fMax2 = fabs (fMAG[iCount]);
    if (fabs (fPHASE[iCount]) > fMax3)
        fMax3 = fabs (fPHASE[iCount]);
    }

for (iCount = 1; iCount <= DIM; iCount++)
    {
    fX[iCount] /= fMax1;
    fMAG[iCount] /= fMax2;
    fPHASE[iCount] /= fMax3;
    }
```

```
/*
 ********PLOT SIGNAL, SPECTRAL MAGNITUDE AND PHASE *****
 */
for (iCount = 1; iCount <= DIM; iCount++)
    {
    _moveto (5 * (iCount - 1), 50);
    _lineto (5 * (iCount - 1), (int)(50 - fX[iCount] *
        25.0 + 0.5));
    _moveto (5 * (iCount - 1), 120);
    _lineto (5 * (iCount - 1), (int)(120 - fMAG[iCount] *
        30.0 + 0.5));

        _moveto (5 * (iCount - 1), 160);
        _lineto (5 * (iCount - 1), (int)(160 - fPHASE[iCount] *
            20.0 + 0.5));
        }

    getch();
    _setvideomode( _DEFAULTMODE );
    }
/*
 *************************************************************
 */
}
```

```
#include <graph.h>
#include <conio.h>
#include <stdio.h>
#include <stdlib.h>
#include <math.h>

#define DIM     512

float   fXR[DIM + 1], fXI[DIM + 1];

/*
*** WHEN COMPARING WITH THE TEXT, NOTE THAT IN THE C PROGRAMS **
******* FLOATING-POINT VARIABLES ARE PREFIXED WITH AN f, *******
************** INTEGER VARIABLES WITH AN i *********************
*/

void main (void)
    {
    /*
     ********* PROGRAM NO. 24   FAST FOURIER TRANSFORM (FFT) ***
     */
    int     iCount, iCount2, iCount3, iN = DIM, iM = 9,
            iT, iLim1, iLim2, iLim3, iJ, iL, iY;
    float   fMax, fMag, fB1, fB2, fC1, fC2, fD,
            fPI = 3.141593, fX1, fX2, fARG, fCos1, fSin1;

    _clearscreen (_GCLEARSCREEN);
    printf ("\n\nPROGRAM 24: Fast Fourier Transform (FFT)\n\n");
    printf ("Press any key to start...\n\n");
    getch();
```

```
for (iCount = 1; iCount <= iN; iCount++)
    fXR[iCount] = fXI[iCount] = 0.0;

for (iCount = 1; iCount <= 32; iCount++)
    fXR[iCount] = 1.0;

/*
 ******************* PLOT INPUT DATA *********************
 */
 (iY=1,iY=1);
 _setvideomode( _MAXRESMODE );

iY = 60;
for (iCount = 1; iCount <= iN; iCount++)
    {
    _moveto (iCount, iY);
    _lineto (iCount, (int)(iY - fXR[iCount] * 40.0 + 0.5));
    }

/*
 **** SELECT TRANSFORM/INVERSE TRANSFORM AND PLOT OFFSET ***
 */
for (iT = 1; iT > -2; iT -= 2)
    {
    iY += 60;
    if (iT < 0)
        fD = (float) iN;
    else
        fD = 1.0;

    /*
     **************** SHUFFLE INPUT DATA ******************
     */
    iLim1 = iN - 1;
    iLim2 = iN / 2;
    iJ = 1;

    for (iCount = 1; iCount <= iLim1; iCount++)
        {

        if ((float) iCount <= ((float) iJ - 0.01))
            {
            fX1 = fXR[iJ];
            fX2 = fXI[iJ];
            fXR[iJ] = fXR[iCount];
            fXI[iJ] = fXI[iCount];
            fXR[iCount] = fX1;
            fXI[iCount] = fX2;
            }
        iL = iLim2;
        while (1)
            {
            if ((float)iL > ((float)(iJ) - 0.01))
                 break;
            iJ -= iL;
            iL /= 2;
            }
        iJ += iL;
        }

  /*
   ************ IN PLACE TRANSFORMATION ****************
   */
  for (iCount = 1; iCount <= iM; iCount++)
      {
      iLim1 = (int) pow (2, iCount-1);
      iLim2 = (int) pow (2, (iM - iCount));
```

```
            for (iCount2 = 1; iCount2 <= iLim2; iCount2++)
                {
                for (iCount3 = 1; iCount3 <= iLim1; iCount3++)
                    {
                    iLim3 = (iCount3 - 1) + (iCount2 - 1) *
                        2 * iLim1 + 1;
                    fB1 = fXR[iLim3];
                    fB2 = fXI[iLim3];
                    fC1 = fXR[iLim3 + iLim1];
                    fC2 = fXI[iLim3 + iLim1];

                    fARG = 2.0 * fPI * (float) (iCount3 - 1) *
                        (float) iLim2 / (float) iN;
                    fCos1 = cos1 (fARG);
                    fSin1 = sin1 (fARG);
                    fX1 = fC1 * fCos1 + fC2 * fSin1 *
                        (float) iT;
                    fX2 = (0.0 - fC1) * fSin1 * (float) iT +
                        fC2 * fCos1;
                    fXR[iLim3] = fB1 + fX1;
                    fXI[iLim3] = fB2 + fX2;
                    fXR[iLim3 + iLim1] = fB1 - fX1;
                    fXI[iLim3 + iLim1] = fB2 - fX2;
                    }
                }
            }

        for (iCount = 1; iCount <= iN; iCount++)
            {
            fXR[iCount] /= fD;
            fXI[iCount] /= fD;
            }

        /*
         ********** PLOT OUTPUT MAGNITUDE ********************
         */
        fMax = 0.0;
        for (iCount = 1; iCount <= 512 /* iN */; iCount++)
            {
            fMag = sqrt (pow (fXR[iCount], 2) +
                pow (fXI[iCount], 2));
            if (fMag > fMax)

                fMax = fMag;
            }

        for (iCount = 1; iCount <= 512 /*iN*/; iCount++)
            {
            _moveto (iCount+1, iY);
            _lineto (iCount+1, (int)(iY -
                (sqrt (pow (fXR[iCount], 2) +
                pow (fXI[iCount], 2)) * 40.0 / fMax + 0.5)));
            }
        }

getch();
_setvideomode( _DEFAULTMODE );

/*
 ***********************************************************
 */
}
```

```c
#include <graph.h>
#include <conio.h>
#include <stdio.h>
#include <stdlib.h>
#include <math.h>

#define DIM      128

float    fXR[DIM + 1], fXI[DIM + 1], fVR[DIM + 1], fVI[DIM + 1],
         fWR[DIM + 1], fWI[DIM + 1], fX[DIM + 1], fH[DIM + 1],
         fY[DIM + 1];

/*
*** WHEN COMPARING WITH THE TEXT, NOTE THAT IN THE C PROGRAMS **
******* FLOATING-POINT VARIABLES ARE PREFIXED WITH AN f, *******
************** INTEGER VARIABLES WITH AN i *********************
*/

void main (void)
    {
    /*
     **** PROGRAM NO. 25    FAST CONVOLUTION *******************
     */
    int    iCount, iCount2, iCount3, iXX, iYY, iM = 7,
           iP, iT, iLim1, iLim2, iLim3, iJ, iL;
    float  fMax, fXHMAG, fMag, fXT, fB1, fB2, fC1, fC2, fD,
           fPI = 3.141593, fX1, fX2, fARG, fCos1, fSin1;

    _clearscreen  (_GCLEARSCREEN);
    printf ("\n\nPROGRAM 25: Fast convolution\n\n");
    printf ("Press any key to start...\n\n");
    getch();

    /*
     *** DEFINE AND PLOT INPUT SIGNAL AND IMPULSE RESPONSE *****
     */
    for (iCount = 1; iCount <= 64; iCount++)
        fX[iCount] = 1.0;

    fH[1] = 1.0;
    fH[2] = 1.5;

    for (iCount = 3; iCount <= 64; iCount++)
        fH[iCount] = 1.5 * fH[iCount-1] - 0.85 * fH[iCount-2];

    _setvideomode( _MRES16COLOR );

    for (iCount = 1; iCount <= DIM; iCount++)
        {
        _moveto (iCount + 1, 70);
        _lineto (iCount + 1, (int)(70 - fX[iCount] * 25.0 + 0.5));
        }

    for (iCount = 1; iCount < DIM; iCount++)
        {
        _moveto (iCount + 190, 70);
        _lineto (iCount + 190, (int)(70 - fH[iCount] * 25.0 + 0.5));
        }

    /*
     ********** CONVOLVE IN TIME DOMAIN THEN PLOT *************
     */
    for (iCount = 1; iCount <= DIM; iCount++)
        for (iCount2 = 1; iCount2 < iCount; iCount2++)
            fY[iCount] += fX[iCount2] * fH[iCount - iCount2 + 1];

    for (iCount = 1; iCount <= DIM; iCount++)
        {
        _moveto (iCount + 100, 160);
        _lineto (iCount + 100, (int)(160 - fY[iCount] * 8.0 + 0.5));
```

```
        }

    printf ("Press any key to start fast convolution");
    getch();

    _clearscreen  (_GCLEARSCREEN);
    _setvideomode( _MRES16COLOR );

    for (iP = 1; iP <= 3; iP++)
        {
        if (iP == 1)
            {
            iT = 1;
            fD = 1.0;
            for (iCount = 1; iCount <= DIM; iCount++)
                {
                fXR[iCount] = fX[iCount];
                fXI[iCount] = 0.0;
                }
            }

        if (iP == 2)
            {
            iT = 1;
            fD = 1.0;
            for (iCount = 1; iCount <= DIM; iCount++)
                {
                fXR[iCount] = fH[iCount];
                fXI[iCount] = 0.0;
                }
            }

        if (iP == 3)
            {
            iT = -1;
            fD = (float) DIM;
            for (iCount = 1; iCount <= DIM; iCount++)
                {
                fXR[iCount] = fVR[iCount] * fWR[iCount] -
                              fVI[iCount] * fWI[iCount];
                fXI[iCount] = fVI[iCount] * fWR[iCount] +
                              fVR[iCount] * fWI[iCount];
                fXHMAG = pow ((pow (fVR[iCount], 2) +
                               pow (fVI[iCount], 2)) *
                              (pow (fWR[iCount], 2) +
                               pow (fWI[iCount], 2)), 0.5);
                _moveto (iCount, 160);
                _lineto (iCount, (int)(160 - fXHMAG / 4.0 + 0.5));
                }
            }
    /*
     **************** SHUFFLE INPUT DATA ******************
     */
    iLim1 = DIM - 1;
    iLim2 = DIM / 2;
    iJ = 1;

    for (iCount = 1; iCount <= iLim1; iCount++)
        {
        if ((float) iCount <= ((float) iJ - 0.01))
            {
            fX1 = fXR[iJ];
            fX2 = fXI[iJ];
            fXR[iJ] = fXR[iCount];
            fXI[iJ] = fXI[iCount];
            fXR[iCount] = fX1;
            fXI[iCount] = fX2;
            }
        iL = iLim2;
        while (1)
```

```
                     {
                     if ((float)iL > ((float)(iJ) - 0.01))
                         break;
                     iJ -= iL;
                     iL /= 2;
                     }
                 iJ += iL;
                 }

    /*
     *************** IN PLACE TRANSFORMATION **************
     */
    for (iCount = 1; iCount <= iM; iCount++)
         {
         iLim1 = (int) pow (2, iCount - 1);
         iLim2 = (int) pow (2, (iM - iCount));

         for (iCount2 = 1; iCount2 <= iLim2; iCount2++)
             {
             for (iCount3 = 1; iCount3 <= iLim1; iCount3++)
                 {
                 iLim3 = (iCount3 - 1) + (iCount2 - 1) * 2 *
                     iLim1 + 1;
                 fB1 = fXR[iLim3];
                 fB2 = fXI[iLim3];
                 fC1 = fXR[iLim3 + iLim1];
                 fC2 = fXI[iLim3 + iLim1];

                 fARG = 2.0 * fPI * (float) (iCount3 - 1) *
                     (float) iLim2 / (float) DIM;
                 fCos1 = cosl (fARG);
                 fSin1 = sinl (fARG);
                 fX1 = fC1 * fCos1 + fC2 * fSin1 *
                     (float) iT;
                 fX2 = (0.0 - fC1) * fSin1 * (float) iT +
                     fC2 * fCos1;
                 fXR[iLim3] = fB1 + fX1;
                 fXI[iLim3] = fB2 + fX2;
                 fXR[iLim3 + iLim1] = fB1 - fX1;
                 fXI[iLim3 + iLim1] = fB2 - fX2;
                 }
             }
         }

    for (iCount = 1; iCount <= DIM; iCount++)
         {
         fXR[iCount] /= fD;
         fXI[iCount] /= fD;
         }

    /*
     * SAVE TRANSFORMS OF INPUT SIGNAL & IMPULSE RESPONSE **
     */
    if (iP == 1)
         {
         for (iCount = 1; iCount <= DIM; iCount++)
             {
             fVR[iCount] = fXR[iCount];
             fVI[iCount] = fXI[iCount];
             }
         }

    if (iP == 2)
         {
         for (iCount = 1; iCount <= DIM; iCount++)
             {
             fWR[iCount] = fXR[iCount];
             fWI[iCount] = fXI[iCount];
             }
         }
```

```
        /*
         ************** FIND PEAK VALUE FOR PLOT **************
         */
        fMax = 0.0;
        for (iCount = 1; iCount <= DIM; iCount++)
            {
            fMag = pow (pow (fXR[iCount], 2) +
                pow (fXI[iCount], 2), 0.5);
            if (fMag > fMax)
                fMax = fMag;
            }


        /*
         **************** SET PLOT OFFSETS ********************
         */
        switch (iP)
            {
            case 1 :
                iXX = 0;
                iYY = 70;
                break;

            case 2 :
                iXX = 190;
                iYY = 70;
                break;

            case 3 :
                iXX = 190;
                iYY = 160;
                break;
            }

        for (iCount = 1; iCount <= DIM; iCount++)
            {
            if (iP == 3)
                fXT = fXR[iCount];
            else
                fXT = pow (pow (fXR[iCount], 2) +
                    pow (fXI[iCount], 2), 0.5);

            _moveto (iCount + iXX, iYY);
            _lineto (iCount + iXX, (int)(iYY -
                fXT * 40.0 / fMax + 0.5));
            }
        }

    getch();
    _setvideomode( _DEFAULTMODE );

    /*
     ***************************************************************
     */
    }
```

```c
/********* PROGRAM NO.26   PROBABILITY MASS FUNCTIONS  ********/
#include <stdio.h>
#include <stdlib.h>
#include <float.h>
#include <graphics.h>
#include <math.h>
#include <conio.h>
#include "graph3.h"


double round(float X);

void main(void) {
    double X[301];
    double Y[301];
    int K,N,B,D;
    double C;
    /******************** UNIFORM DISTRIBUTION  ******************/
    for (K=0; K<=300; K++)
        {
        X[K]=0; Y[K]=0;
        }

    randomize();
    graphmode();

    for (K=1; K<=3000; K++)
        {
        B=(random(10000)/(float)10000)*300;
        X[B]=X[B]+1;
        draw(B,50,B,round(50-X[B]),1);
        }

    /************ GAUSSIAN DISTRIBUTION (APPROXIMATE)  ***********/
    for (K=1; K<=3000; K++)
        {
        C=0;
        for (N=1; N<=30; N++)
            C=C+(random(10000)/(float)10000);
        D=(int)(C*10);
        Y[D]=Y[D]+1;
        draw(D,180,D,round(180-Y[D]),1);
        }

    getch();
    closegraph();
    }


    /***********************************************************/

    double round(float X) {
    if (fmod(X,1)>= 0.5 )
        return ceil(X);
    else
        return floor(X);
    }
```

```
/******* PROGRAM NO.27  ESTIMATES OF MEAN AND VARIANCE  *******/
#include <stdio.h>
#include <stdlib.h>
#include <float.h>
#include <graphics.h>
#include <math.h>
#include <conio.h>
#include "graph3.h"

double round(float X);

void main(void)
    {
    double X[321];
    double M[17];
    double V[17];
    int J,K,N;
    double A,B,C,RM,RV,Z;

    randomize();
    graphmode();
    for (N=1; N<=16; N++)
        {
        M[N]=0; V[N]=0;
        }

    /******* FORM 320 VALUES OF GAUSSIAN SEQUENCE, AND PLOT  ******/
    for (N=1; N<=320; N++)
        {
        C=0;
        for (K=1; K<=12; K++)
            C=C+(random(10000)/(float)10000)-0.5;
        X[N]=C;
        draw(N,50,N,round(50-X[N]*10),1);
        }

    /************  ESTIMATE SAMPLE MEANS AND VARIANCES  ************/
    for (J=1; J<=16; J++)
        {
        for (K=1; K<=20; K++)
            {
            N=K+20*(J-1)-1;
            M[J]=M[J]+X[N]/20;
            V[J]=V[J]+X[N]*X[N]/20;
            }
        draw(N,100,N,round(100-M[J]*50),1);
        draw(N,160,N,round(160-V[J]*20),1);
        }

    printf("PRESS ANY KEY TO CONTINUE ");
    getch();
    cleardevice();
    /*************  ESTIMATE RUNNING MEAN AND VARIANCE  ************/
    A=0;B=0;
    for (N=1; N<=320; N++)
        {
        A=A+X[N];RM=A/N;B=B+X[N]*X[N];RV=B/N;
        draw(N,80,N,round(80-RM*50),1);
        draw(N,150,N,round(150-RV*20),1);
        }

    getch();
    closegraph();
    }
    /***************************************************************/

    double round(float X) {
    if (fmod(X,1)>= 0.5 )
        return ceil(X);
    else
      return floor(X);
    }
```

```c
/*******  PROGRAM N0.28        PROCESSING RANDOM SEQUENCES  *******/
#include <stdio.h>
#include <stdlib.h>
#include <float.h>
#include <graphics.h>
#include <math.h>
#include <conio.h>
#include "graph3.h"

double round(float X);

void main(void)
    {
    double GN[1201];
    double X[1201];
    double Y[1201];
    double H[11];
    double J[20];
    double XACF[102];
    double YACF[102];
    double EST[102];
    int N,K,L,M,P,Q;
    double C,SUM,MY,A,B,Z,MYE,VARE;

    randomize();
    graphmode();
    for (N=1; N<=1200; N++)
        Y[N]=0;
    for (N=1; N<=19; N++)
        J[N]=0;
    for (N=1; N<=101; N++)
        EST[N]=0;
    /***********  ENTER PROCESSOR'S IMPULSE RESPONSE  ************/
    H[1]=1;H[2]=2;H[3]=1;H[4]=-1;H[5]=-2;
    H[6]=-1;H[7]=1;H[8]=2;H[9]=1;H[10]=0;
    /*******  FORM GAUSSIAN NOISE, ZERO MEAN, UNIT VARIANCE  *******/
    cleardevice();
    for (N=1; N<=1200; N++)
        {
        C=0;
        for (K=1; K<=12; K++)
        C=C+(random(10000)/(float)10000)-0.5;
        GN[N]=C;
        }

    /********  LOW-PASS FILTER & ADD UNIT MEAN TO FORM X[N]  ********/
    for (N=4; N<=1200; N++)
        X[N]=(GN[N]+GN[N-1]+GN[N-2]+GN[N-3])*0.5+1;

    /*************   PLOT 320 VALUES OF X[N] ON SCREEN  *************/
    for (N=1; N<=320; N++)
        draw(N,50,N,round(50-X[N+100]*6),1);

    /*****************  DEFINE AND PLOT INPUT ACF  *****************/
    for (K=1; K<=101; K++)
        XACF[K]=1;
    XACF[48]=1.25;XACF[49]=1.5;XACF[50]=1.75;XACF[51]=2;
    XACF[52]=XACF[50];XACF[53]=XACF[49];XACF[54]=XACF[48];
    for (N=1; N<=83; N++)
        draw(N,180,N,round(180-XACF[N+9]*25),1);

    /**********  FIND ACF OF PROCESSOR'S IMPULSE RESPONSE **********/
    for (K=1; K<=10; K++)
    for (L=1; L<=K; L++)
        J[K]=J[K]+H[L]*H[10-K+L];
    for (K=11; K<=19; K++)
        J[K]=J[20-K];
```

```
/***************   PREDICT MEAN VALUE OF OUTPUT  ***************/
SUM=0;
for (K=1; K<=10; K++)
    SUM=SUM+H[K];
MY-1*SUM;

/***************   PREDICT AND PLOT OUTPUT ACF  ***************/
for (M=10; M<=92; M++)
    {
    SUM=0;
    for (K=1; K<=19; K++)
        SUM=SUM+XACF[M-10+K]*J[K];
    YACF[M]=SUM;
    }

for (N=1; N<=83; N++)
    draw(N+120,180,N+120,round(180-YACF[N+9]*50/YACF[51]),1);

/***********   CONVOLVE X[N] AND H[N] TO FORM Y[N]   ***********/
for (N=15; N<=1200; N++)
    for (K=1; K<=10; K++)
        Y[N]=Y[N]+H[K]*X[N-K];

/************   PLOT 320 VALUES OF Y[N] ON SCREEN  ***********/
A=9/sqrt(YACF[51]);
for (N=1; N<=320; N++)
    draw(N,100,N,round(100-Y[N+100]*A),1);

/***************   ESTIMATE AND PLOT OUTPUT ACF  ***************/
B=20*YACF[51];
for (K=1; K<=100; K++)
    {
    for (M=10; M<=92; M++)
        {
        P=M-51;
        for (L=1; L<=10; L++)
            {
            Q=10*(K-1)+L+60;
            EST[M]=EST[M]+Y[Q]*Y[Q+P];
            }
        draw(M+227,180,M+227,round(180-EST[M]/B),1);
        }

    printf("PRESS ANY KEY TO CONTINUE\n");
    getch();
    cleardevice();

/******   PREDICT AND ESTIMATE OUTPUT MEAN AND VARIANCE   ******/
SUM=0;
for (N=15; N<=1200; N++)
    SUM=SUM+Y[N];
MYE=SUM/1185;
printf("OUTPUT MEAN (PREDICTED)=%f\n", MY);
printf("OUTPUT MEAN (ESTIMATED)=%f\n",MYE);
SUM=0;
for (N=15; N<=1200; N++)
    SUM=SUM+(Y[N]-MY)*(Y[N]-MY);
VARE=SUM/1185;
printf("OUTPUT VARIANCE (PREDICTED)=%f\n",YACF[51]-MY*MY);
printf("OUTPUT VARIANCE (ESTIMATED)=%f\n",VARE);
getch();
closegraph();
}

/****************************************************************/

double round(float X)
{
```

```c
        if (fmod(X,1)>= 0.5 )
            return ceil(X);
        else
            return floor(X);
    }




/*********** PROGRAM NO.29      MATCHED FILTERING  ***********/
#include <stdio.h>
#include <stdlib.h>
#include <float.h>
#include <graphics.h>
#include <math.h>
#include <conio.h>
#include "graph3.h"

double round(float X);

void main(void)
    {
    double X[421];
    double Y[421];
    double H[101];
    double S[101];
    int N,K;
    double NV,SUM,MX,MY;

    for (N=1; N<=420; N++)
        {
        X[N]=0; Y[N]=0;
        }

    randomize();
    graphmode();
/***************** DEFINE INPUT SIGNAL  ******************/
    S[1]=1.0;S[2]=0.575;
    for (N=3; N<=60; N++)
        S[N]=1.575*S[N-1]-0.9025*S[N-2];

/******** LOAD TWO VERSIONS OF SIGNAL INTO INPUT ARRAY  *******/
    for (N=1; N<=60; N++)
        {
        X[N+110]=S[N];
        X[N+280]=X[N+280]+S[N];
        }

/******** ADD REQUIRED AMOUNT OF WHITE GAUSSIAN NOISE  ********/
    printf("ENTER NOISE VARIANCE: ");
    scanf("%lf",&NV);
    for (N=1; N<=420; N++)
        {
        SUM=0;
        for (K=1; K<=12; K++)
            SUM=SUM+(random(10000)/(float)10000)-0.5;
        X[N]=X[N]+NV*SUM;
        }

/***************** DEFINE IMPULSE RESPONSE  *****************/
    for (N=1; N<=60; N++)
        H[N]=S[61-N];

/************ CONVOLVE INPUT AND IMPULSE RESPONSE  ***********/
    for (N=101; N<=420; N++)
    for (K=1; K<=100; K++)
        Y[N]=Y[N]+H[K]*X[N-K];
```

```
/******  NORMALISE ARRAYS TO MAX VALUE OF UNITY FOR PLOT  ******/
MX=0;MY=0;
for (N=101; N<=420; N++)
    {
    if (fabs(X[N])>MX)
    MX=fabs(X[N]);
    if (fabs(Y[N])>MY)
    MY=fabs(Y[N]);
    }
for (N=1; N<=420; N++)
    {
    X[N]=X[N]/MX;
    Y[N]=Y[N]/MY;
    }

/***********  PLOT INPUT AND OUTPUT SIGNAL ARRAYS  ************/
cleardevice();
for (N=101; N<=420; N++)
    draw(N-100,50,N-100,round(50-X[N]*22),1);

for (N=101; N<=420; N++)
    draw(N-100,150,N-100,round(150-Y[N]*22),1);

getch();
closegraph();
}
/**********************************************************************/

double round(float X)
{
if (fmod(X,1)>= 0.5 )
    return ceil(X);
else
    return floor(X);
}
```

A1.3   PASCAL PROGRAMS

```
(*********   PROGRAM NO.1    DIGITAL BANDPASS FILTER  *********)
program DSP01;
uses graph3;
var X: array[1..320] of real;
var Y: array[1..320] of real;
var N: integer;
(*******   GENERATE INPUT SIGNAL, AND LOAD INTO ARRAY X  *******)
begin
  for N:=1 to 320 do
    begin
      X[N]:=sin(3.141593*N*(0.05+(0.0005*N)));
    end;
(********   ESTIMATE OUTPUT SIGNAL OF RECURSIVE FILTER  ********)
  Y[1]:=0;
  Y[2]:=0;
    for N:=3 to 320 do
      begin
        Y[N]:=1.5*Y[N-1]-0.85*Y[N-2]+X[N];
      end;
(*********   PLOT INPUT AND OUTPUT SIGNALS ON SCREEN  *********)
  graphmode;
    for N:=1 to 320 do
      begin
        draw(N,50,N,round(50-X[N]*25),1);
        draw(N,150,N,round(150-Y[N]*4),1);
      end;
end.
(***************************************************************)
```

```
(*****   PROGRAM NO.2    SAMPLED EXPONENTIALS AND SINUSOIDS  ****)
program DSP02;
uses graph3;
var X: array[1..100] of real;
var N: integer;
var B,W: real;
begin
  writeln('Value of beta-zero?');
  readln(B);
  writeln('Value of omega?');
  readln(W);
    begin
      for N:=1 to 100 do
        begin
          X[N]:=exp(B*N)*cos(W*N);
        end;
(*************   DRAW AXES, THEN SIGNAL SAMPLES  **************)
      graphmode;
      draw(1,100,300,100,1);
      draw(1,0,1,200,1);
      for N:=1 to 100 do
        begin
          draw(3*N,100,3*N,round(100-X[N]*30),1);
        end;
    end;
end.
(***************************************************************)
```

```
(******  PROGRAM NO.3   AMBIGUITY IN DIGITAL SINUSOIDS  ******)
program DSP03;
uses graph3;
var X: array[1..320] of real;
var N,M,K: integer;
begin
(****  GENERATE SIGNAL WITH TEN EQUAL FREQUENCY INCREMENTS  ****)
  for M:=0 to 9 do
    begin
      for K:=1 to 32 do
        begin
          N:=K+32*M;
          X[N]:=cos(N*2*3.141593*M/8);
        end;
    end;
(************************* PLOT ***************************)
  graphmode;
  for N:=1 to 320 do
    begin
      draw(N,50,N,round(50-X[N]*25),1);
    end;
end.
(*************************************************************)
```

```pascal
(******  PROGRAM NO.4  IMPULSE RESPONSE OF LTI PROCESSOR  ******)
program DSP04;
uses graph3;
var H: array[1..120] of real;
var X: array[1..120] of real;
var N: integer;
var A1,A2,A3,B0,B1,B2,M: real;
begin
(**************  LOAD IMPULSE INTO INPUT ARRAY  **************)
  for N:=1 to 120 do
    begin
      X[N]:=0;H[N]:=0
    end;
  X[10]:=1.0;
  writeln('ENTER 3 RECURSIVE MULTIPLIERS');
  readln(A1,A2,A3);
  writeln('ENTER 3 NONRECURSIVE MULTIPLIERS');
  readln(B0,B1,B2);
(*****  FIND IMPULSE RESPONSE, TAKING N=10 AS TIME ORIGIN ******)
  for N:=10 TO 120 do
    begin
      H[N]:= A1*H[N-1]+A2*H[N-2]+A3*H[N-3]+B0*X[N]+B1*X[N-1]
                                                +B2*X[N-2];
    end;
(*********  NORMALISE H[N] TO A MAXIMUM VALUE OF UNITY  ********)
  M:=0;
  for N:=1 to 120 do
    begin
      if abs(H[N])>M then M:=abs(H[N]);
    end;
  for N:=1 TO 120 do
    begin
      H[N]:=H[N]/M;
    end;
(**********  PLOT HORIZONTAL AXIS AND IMPULSE INPUT  **********)
  graphmode;
  draw(0,50,300,50,1);
  for N:=1 to 100 do
    begin
      draw(3*N,50,3*N,round(50-X[N+9]*25),1);
    end;
(*********  PLOT HORIZONTAL AXIS AND IMPULSE RESPONSE  *********)
  draw (0,120,300,120,1);
  for N:=1 to 100 do
    begin
      draw(3*N,120,3*N,round(120-H[N+9]*60),1);
    end;
end.
(***********************************************************)
```

```
(************  PROGRAM NO.5    DIGITAL CONVOLUTION  ************)
program DSP05;
uses graph3;
var X: array[1..320] of real;
var Y: array[1..320] of real;
var H: array[1..60]  of real;
var N,K: integer;
var MX,MY: real;
(**********  CLEAR ARRAYS AND DEFINE INPUT SIGNAL  ***********)
begin
  for N:=1 to 320 do
    begin
      X[N]:=0; Y[N]:=0;
    end;
  for N:=60 to 320 do
    begin
      X[N]:=sin(2*3.141593*N/60)+sin(2*3.141593*N/10);
    end;
(******  DEFINE IMPULSE RESPONSE (MAX NO. OF TERMS=60)  ******)
  for N:=1 to 60 do
    begin
      H[N]:=0;
    end;
  for N:=1 TO 10 do
    begin
      H[N]:=0.1;
    end;
(************  CONVOLVE INPUT AND IMPULSE RESPONSE  ************)
  for N:=61 to 320 do
    begin
      for K:=1 to 60 do
        begin
          Y[N]:=Y[N]+H[K]*X[N-K];
        end;
    end;
(******  NORMALISE SIGNALS TO MAX VALUE OF UNITY FOR PLOT ******)
  MX:=0;
  MY:=0;
  for N:=1 to 320 do
    begin
      if abs(X[N])>MX then MX:=abs(X[N]);
      if abs(Y[N])>MY then MY:=abs(Y[N]);
    end;
  for N:=1 to 320 do
    begin
      X[N]:=X[N]/MX;Y[N]:=Y[N]/MY;
    end;
(**********  PLOT HORIZONTAL AXIS AND INPUT SIGNAL  ***********)
  graphmode;
  draw(0,50,320,50,1);
  for N:=1 to 320 do
    begin
      draw(N,50,N,round(50-X[N]*30),1);
    end;
(**********  PLOT HORIZONTAL AXIS AND OUTPUT SIGNAL  *********)
  draw(0,150,320,150,1);
  for n:=1 to 320 do
    begin
      draw(N,150,N,round(150-Y[N]*30),1);
    end;
end.
(*************************************************************)
```

```
(******   PROGRAM NO.6       START-UP AND STOP TRANSIENTS  *******)
program DSP06;
uses graph3;
var X: array[1..320] qf real;
var Y: array[1..320] of real;
var N: integer;
var MX,MY: real;
(********************   DEFINE INPUT SIGNAL   *******************)
begin
  for N:=1 to 320 do
    begin
      X[N]:=0; Y[N]:=0;
    end;
  for N:=20 to 200 do
    begin
      X[N]:=cos(2*3.141593*N/20);
    end;
(**********  ESTIMATE OUTPUT FROM BANDSTOP FILTER  ***********)
  for N:=3 TO 320 do
    begin
      Y[N]:=1.8523*Y[N-1]-0.94833*Y[N-2]+X[N]-1.9021*X[N-1]
                                                      +X[N-2];
    end;
(******  NORMALISE SIGNALS TO MAX VALUE OF UNITY FOR PLOT ******)
  MX:=0;
  MY:=0;
  for N:=1 to 320 do
    begin
      if abs(X[N])>MX then MX:=abs(X[N]);
      if abs(Y[N])>MY then MY:=abs(Y[N]);
    end;
  for N:=1 to 320 do
    begin
      X[N]:=X[N]/MX;Y[N]:=Y[N]/MY;
    end;
(**********  PLOT HORIZONTAL AXIS AND INPUT SIGNAL  ***********)
  graphmode;
  draw(0,50,320,50,1);
  for N:=1 to 320 do
    begin
      draw(N,50,N,round(50-X[N]*35),1);
    end;
(*********  PLOT HORIZONTAL AXIS AND OUTPUT SIGNAL  **********)
  draw(0,150,320,150,1);
  for N:=1to 320 do
    begin
      draw(N,150,N,round(150-Y[N]*35),1);
    end;
end.
(*************************************************************)
```

```
(******** PROGRAM NO.7   DISCRETE-TIME FOURIER SERIES ********)
(************* FOR SIGNAL WITH 7 SAMPLE VALUES *************)
program DSP07;
var X: array[1..7] of real;
var AR: array[1..7] of real;
var AI: array[1..7] of real;
var N,K: integer;
begin
  for N:=1 to 7 do
    begin
      AR[N]:=0; AI[N]:=0;
    end;
  X[1]:=2; X[2]:=1; X[3]:=-2; X[4]:=3;
  X[5]:=-1; X[6]:=-1; X[7]:=1;
  for K:=1 to 7 do
    begin
      for N:=1 to 7 do
        begin
          AR[K]:=AR[K]+(1/7)*X[N]*cos(2*3.141593*(K-1)*(N-1)/7);
          AI[K]:=AI[K]-(1/7)*X[N]*sin(2*3.141593*(K-1)*(N-1)/7);
        end;
    end;
  for N:=1 to 7 do
    begin
      writeln(AR[N],AI[N]);
    end;
end.
(****************************************************************)
```

```pascal
(********  PROGRAM NO.8   DISCRETE-TIME FOURIER SERIES   ********)
(*****  64 REAL SAMPLE VALUES, SPECTRAL MAGNITUDE & PHASE  *****)
program DSP08;
uses graph3;
var X: array[1..64] of real;
var AR: array[1..64] of real;
var AI: array[1..64] of real;
var MAG: array[1..64] of real;
var PHASE: array[1..64] of real;
var N,K: integer;
var B,M1,M2,M3: real;
begin
  for N:=1 to 64 do
    begin
      AR[N]:=0;AI[N]:=0;
      B:=2*3.141593*(N-1);
      X[N]:=sin(B/64)+cos(B/16)+0.6*cos(B/8)+0.5*sin(B/4);
    end;
  for K:=1 to 64 do
    begin
      for N:=1 to 64 do
        begin
          AR[K]:=AR[K]+(1/64)*X[N]*cos(2*3.141593*(K-1)*(N-1)/64);
          AI[K]:=AI[K]-(1/64)*X[N]*sin(2*3.141593*(K-1)*(N-1)/64);
        end;
    end;
(**********  CONVERT TO SPECTRAL MAGNITUDE & PHASE  **********)
  for K:=1 to 64 do
    begin
      if abs(AI[K])<0.001 then AI[K]:=0;
      if abs(AR[K])<0.00001 then AR[K]:=0.00001;
      MAG[K]:=sqrt(AR[K]*AR[K]+AI[K]*AI[K]);
      PHASE[K]:=arctan(AI[K]/AR[K]);
    end;
(*******  NORMALISE TO MAXIMUM VALUE OF UNITY FOR PLOT  ********)
  M1:=0.001;M2:=0.001;M3:=0.001;
    for N:=1 to 64 do
      begin
        if abs(X[N])>M1 then M1:=abs(X[N]);
        if abs(MAG[N])>M2 then M2:=abs(MAG[N]);
        if abs(PHASE[N])>M3 then M3:=abs(PHASE[N]);
      end;
  for N:=1 to 64 do
    begin
      X[N]:=X[N]/M1;MAG[N]:=MAG[N]/M2;
      PHASE[N]:=PHASE[N]/M3;
    end;
(*******  PLOT SIGNAL, SPECTRAL MAGNITUDE, AND PHASE  ********)
  graphmode;
  for N:=1 to 64 do
    begin
      draw(5*(N-1),50,5*(N-1),round(50-X[N]*25),1);
      draw(5*(N-1),120,5*(N-1),round(120-MAG[N]*30),1);
      draw(5*(N-1),160,5*(N-1),round(160-PHASE[N]*20),1);
    end;
end.
(***************************************************************)
```

```
(****  PROGRAM NO.9  FREQUENCY RESPONSE OF BANDPASS FILTER *****)
program DSP09;
uses graph3;
var N: integer;
var W,A,B,MAG,PHASE: real;
begin
  graphmode;
  for N:=1 to 320 do
    begin
      W:=3.141593*(N-1)/320;
      A:=1-1.5*cos(W)+0.85*cos(2*W);
      B:=1.5*sin(W)-0.85*sin(2*W);
      MAG:=1/sqrt(A*A+B*B);
      PHASE:=arctan(B/A);
      draw(N,100,N,round(100-5*MAG),1);
      draw(N,160,N,round(160-8*PHASE),1);
    end;
end.
(**************************************************************)
```

```
(**********  PROGRAM NO.10      INVERSE Z-TRANSFORM  **********)
program DSP10;
uses graph3;
var X: array[1..320] of real;
var Y: array[1..320] of real;
var N: integer;
begin
  for N:=1 to 320 do
    begin
      X[N]:=0;Y[N]:=0;
    end;
  X[5]:=1;
  graphmode;
  for N:=5 to 320 do
    begin
      Y[N]:=-0.54048*Y[N-1]+0.62519*Y[N-2]+0.66354*Y[N-3]
            -0.60317*Y[N-4]-0.69341*Y[N-5]
            +X[N]-X[N-1]+X[N-2]-X[N-3];
      draw(N,100,N,round(100-20*Y[N]),1);
    end;
end.
(**************************************************************)
```

```
(****   PROGRAM NO.11   CHARACTERISTICS OF 2ND-ORDER SYSTEMS   ****)
program DSP11;
uses graph3;
var I: array[1..320] of real;
var F: array[1..320] of real;
var R,THETA,T,A,B,GAIN,C,D,W,M: real;
var N: integer;
(***  I=IMPULSE RESPONSE, F=FREQUENCY RESPONSE (MAGNITUDE)  ****)
begin
  writeln('ENTER RADIUS, ANGLE (DEGREES) OF Z-PLANE POLES ');
  readln(R,THETA);
  T:=THETA*3.141593/180;
(******  LOAD IMPULSE RESPONSE, STARTING AT LOCATION I(2)  *****)
  I[1]:=0; I[2]:=1;
  for N:=3 TO 320 do
    begin
      I[N]:=2*R*cos(T)*I[N-1]-R*R*I[N-2];
    end;
(******   NORMALISE TO PEAK VALUE OF UNITY, THEN  PLOT  ********)
  M:=0;
  for N:=1 to 320 do
    begin
      if abs(I[N])>M then M:=abs(I[N]);
    end;
  for N:=1 to 320 do
    begin
      I[N]:=I[N]/M;
    end;
  graphmode;
  for N:=1 to 320 do
    begin
      draw(N,70,N,round(70-I[N]*30),1);
    end;
(********************   ESTIMATE PEAK GAIN   *******************)
  A:=1-2*R*cos(T)*cos(T)+R*R*cos(2*T);
  B:=2*R*cos(T)*sin(T)-R*R*sin(2*T);
  GAIN:=1/sqrt(A*A+B*B);
(******  FIND FREQUENCY RESPONSE (NORMALISED), THEN PLOT  *****)
  for N:=1 to 320 do
    begin
      W:=(N-1)*3.141593/320;
      C:=1-2*R*cos(T)*cos(W)+R*R*cos(2*W);
      D:=2*R*cos(T)*sin(W)-R*R*sin(2*W);
      F[N]:=1/(sqrt(C*C+D*D)*GAIN);
    end;
  for N:=1 to 320 do
    begin
      draw(N,170,N,round(170-F[N]*60),1);
    end;
end.
(****************************************************************)
```

```
(****    PROGRAM NO.12    F-RESPONSE OF MOVING-AVERAGE FILTER   ****)
program DSP12;
uses graph3;
var J,W,H: real;
var N,M,K: integer;
begin
  writeln('Enter value of M');
  readln(M);
  J:=1/(2*M+1);
  graphmode;
  for N:=1 to 320 do
    begin
      W:=3.141593*(N-1)/320;H:=J;
        for K:=1 to M do
          begin
            H:=H+2*J*cos(K*W);
          end;
        draw(N,100,N,round(100-50*abs(H)),1);
    end;
end.
(***************************************************************)


(********************** PROGRAM NO.13 **********************)
(***  NONRECURSIVE FILTER DESIGN BY FOURIER TRANSFORM METHOD  **)
(*******  IMPULSE RESPONSE TRUNCATED TO (2M+1) TERMS  ********)
program DSP13;
uses graph3;
var H: array[1..320] of real;
var A,B,W,W0,W1,H0,HF: real;
var M,N,K,X: integer;
begin
  writeln('ENTER CENTER-FREQUENCY (IN DEGREES)');readln(A);
  W0:=A*3.141593/180;
  writeln('ENTER FILTER BANDWIDTH (IN DEGREES)');readln(B);
  W1:=B*0.5*3.141593/180;
  writeln('ENTER VALUE OF M');readln(M);
  writeln('IMPULSE RESPONSE VALUES H(0) TO H(M) ARE');
  H0:=W1/3.141593;writeln(H0);
  for N:=1 to M do
    begin
      H[N]:=(1/(N*3.141593))*sin(N*W1)*cos(N*W0);
      writeln(H[N]);
    end;
  writeln('ENTER ANY INTEGER TO BEGIN F-RESPONSE PLOT');
  readln(X);
  graphmode;
  for N:=1 to 320 do
    begin
      W:=3.141593*(N-1)/320;HF:=H0;
        for K:=1 to M do
          begin
            HF:=HF+2*H[K]*cos(K*W);
          end;
        draw(N,100,N,round(100-70*abs(HF)),1);
    end;
end.
(***************************************************************)
```

```pascal
(********************** PROGRAM NO.14 ************************)
(*************** TIME-DOMAIN PERFORMANCE OF *****************)
(********** 21-TERM LINEAR-PHASE BANDPASS FILTER ***********)
program DSP14;
uses graph3;
var X: array[1..320] of real;
var Y: array[1..320] of real;
var H: array[0..20] of real;
var N,K: integer;
begin
  for N:=1 to 320 do
    begin
      X[N]:=0;Y[N]:=0;
    end;
(***  DEFINE IMPULSE RESPONSE VALUES, STARTING WITH CENTRAL  ***)
(********  VALUE AND WORKING TOWARDS ONE OF THE "TAILS"  ******)
  H[10]:=0.083333;H[11]:=0.041192;H[12]:=-0.039789;
  H[13]:=-0.075026;H[14]:=-0.034458;H[15]:=0.030746;
  H[16]:=0.053052;H[17]:=0.021962;H[18]:=-0.017229;
  H[19]:=-0.025009;H[20]:=-0.007958;
  for N:=10 to 20 do
    begin
      H[20-N]:=H[N];
    end;
(********  LOAD INPUT ARRAY WITH TWO SIGNAL COMPONENTS  *******)
  for N:=108 to 180 do
    begin
      X[N]:=cos(N*5*3.141593/18)+cos(N*7*3.141593/18);
    end;
(************************* PLOT ***************************)
  graphmode;
  for N:=1 to 320 do
    begin
      draw(N,30,N,round(30-10*X[N]),1);
    end;
(******  ADD TWO OUT-OF-BAND COMPONENTS TO INPUT SIGNAL  ******)
  for N:=1 to 320 do
    begin
      X[N]:=X[N]+sin(N*3.141593/8)+sin(N*3.141593/1.8);
    end;
(************************* PLOT ***************************)
  for N:=1 to 320 do
    begin
      draw(N,90,N,round(90-10*X[N]),1);
    end;
(****  CONVOLVE INPUT SIGNAL AND IMPULSE RESPONSE, AND PLOT  ***)
  for N:=22 to 320 do
    begin
      Y[N]:=0;
        for K:=0 to 20 do
          begin
            Y[N]:=Y[N]+H[K]*X[N-K];
          end;
        draw(N,150,N,round(150-25*Y[N]),1);
    end;
end.
(*************************************************************)
```

```pascal
(*********************** PROGRAM NO.15 **********************)
(*** SPECTRAL PROPERTIES OF RECTANGULAR WINDOW (LOG PLOT) ****)
program DSP15;
uses graph3;
var FREQ,H,DB: real;
var N,M,K: integer;
begin
  writeln('ENTER VALUE OF M');
  readln(M);
  graphmode;
(************ DRAW RECTANGULAR GRID FOR PLOT ***************)
  for K:=0 to 5 do
    begin
      draw(0,30+20*K,320,30+20*K,1);
      draw(64*K,30,64*K,130,1);
    end;
  draw(319,30,319,130,1);
(***** ESTIMATE SPECTRUM, CONVERT TO DECIBELS, AND PLOT ******)
  for N:=1 to 320 do
    begin
      FREQ:=3.141593*(N-1)/320;H:=1;
        for K:=1 to M do
          begin
            H:=H+2*cos(K*FREQ);
          end;
        H:=H/(1+2*M);
        DB:=20*ln(abs(H))*0.4343;
        if (DB)<-50. then DB:=-50;
        draw(N,130,N,round(30-2*DB),1);
    end;
end.
(****************************************************************)
```

```
(*********************   PROGRAM NO.16   *********************)
(*****   SPECTRAL PROPERTIES OF VON HANN & HAMMING WINDOWS *****)
program DSP16;
uses graph3;
var SEE: array[1..200] of real;
var W: array[1..320] of real;
var A,B,C,DB,FREQ,PEAK: real;
var X,K,N,M: integer;
label 1,2;
begin
  writeln('SELECT WINDOW TYPE: 1= VON HANN; 2= HAMMING');
  readln(X);
  writeln('ENTER VALUE OF M'); readln(M);
(*****   LOAD ONE HALF OF WINDOW (EXCLUDING CENTRAL VALUE)   *****)
(*********************   INTO ARRAY SEE   *********************)
  if (X)=1   then goto 1;
      A:=0.54;B:=0.46;C:=M; goto 2;
    1: A:=0.5;B:=0.5;C:=M+1;
    2: for K:=1 to M do
          begin
            SEE[K]:=A+B*cos(K*3.141593/C);
          end;
(*************   DRAW RECTANGULAR GRID FOR PLOT   **************)
  graphmode;
  for K:=0 to 5 do
    begin
      draw(0,30+20*K,320,30+20*K,1);
      draw(64*K,30,64*K,130,1)
    end;
  draw(319,30,319,130,1);
(********************   ESTIMATE SPECTRUM   *******************)
  for N:=1 to 320 do
    begin
      FREQ:=3.141593*(N-1)/320;W[N]:=1;
        for K:=1 to M do
          begin
            W[N]:=W[N]+2*SEE[K]*cos(K*FREQ);
          end;
(***   NORMALISE TO PEAK VALUE, CONVERT TO DECIBELS, AND PLOT   **)
            if (N)=1 then PEAK:=W[N];
            W[N]:=W[N]/PEAK;
            DB:=20*ln(abs(W[N]))*0.4343;
            if (DB)<-50 then DB:=-50;
            draw(N,130,N,round(30-2*DB),1);
    end;
end.
(**************************************************************)
```

```
(******************* PROGRAM NO.17 ***************************)
(**  NONRECURSIVE FILTER DESIGN BY FOURIER TRANSFORM METHOD  ***)
(******* WITH RECTANGULAR, VON HANN, OR HAMMING WINDOW *******)
(****** AND DECIBEL PLOT OF FREQUENCY RESPONSE MAGNITUDE ******)
(*********** (IMPULSE RESPONSE HAS (2M+1) TERMS) ************)
program DSP17;
uses graph3;
var SEE: array[1..200] of real;
var H: array[1..200] of real;
var W: array[1..320] of real;
var H0,A,B,C,FREQ,MAX,DB,W0,W1: real;
var M,X,K,N,Y: integer;
label 1;
begin
  writeln('ENTER CENTER-FREQUENCY (IN DEGREES)'); readln(A);
  W0:=A*3.141593/180;
  writeln('ENTER FILTER BANDWIDTH (IN DEGREES)'); readln(B);
  W1:=B*0.5*3.141593/180;
  writeln('ENTER VALUE OF M'); readln(M);
  writeln('SELECT WINDOW:');
  writeln('0=RECTANGULAR; 1=VON HANN; 2=HAMMING'); readln(X);
(******************* COMPUTE WINDOW VALUES *****************)
    if X=0 then begin A:=1;B:=0;C:=1;goto 1 end;
    if X=1 then begin A:=0.5;B:=0.5;C:=M+1;goto 1 end;
      A:=0.54;B:=0.46;C:=M;
  1: for N:=1 to M do begin
        SEE[N]:=A+B*cos(N*3.141593/C);end;
(************* COMPUTE IMPULSE RESPONSE VALUES **************)
  H0:=W1/3.141593;
    for N:=1 to M do begin
    H[N]:=(1/(N*3.141593))*sin(N*W1)*cos(N*W0)*SEE[N];end;
(************* DRAW RECTANGULAR GRID FOR PLOT ***************)
  graphmode;
    for K:=0 to 5 do begin
      draw(0,30+20*K,320,30+20*K,1);
      draw(64*K,30,64*K,130,1);end;
    draw(319,30,319,130,1);
(************** COMPUTE FREQUENCY RESPONSE *****************)
  for N:=1 to 320 do begin
    FREQ:=3.141593*(N-1)/320;W[N]:=H0;
      for K:=1 to M do begin
      W[N]:=W[N]+2*H[K]*COS(K*FREQ);
      end;
  end;
(***** NORMALISE TO UNITY, CONVERT TO DECIBELS, AND PLOT *****)
  MAX:=0;
    for N:=1 to 320 do begin
      if abs(W[N])>MAX then MAX:=abs(W[N]);
    end;
    for N:=1 to 320 do begin
    DB:=20*ln(abs(W[N])/MAX)*0.4343;if DB<-50 then DB:=-50;
    draw(N,130,N,round(30-2*DB),1);
    end;
  writeln('ENTER ANY INTEGER TO PRINT h[n] VALUES');
  readln (Y);
  clearscreen;
  writeln('VALUES h[0] TO h[M]:');
  writeln ('(CORRECTED FOR UNITY MAXIMUM GAIN)');
  writeln ( H0/MAX);
    for N:=1 to M do begin writeln(H[N]/MAX);end;
end.
(***************************************************************)
```

```
(******************** PROGRAM NO.18 *************************)
(*****  NONRECURSIVE FILTER DESIGN WITH THE KAISER WINDOW  *****)
program DSP18;
uses graph3;
var SEE: array[0..200] of real;
var H: array[0..200] of real;
var W: array[0..320] of real;
var P,Q,R,S,T,A,I,B,C,D: real;
var ALPHA,BETA,A0,H0,W0,W1,FREQ,MAX,DB,MREAL: real;
var X,N,K,M,Y: integer;
label 1,2;
begin
  1:writeln('ENTER CENTER-FREQUENCY (IN DEGREES)'); readln(P);
    W0:=P*3.141593/180;
    writeln('ENTER FILTER BANDWIDTH (IN DEGREES)'); readln(Q);
    W1:=Q*0.5*3.141593/180;
    writeln('ENTER KAISER DESIGN PARAMETERS:');
    writeln('RIPPLE (AS A FRACTION):'); readln(R);
    writeln('TRANSITION WIDTH (DEGREES):'); readln(S); T:=S/360;
(**************  ESTIMATE KAISER PARAMETERS  *****************)
  A:=-20*ln(R)*0.4343;MREAL:=(A-7.95)/(28.72*T);M:=trunc(MREAL+1);
  writeln('NO. OF IMPULSE RESPONSE TERMS=',2*M+1);
  writeln('CHANGE  PARAMETERS, OR CONTINUE (0/1)?');
  readln(X); if X=0 then GOTO 1;
    if A>49 then begin ALPHA:=0.1102*(A-8.7);goto 2 end;
    if A>21 then begin
     ALPHA:=(0.5842*exp(0.4*ln(A-21))+(0.07886*(A-21)));goto 2
    end;ALPHA:=0;
  2:writeln('FOR THIS WINDOW, ALPHA=',ALPHA);
    writeln('ENTER ANY INTEGER TO START PLOT'); readln(X);
(*******************  COMPUTE WINDOW VALUES  ******************)
  for N:=0 to M do begin
    BETA:=ALPHA*sqrt(1-(N/M)*(N/M));I:=1;B:=1;
      for K:=1 to 20 do begin
        C:=B*BETA*BETA/4;D:=C/(K*K);I:=I+D;B:=D;end;SEE[N]:=I;
  end;A0:=SEE[0]; for N:=0 to M do begin
                  SEE[N]:=SEE[N]/A0;end;
(*************  COMPUTE IMPULSE RESPONSE VALUES  *************)
  H0:=W1/3.141593;
    for N:=1 to M do begin
    H[N]:=(1/(N*3.141593))*sin(N*W1)*cos(N*W0)*SEE[N];end;
(*************  DRAW RECTANGULAR GRID FOR PLOT  *************)
  graphmode;
    for K:=0 to 5 do begin draw(0,30+20*K,320,30+20*K,1);
              draw(64*K,30,64*K,130,1);end;draw(319,30,319,130,1);
(**************  COMPUTE FREQUENCY RESPONSE  ****************)
  for N:=1 to 320 do begin
    FREQ:=3.141593*(N-1)/320;W[N]:=H0;
      for K:=1 to M do begin
      W[N]:=W[N]+2*H[K]*cos(K*FREQ);end;end;
(*****  NORMALISE TO UNITY, CONVERT TO DECIBELS, AND PLOT  *****)
  MAX:=0;
  for N:=1 to 320 do begin
    if abs(W[N])>MAX then MAX:=abs(W[N]);end;
      for N:=1 to 320 do begin
        DB:=20*ln(abs(W[N])/MAX)*0.4343;if DB<-50 then DB:=-50;
        draw(N,130,N,round(30-2*DB),1);end;
    writeln('ENTER ANY INTEGER TO PRINT h[n] VALUES');readln(Y);
  clearscreen;
    writeln('VALUES h[0] TO h[M]:');
    writeln('(CORRECTED FOR UNITY MAXIMUM GAIN)');
    writeln(H0/MAX);
  for N:=1 to M do begin writeln(H[N]/MAX);end;
end.
(***********************************************************)
```

```
(********   PROGRAM NO.19      DIGITAL DIFFERENTIATOR  **********)
program DSP19;
uses graph3;
var SEE: array[1..200] of real;
var H: array[1..200] of real;
var A,B,W,HF,C: real;
var M,N,K,X,Y: integer;
label 1,2,3;
begin
  writeln('ENTER VALUE OF M');
  readln(M);
  writeln('SELECT WINDOW:');
  writeln('1=RECTANGULAR; 2=HAMMING');
  readln(X);
(*****************   COMPUTE WINDOW VALUES  ******************)
    if (X=1) then begin A:=1;B:=0;C:=1;goto 1 end
             else begin A:=0.54;B:=0.46;C:=M end;
  1: for N:=1 to M do
        begin
          SEE[N]:=A+B*cos(N*3.141593/C);
        end;
(****************   COMPUTE IMPULSE RESPONSE  *****************)
      for N:=1 to M do
        begin
          if odd(N) then goto 2;
            H[N]:=(1/N)*SEE[N];goto 3;
  2: H[N]:=-(1/N)*SEE[N];
  3: end;
(*****   COMPUTE AND PLOT FREQUENCY RESPONSE (LINEAR SCALE) *****)
  graphmode;
  for N:=1 to 320 do
    begin
      W:=(N-1)*3.141593/320;HF:=0;
        for K:=1 to M do
          begin
            HF:=HF-2*H[K]*sin(K*W);
          end;
        draw(N,150,N,round(150-20*HF),1);
    end;
  writeln('ENTER ANY INTEGER TO PRINT h[n]');
  writeln('  (REMEMBER THAT h(0) IS ZERO!)'); readln(Y);
  clearscreen;
  writeln('VALUES H(1) TO H(M) ARE:');
    for N:=1 to M do
      begin
        writeln (H[N]);
      end;
end.
(****************************************************************)
```

```
(*********  PROGRAM NO.20    FILTER MAGNITUDE RESPONSE *********)
(***********  FROM Z-PLANE POLE AND ZERO LOCATIONS  ***********)
program DSP20;
uses graph3;
var RP: array[1..20,1..2] of real;
var RZ: array[1..20,1..2] of real;
var CP: array[1..20,1..2] of real;
var CZ: array[1..20,1..2] of real;
var H: array[1..320] of real;
var A,B,W,C1,C2,S1,S2,D,R,T,E,MAX,DB,BASE,EX: real;
var NRP,NRZ,N,NCP,NCZ,K: integer;
label 1,2,3,4,5,6,7,8;
begin
    writeln('NO. OF SEPARATE REAL POLES'); readln(NRP);
      if NRP=0 then goto 1;
    writeln('ENTER VALUE, ORDER, OF EACH IN TURN');
    writeln('    FOLLOWED BY <RETURN>');
      for N:=1 to NRP do begin
        readln(RP[N,1],RP[N,2]); end;
  1: writeln('NO. OF SEPARATE REAL ZEROS'); readln(NRZ);
      if NRZ=0 then goto 2;
    writeln('ENTER VALUE, ORDER, OF EACH IN TURN');
    writeln('    FOLLOWED BY <RETURN>');
      for N:=1 to NRZ do begin
        readln(RZ[N,1],RZ[N,2]); end;
  2: writeln('NO. OF COMPLEX-CONJUGATE POLE PAIRS'); readln(NCP);
      if NCP=0 then goto 3;
    writeln('ENTER RADIUS, ANGLE (DEGREES), OF EACH');
    writeln('   IN TURN, FOLLOWED BY <RETURN>');
      for N:=1 to NCP do begin
        readln(CP[N,1],CP[N,2]); end;
  3: writeln('NO. OF COMPLEX-CONJUGATE ZERO PAIRS'); readln(NCZ);
      if NCZ=0 then goto 4;
    writeln('ENTER RADIUS, ANGLE (DEGREES), OF EACH');
    writeln('   IN TURN, FOLLOWED BY <RETURN>');
    for N:=1 to NCZ do begin
        readln(CZ[N,1],CZ[N,2]);end;
(***********  COMPUTE FREQUENCY RESPONSE MAGNITUDE  ***********)
  4: for N:=1 to 320 do begin
      W:=3.141593*(N-1)/320;
      C1:=cos(W);C2:=cos(2*W);S1:=sin(W);S2:=sin(2*W);H[N]:=1;
        if NRP=0 then goto 5;
          for K:=1 to NRP do begin
            A:=RP[K,1];B:=RP[K,2];BASE:=1-2*A*C1+A*A;EX:=B/2;
            H[N]:=H[N]/(exp(EX*ln(BASE))); end;
  5:      if NRZ=0 then goto 6;
          for K:=1 to NRZ do begin
            A:=RZ[K,1];B:=RZ[K,2];BASE:=1-2*A*C1+A*A;EX:=B/2;
            H[N]:=H[N]*(exp(EX*ln(BASE))); end;
  6:      if NCP=0 then goto 7;
          for K:=1 to NCP do begin
            R:=CP[K,1];T:=CP[K,2]*3.141593/180;
            D:=C2-2*R*cos(T)*C1+R*R;E:=S2-2*R*cos(T)*S1;
            H[N]:=H[N]/(exp(0.5*ln(D*D+E*E))); end;
  7:      if NCZ=0 then goto 8;
          for K:=1 to NCZ do begin
            R:=CZ[K,1];T:=CZ[K,2]*3.141593/180;
            D:=C2-2*R*cos(T)*C1+R*R;E:=S2-2*R*cos(T)*S1;
            H[N]:=H[N]*(exp(0.5*ln(D*D+E*E))); end;
  8: end;
(************  FIND MAXIMUM VALUE AND PRINT OUT  *************)
  graphmode;
```

```
      MAX:=0;
      for N:=1 to 320 do begin
        if abs(H[N])>MAX then MAX:=abs(H[N]); end;
      writeln('MAXIMUM GAIN=',20*ln(MAX)*0.4343,' dB');
(************** DRAW RECTANGULAR GRID FOR PLOT **************)
      for K:=0 to 5 do begin
        draw(0,30+20*K,320,30+20*K,1);
        draw(64*K,30,64*K,130,1);end;   draw(319,30,319,130,1);
(****  NORMALISE RESPONSE TO UNITY, CONVERT TO DB, AND PLOT  ***)
      for N:=1 to 320 do begin
        if abs(H[N])<0.000001 then H[N]:=0.000001;
        DB:=20*ln(abs(H[N])/MAX)*0.4343;if DB<-50 then DB:=-50;
        draw(N,130,N,round(30-2*DB),1); end;
end.
(*************************************************************)
```

```
(************ PROGRAM NO.21    POLES AND ZEROS OF ************)
(********* BUTTERWORTH AND CHEBYSHEV DIGITAL FILTERS *********)
program DSP21;
var PR: array[1..50] of real;
var PI: array[1..50] of real;
var FF,FT,N,K,J,IR,M,M1,N1,N2,ST: integer;
var F1,D,W1,C1,C2,B1,B2,E,X,A,Y,B: real;
var T,C3,C4,C5,R,I,B3,F2,F3,F4,F5,FR,FI,GR,AR: real;
var GI,AI,SR,SI,H1,H2,P1R,P2R,P1I,P2I,TH: real;
label 1,2,3,4,5,6,7,8,9,10,11,12,13,14,15,16;
begin
      for K:=1 to 50 do begin PR[K]:=0;PI[K]:=0;end;
      writeln('CHOOSE FILTER FAMILY: 1=BUTTERWORTH; 2=CHEBYSHEV');
      readln(FF);
      writeln('CHOOSE FILTER TYPE:');
      writeln('  1=LOWPASS; 2=HIGHPASS; 3=BANDPASS'); readln(FT);
      writeln('FILTER ORDER');readln(N);if FT=3 then begin
        N:=trunc(N/2);goto 1 end;
      writeln('CUT-OFF FREQUENCY (DEGREES)');readln(F1);ST:=2;
        if FT>1 then F1:=180-F1;goto 2;
 1: writeln('LOWER,UPPER CUT-OFF FREQUENCIES (DEGREES)');
      readln(F2,F3);F1:=F3-F2;ST:=1;
 2: if FF=2 then begin writeln('PASSBAND RIPPLE (FRACTION)');
      readln(D) end;
(*********** FIND BUTTERWORTH/CHEBYSHEV PARAMETERS ***********)
      IR:=N mod 2;N1:=N+IR;N2:=trunc((3*N+IR)/2-1);
      W1:=3.141593*F1/360;C1:=sin(W1)/cos(W1);B1:=2*C1;C2:=C1*C1;
      B2:=0.25*B1*B1; if FF=1 then goto 3;
      E:=1/sqrt(1/((1-D)*(1-D))-1);
      X:=exp(1/N*ln(sqrt(E*E+1)+E));
      Y:=1/X;A:=0.5*(X-Y);B:=0.5*(X+Y);
(****** FIND REAL AND IMAGINARY PARTS OF LOW-PASS POLES ******)
 3: for K:=N1 to N2 do begin
      T:=3.141593*(2*K+1-IR)/(2*N);if FF=1 then goto 4;
```

```
           C3:=A*Cl*COS(T);C4:=B*Cl*SIN(T);C5:=(1-C3)*(1-C3)+C4*C4;
           R:=2*(1-C3)/C5-1;I:=2*C4/C5; goto 5;
    4:     B3:=1-Bl*cos(T)+B2;R:=(1-B2)/B3;I:=Bl*sin(T)/B3;
    5:     M:=(N2-K)*2+1;
           PR[M+IR]:=R;PI[M+IR]:=I;PR[M+IR+1]:=R;PI[M+IR+1]:=-I;end;
         if IR=0 then goto 8;
           if FF=1 then goto 6;
             R:=2/(1+A*Cl)-1;goto 7;
    6:       R:=(1-B2)/(1+Bl+B2);
    7:       PR[1]:=R;PI[1]:=0;
    8:         if FT=3 then goto 10;
(************  PRINT OUT Z-PLANE ZERO LOCATIONS  *************)
      if FT>1 then goto 9;
      writeln('REAL ZERO, OF ORDER ',N,', AT Z=-1');goto 13;
   9: writeln('REAL ZERO, OF ORDER ',N,', AT Z=1');
      for M:=1 to N do begin
      PR[M]:=-PR[M];end;goto 13;
   10: writeln('REAL ZEROS, OF ORDER ',N,', AT Z=1 AND AT Z=-1');
(***********  LOW-PASS TO BANDPASS TRANSFORMATION  *************)
      F4:=F2*3.141593/360;F5:=F3*3.141593/360;
      A:=cos(F4+F5)/cos(F5-F4);
        for M1:=0 to 24 do begin M:=1+2*M1;
          AR:=PR[M];AI:=PI[M];
            if abs(AI)<0.0001 then goto 11;
              FR:=A*0.5*(1+AR);FI:=A*0.5*AI;
              GR:=FR*FR-FI*FI-AR;GI:=2*FR*FI-AI;
              SR:=sqrt(abs(GR+sqrt(GR*GR+GI*GI))/2);SI:=GI/(2*SR);
              PlR:=FR+SR;PlI:=FI+SI;P2R:=FR-SR;P2I:=FI-SI;goto 12;
   11: H1:=A*(1+AR)/2;H2:=H1*H1-AR;
        if H2>0 then begin PlR:=H1+sqrt(H2);P2R:=H1-sqrt(H2);
        PlI:=0;P2I:=0;goto 12 end;
      PlR:=H1;P2R:=H1;PlI:=sqrt(abs(H2));P2I:=-PlI;
   12: PR[M]:=PlR;PR[M+1]:=P2R;PI[M]:=PlI;PI[M+1]:=P2I;end;
(*************  PRINT OUT Z-PLANE POLE LOCATIONS  *************)
   13: writeln('RADII, ANGLES OF Z-PLANE POLES:');
       writeln('        R','                  THETA');
         for J:=1 to N do begin
           if ST=1 then goto 14;
           if J mod ST=0 then goto 16;
   14: M:=J;if IR=0 then goto 15;
           if J=2 then M:=N+1;
   15: R:=sqrt(PR[M]*PR[M]+PI[M]*PI[M]);
       TH:=arctan(abs(PI[M])/abs(PR[M]))*180/3.141593;
         if PR[M]<0 then TH:=180-TH;writeln(R,TH);
   16:   end;
end.
(***************************************************************)
```

```
(********  PROGRAM NO.22    FREQUENCY-SAMPLING FILTER  *********)
program DSP22;
uses graph3;
var X: array[1..320] of real;
var Y: array[1..320] of real;
var W: array[1..320] of real;
var P: array[1..320] of real;
var Q: array[1..320] of real;
var R: array[1..320] of real;
var S: array[1..320] of real;
var T: array[1..320] of real;
var U: array[1..320] of real;
var V: array[1,.320] of real;
var N,M,K: integer;
var FREQ,HR,HI,H: real;
begin
  for N:=1 to 320 do
    begin
      X[N]:=0;Y[N]:=0;W[N]:=0;P[N]:=0;Q[N]:=0;
      R[N]:=0;S[N]:=0;T[N]:=0;U[N]:=0;V[N]:=0;
    end;
  graphmode;
(*************  LOAD IMPULSE INTO INPUT ARRAY   ***************)
  X[120]:=1;
    for N:=120 TO 280 do
      begin
        W[N]:=X[N]-0.886867*X[N-120];
        P[N]:=0.5171211*P[N-1]-0.998001*P[N-2]+0.5*W[N];
        Q[N]:=0.415408*Q[N-1]-0.998001*Q[N-2]-W[N];
        R[N]:=0.312556*R[N-1]-0.998001*R[N-2]+W[N];
        S[N]:=0.208848*S[N-1]-0.998001*S[N-2]-W[N];
        T[N]:=0.104567*T[N-1]-0.998001*T[N-2]+W[N];
        U[N]:=-0.998001*U[N-2]-0.666667*W[N];
        V[N]:=-0.104567*V[N-1]-0.998001*V[N-2]+0.333333*W[N];
        Y[N]:=P[N]+Q[N]+R[N]+S[N]+T[N]+U[N]+V[N];
(***************   PLOT IMPULSE RESPONSE VALUE   ***************)
        M:=(N-120)*2;
        draw(M,70,M,round(70-Y[N]*7),1);
      end;
  for N:=1 to 320 do
    begin
      FREQ:=(N-1)*3.141593/320;HR:=0;HI:=0;H:=0;
        for K:=1 to 120 do
          begin
            HR:=HR+Y[119+K]*cos(K*FREQ);
            HI:=HI+Y[119+K]*sin(K*FREQ);
          end;
        H:=sqrt(HR*HR+HI*HI);
(*************   PLOT FREQUENCY RESPONSE VALUE   ***************)
      draw(N,170,N,round(170-0.75*H),1);
    end;
end.
(*************************************************************)
```

```
(*****   PROGRAM NO.23    DISCRETE FOURIER TRANSFORM (DFT)  *****)
(******* BY DIRECT CALCULATION, AND WITH TABLE LOOKUP  ********)
(******  REAL SIGNAL, N=64, SPECTRAL MAGNITUDE & PHASE  ********)
program DSP23;
uses graph3;
var X: array[1..64] of real;
var S: array[1..64] of real;
var C: array[1..64] of real;
var XXR: array[1..64] of real;
var XXI: array[1..64] of real;
var MAG: array[1..64] of real;
var PHASE: array[1..64] of real;
var N,K,D,P,J,L,I: integer;
var B,W0,M1,M2,M3: real;
label 1,2;
(*********************  GENERATE SIGNAL  *********************)
begin
    for N:=1 to 64 do begin
      B:=2*3.141593*(N-1);
      X[N]:=sin(B/64)+cos(B/16)+0.6*cos(B/8)+0.5*sin(B/4);end;
  P:=1;
    for N:=1 to 64 do begin XXR[N]:=0;XXI[N]:=0;end;
   writeln('ENTER ANY INTEGER TO START FIRST DFT');readln(D);
     for K:=1 to 64 do begin
      for N:=1 to 64 do begin
       XXR[K]:=XXR[K]+X[N]*cos(2*3.141593*(K-1)*(N-1)/64);
       XXI[K]:=XXI[K]-X[N]*sin(2*3.141593*(K-1)*(N-1)/64);
     end;end;
  if P=1 then goto 2;
1:  for J:=1 to 64 do begin XXR[J]:=0;XXI[J]:=0;end;
   writeln('ENTER ANY INTEGER TO START SECOND DFT');readln(D);
(*******************  FORM TABLE LOOKUP ' *********************)
W0:=2*3.141593/64;
   for J:=1 to 64 do begin
    C[J]:=cos(W0*(J-1));S[J]:=sin(W0*(J-1));end;
     for K:=1 to 64 do begin
      for N:=1 to 64 do begin
        I:=(K-1)*(N-1);L:=I mod 64;
        XXR[K]:=XXR[K]+X[N]*C[L+1];
        XXI[K]:=XXI[K]-X[N]*S[L+1];
     end;end;
(************  CONVERT TO SPECTRAL MAGNITUDE & PHASE  **********)
2: for K:=1 to 64 do begin
     if abs(XXI[K])<0.001 then XXI[K]:=0;
     if abs(XXR[K])<0.00001 then XXR[K]:=0.00001;
     MAG[K]:=sqrt(XXR[K]*XXR[K] + XXI[K]*XXI[K]);
     PHASE[K]:=arctan(XXI[K]/XXR[K]);end;
(*******  NORMALISE TO MAXIMUM VALUE OF UNITY FOR PLOT  ********)
   M1:=0.001;M2:=0.001;M3:=0.001;
     for N:=1 to 64 do begin
       if abs(X[N])>M1 then M1:=abs(X[N]);
       if abs(MAG[N])>M2 then M2:=abs(MAG[N]);
       if abs(PHASE[N])>M3 then M3:=abs(PHASE[N]);end;
     for N:=1 to 64 do begin
       X[N]:=X[N]/M1;MAG[N]:=MAG[N]/M2;PHASE[N]:=PHASE[N]/M3;end;
(*******  PLOT SIGNAL, SPECTRAL MAGNITUDE, AND PHASE  *********)
  graphmode;
    for N:=1 to 64 do begin
       draw (5*(N-1),50,5*(N-1),round(50-X[N]*25),1);
       draw (5*(N-1),120,5*(N-1),round(120-MAG[N]*30),1);
       draw (5*(N-1),160,5*(N-1),round(160-PHASE[N]*20),1);end;
  P:=P+1;if P=2 then goto 1;
end.
(************************************************************)
```

```
(******   PROGRAM NO.24      FAST FOURIER TRANSFORM   **********)
program DSP24;
uses graph3;
var XR: array[1..512] of real;
var XI: array[1..512] of real;
var LIM1,LIM2,LIM3,L,R,M,K,N,Y,I,J: integer;
var T,D,X1,X2,PI,B1,B2,C1,C2,ARG,SIN1,COS1,MAG,MAX: real;
label 1,2,3,4,5;
(******   DEFINE TRANSFORM LENGTH AND ENTER INPUT DATA   ********)
begin
   N:=512;M:=9;
   for K:=1 to N do begin
     XR[K]:=0;XI[K]:=0;end;
   for K:=1 to 32 do begin
     XR[K]:=1;end;
(******   PLOT INPUT DATA USING HIGH RESOLUTION GRAPHICS   *******)
   hires;
   Y:=60; for K:=1 to N do begin
     draw(K,Y,K,round(Y-XR[K]*40),1);end;
(*****   SELECT TRANSFORM/INVERSE TRANSFORM AND PLOT OFFSET   ****)
   T:=1;D:=1;
 1: Y:=Y+60;IF T<0 THEN D:=N;
(********************   SHUFFLE INPUT DATA   ******************)
    LIM1:=N-1;LIM2:=trunc(N/2);J:=1;
    for I:=1 to LIM1 do begin
      if I>J-1 then goto 2;
      X1:=XR[J];X2:=XI[J];
      XR[J]:=XR[I];XI[J]:=XI[I];XR[I]:=X1;XI[I]:=X2;
 2: L:=LIM2;
 3: if L>J-1 then goto 4;
    J:=J-L;L:=trunc(L/2);goto 3;
 4: J:=J+L; end;
(*******************   IN-PLACE TRANSFORMATION   *****************)
    PI:=3.141593;
    for I:=1 to M do begin
      LIM1:=trunc(exp((I-1)*ln(2)));
      LIM2:=trunc(exp((M-I)*ln(2)));
        for L:=1 to LIM2 do begin
          for R:=1 to LIM1 do begin
            LIM3:=(R-1)+(L-1)*2*LIM1+1;
            B1:=XR[LIM3];B2:=XI[LIM3];C1:=XR[LIM3+LIM1];
            C2:=XI[LIM3+LIM1];
            ARG:=2*PI*(R-1)*LIM2/N;COS1:=cos(ARG);SIN1:=sin(ARG);
            X1:=C1*COS1+C2*SIN1*T;X2:=-C1*SIN1*T+C2*COS1;
            XR[LIM3]:=B1+X1;XI[LIM3]:=B2+X2;
            XR[LIM3+LIM1]:=B1-X1;XI[LIM3+LIM1]:=B2-X2;
   end;end;end;
   for K:=1 to N do begin
     XR[K]:=XR[K]/D;XI[K]:=XI[K]/D;end;
(*****************   PLOT OUTPUT (MAGNITUDE)   *****************)
  MAX:=0;
    for K:=1 to N do begin
      MAG:=sqrt(XR[K]*XR[K]+XI[K]*XI[K]);
      if MAG>MAX then MAX:=MAG;end;
    for K:=1 to N do begin
      draw(K,Y,K,round(Y-sqrt(XR[K]*XR[K]+XI[K]*XI[K])*40/MAX),1);
    end;
    if T<0 then GOTO 5;
    T:=-1;goto 1;
 5: end.
 (***************************************************************)
```

```
(*********  PROGRAM NO.25        FAST CONVOLUTION  *************)
program DSP25;
uses graph3;
var X: array[1..128] of real;
var H: array[1..128] of real;
var Y: array[1..128] of real;
var XR: array[1..128] of real;
var XI: array[1..128] of real;
var VR: array[1..128] of real;
var VI: array[1..128] of real;
var WI: array[1..128] of real;
var WR: array[1..128] of real;
var A,K,N,M,L,P,T,D,LIM1,LIM2,J,I,R,LIM3,XX,YY: integer;
var XHMAG,X1,X2,PI,B1,B2,C1,C2,ARG,COS1,SIN1,MAX,MAG,XT: real;
label 1,2,3,4;
begin
    for K:=1 to 128 do begin X[K]:=0;H[K]:=0;Y[K]:=0;end;
    N:=128;M:=7;
(*****  DEFINE AND PLOT INPUT SIGNAL AND IMPULSE RESPONSE  *****)
    for K:=1 to 64 do begin X[K]:=1;end;
    H[1]:=1;H[2]:=1.5;
    for K:=3 to 64 do begin H[K]:=1.5*H[K-1]-0.85*H[K-2];end;
 graphmode;
    for K:=1 to N do begin
    draw(K,70,K,round(70-25*X[K]),1);end;
    for K:=1 to N do begin
    draw(K+190,70,K+190,round(70-25*H[K]),1);end;
(*************  CONVOLVE IN TIME DOMAIN, THEN PLOT  ************)
    for K:=1 to 128 do begin
    for L:=1 to K do begin
        Y[K]:=Y[K]+X[L]*H[K-L+1];
    end;end;
    for K:=1 to N do begin
        draw(K+100,160,K+100,round(160-8*Y[K]),1);end;
    writeln('ENTER ANY INTEGER TO START FAST CONVOLN');
    readln(A);
 clearscreen;
  for P:=1 to 3 do begin
        if P=1 then begin T:=1;D:=1;
        for K:=1 to N do begin
            XR[K]:=X[K];XI[K]:=0;end end;
        if P=2 then begin T:=1; D:=1;
        for K:=1 to N do begin
            XR[K]:=H[K];XI[K]:=0;end end;
        if P<3 then goto 1;
        T:=-1;D:=N;
        for K:=1 to N do begin
            XR[K]:=VR[K]*WR[K]-VI[K]*WI[K];
            XI[K]:=VI[K]*WR[K]+VR[K]*WI[K];
            XHMAG:=sqrt((VR[K]*VR[K]+VI[K]*VI[K])*
                                    (WR[K]*WR[K]+WI[K]*WI[K]));
            draw(K,160,K,round(160-XHMAG/4),1);
        end;
(*******************  SHUFFLE INPUT DATA  ******************)
 1: LIM1:=N-1;LIM2:=trunc(N/2);J:=1;
    for I:=1 to LIM1 do begin
        if I>J-1 then GOTO 2;
        X1:=XR[J];X2:=XI[J];
```

```
          XR[J]:=XR[I];XI[J]:=XI[I];XR[I]:=X1;XI[I]:=X2;
2: L:=LIM2;
3: if L>J-1 then goto 4;
      J:=J-L;L:=trunc(L/2);goto 3;
4: J:=J+L;end;
(****************** . IN-PLACE TRANSFORMATION   *****************)
    PI:=3.141593;
    for I:=1 to M do begin
      LIM1:=round(exp((I-1)*ln(2)));
      LIM2:=round(exp((M-I)*ln(2)));
      for L:=1 to LIM2 do begin
        for R:=1 to LIM1 do begin
          LIM3:=(R-1)+(L-1)*2*LIM1+1;
          B1:=XR[LIM3];B2:=XI[LIM3];
          C1:=XR[LIM3+LIM1];C2:=XI[LIM3+LIM1];
          ARG:=2*PI*(R-1)*LIM2/N;COS1:=cos(ARG);SIN1:=sin(ARG);
          X1:=C1*COS1+C2*SIN1*T;X2:=-C1*SIN1*T+C2*COS1;
          XR[LIM3]:=B1+X1;XI[LIM3]:=B2+X2;
          XR[LIM3+LIM1]:=B1-X1;XI[LIM3+LIM1]:=B2-X2;
      end;end;end;
    for K:=1 to N do begin
      XR[K]:=XR[K]/D;XI[K]:=XI[K]/D;end;
(*** SAVE TRANSFORMS OF INPUT SIGNAL AND IMPULSE RESPONSE  ****)
    if P=1 then begin
      for K:=1 to N do begin
        VR[K]:=XR[K];VI[K]:=XI[K];
    end end;
    if P=2 then begin
      for K:=1 to N do begin
        WR[K]:=XR[K];WI[K]:=XI[K];
    end end;
(*************** FIND PEAK VALUE FOR PLOT   ******************)
    MAX:=0;
    for K:=1 to N do begin
      MAG:=sqrt(XR[K]*XR[K]+XI[K]*XI[K]);
      if MAG>MAX then MAX:=MAG;
    end;
(******************** SET PLOT OFFSETS  ******************)
    if P=1 then begin XX:=0;YY:=70 end;
    if P=2 then begin XX:=190;YY:=70 end;
    if P=3 then begin XX:=190;YY:=160 end;
(******* PLOT MAGNITUDE OF FFT, OR REAL PART OF IFFT  ********)
    for K:=1 to N do begin
      XT:=sqrt(XR[K]*XR[K]+XI[K]*XI[K]);
      if P=3 then XT:=XR[K];
      draw(K+XX,YY,K+XX,round(YY-XT*40/MAX),1);
    end;
  end;
end.
(*************************************************************)
```

```
(********    PROGRAM NO.26    PROBABILITY MASS FUNCTIONS   ********)
program DSP26;
uses graph3;
var X: array[0..300] of real;
var Y: array[0..300] of real;
var K,N,B,D: integer;
var C: real;
(******************    UNIFORM DISTRIBUTION   ******************)
begin
  for K:=0 to 300 do begin
    X[K]:=0;Y[K]:=0;end;
  randomize;graphmode;
    for K:=1 to 3000 do begin
      B:=trunc(random*300);X[B]:=X[B]+1;
      draw(B,50,B,round(50-X[B]),1);end;
(************   GAUSSIAN DISTRIBUTION (APPROXIMATE)   ***********)
  for K:=1 to 3000 do begin
    C:=0;for N:=1 to 30 do begin
          C:=C+random;end;
    D:=trunc(C*10);Y[D]:=Y[D]+1;
    draw(D,180,D,round(180-Y[D]),1);end;
end.
(*************************************************************)
 .
```

```
(*******   PROGRAM NO.27   ESTIMATES OF MEAN AND VARIANCE   *******)
program DSP27;
uses graph3;
var X: array[1..320] of real;
var M: array[1..16] of real;
var V: array[1..16] of real;
var J,K,N: integer;
var A,B,C,RM,RV,Z: real;
begin
  randomize;
  graphmode;
    for N:=1 to 16 do begin
    M[N]:=0;V[N]:=0;end;
(*******   FORM 320 VALUES OF GAUSSIAN SEQUENCE, AND PLOT   ******)
  for N:=1 to 320 do begin
    C:=0;for K:=1 to 12 do begin
         C:=C+random-0.5;end;
    X[N]:=C;draw(N,50,N,round(50-X[N]*10),1);
  end;
(************   ESTIMATE SAMPLE MEANS AND VARIANCES   ************)
  for J:=1 to 16 do begin
    for K:=1 to 20 do begin
      N:=K+20*(J-1)-1;M[J]:=M[J]+X[N]/20;
      V[J]:=V[J]+X[N]*X[N]/20;end;
    draw(N,100,N,round(100-M[J]*50),1);
    draw(N,160,N,round(160-V[J]*20),1);end;
    writeln('INPUT DUMMY VARIABLE');readln(Z);clearscreen;
(*************   ESTIMATE RUNNING MEAN AND VARIANCE   ************)
  A:=0;B:=0;
    for N:=1 to 320 do begin
    A:=A+X[N];RM:=A/N;B:=B+X[N]*X[N];RV:=B/N;
    draw(N,80,N,round(80-RM*50),1);
    draw(N,150,N,round(150-RV*20),1);end;
end.
(***************************************************************)
```

```
(******* PROGRAM NO.28     PROCESSING RANDOM SEQUENCES *******)
program DSP28;
uses graph3;
var GN: array[1..1200] of real;
var X: array[1..1200] of real;
var Y: array[1..1200] of real;
var H: array[1..10] of real;
var J: array[1..19] of real;
var XACF: array[1..101] of real;
var YACF: array[1..101] of real;
var EST: array[1..101] of real;
var N,K,L,M,P,Q: integer;
var C,SUM,MY,A,B,Z,MYE,VARE: real;
begin
  randomize;
  graphmode;
    for N:=1 to 1200 do begin
      Y[N]:=0;end;
    for N:=1 to 19 do begin
      J[N]:=0;end;
    for N:=1 to 101 do begin
      EST[N]:=0;end;
(************ ENTER PROCESSOR'S IMPULSE RESPONSE ************)
  H[1]:=1;H[2]:=2;H[3]:=1;H[4]:=-1;H[5]:=-2;
  H[6]:=-1;H[7]:=1;H[8]:=2;H[9]:=1;H[10]:=0;
(******* FORM GAUSSIAN NOISE, ZERO MEAN, UNIT VARIANCE *******)
  clearscreen;
  for N:=1 to 1200 do begin
    C:=0;for K:=1 to 12 do begin
          C:=C+random-0.5;
        end;
    GN[N]:=C;
  end;
(******** LOW-PASS FILTER & ADD UNIT MEAN TO FORM X[N] ********)
  for N:=4 to 1200 do begin
    X[N]:=(GN[N]+GN[N-1]+GN[N-2]+GN[N-3])*0.5+1;
  end;
(************ PLOT 320 VALUES OF X[N] ON SCREEN ************)
  for N:=1 to 320 do begin
    draw(N,50,N,round(50-X[N+100]*6),1);
  end;
(**************** DEFINE AND PLOT INPUT ACF ****************)
  for K:=1 to 101 do begin
    XACF[K]:=1;
  end;
  XACF[48]:=1.25;XACF[49]:=1.5;XACF[50]:=1.75;XACF[51]:=2;
  XACF[52]:=XACF[50];XACF[53]:=XACF[49];XACF[54]:=XACF[48];
  for N:=1 to 83 do begin
    draw(N,180,N,round(180-XACF[N+9]*25),1);
  end;
(********** FIND ACF OF PROCESSOR'S IMPULSE RESPONSE **********)
  for K:=1 to 10 do begin
    for L:=1 to K do begin
      J[K]:=J[K]+H[L]*H[10-K+L];
    end;
  end;
    for K:=11 to 19 do begin
      J[K]:=J[20-K];
    end;
(*************** PREDICT MEAN VALUE OF OUTPUT ***************)
  SUM:=0;for K:=1 to 10 do begin
          SUM:=SUM+H[K];
        end;
  MY:=1*SUM;
(*************** PREDICT AND PLOT OUTPUT ACF ***************)
  for M:=10 to 92 do begin
    SUM:=0;for K:=1 to 19 do begin
            SUM:=SUM+XACF[M-10+K]*J[K];
```

```
            end;
        YACF[M]:=SUM;
      end;
    for N:=1 to 83 do begin
      draw(N+120,180,N+120,round(180-YACF[N+9]*50/YACF[51]),1);
    end;
(************  CONVOLVE X[N] AND H[N] TO FORM Y[N]  ************)
    for N:=15 to 1200 do begin
      for K:=1 to 10 do begin
        Y[N]:=Y[N]+H[K]*X[N-K];
      end;
    end;
(*************  PLOT 320 VALUES OF Y[N] ON SCREEN  *************)
    A:=9/SQRT(YACF[51]);
      for N:=1 to 320 do begin
        draw(N,100,N,round(100-Y[N+100]*A),1);
      end;
(***************  ESTIMATE AND PLOT OUTPUT ACF  ***************)
    B:=20*YACF[51];
      for K:=1 to 100 do begin
        for M:=10 to 92 do begin
          P:=M-51;
            for L:=1 to 10 do begin
              Q:=10*(K-1)+L+60;EST[M]:=EST[M]+Y[Q]*Y[Q+P];
            end;
          draw(M+227,180,M+227,round(180-EST[M]/B),1);
        end;
      end;
    writeln('ENTER A DUMMY VARIABLE TO CONTINUE');readln(Z);
    clearscreen;
(******  PREDICT AND ESTIMATE OUTPUT MEAN AND VARIANCE  ******)
    SUM:=0;for N:=15 to 1200 do begin
             SUM:=SUM+Y[N];
           end;
    MYE:=SUM/1185;
    writeln('OUTPUT MEAN (PREDICTED)=');writeln(MY);
    writeln('OUTPUT MEAN (ESTIMATED)=');writeln(MYE);
    SUM:=0;for N:=15 to 1200 do begin
             SUM:=SUM+(Y[N]-MY)*(Y[N]-MY);
           end;
    VARE:=SUM/1185;
    writeln('OUTPUT VARIANCE (PREDICTED)=');
    writeln(YACF[51]-MY*MY);
    writeln('OUTPUT VARIANCE (ESTIMATED)=');writeln(VARE);
end.
(*************************************************************)
```

```pascal
(***********  PROGRAM NO.29     MATCHED FILTERING  ***********)
program DSP29;
uses graph3;
var X: array[1..420] of real;
var Y: array[1..420] of real;
var H: array[1..100] of real;
var S: array[1..100] of real;
var N,K: integer;
var NV,SUM,MX,MY: real;
begin
  for N:=1 to 420 do begin
  X[N]:=0;Y[N]:=0;end;
  randomize;
  graphmode;
(******************** DEFINE INPUT SIGNAL  ******************)
  S[1]:=1;S[2]:=0.575;
    for N:=3 to 60 do begin
      S[N]:=1.575*S[N-1]-0.9025*S[N-2];
    end;
(*******  LOAD TWO VERSIONS OF SIGNAL INTO INPUT ARRAY  ******)
  for N:=1 to 60 do begin
    X[N+110]:=S[N];X[N+280]:=X[N+280]+S[N];
  end;
(*******  ADD REQUIRED AMOUNT OF WHITE GAUSSIAN NOISE  *******)
  writeln('ENTER NOISE VARIANCE');readln(NV);
    for N:=1 to 420 do begin
      SUM:=0;
        for K:=1 to 12 do begin
          SUM:=SUM+random-0.5;
        end;
      X[N]:=X[N]+NV*SUM;
    end;
(****************** DEFINE IMPULSE RESPONSE  ****************)
  for N:=1 to 60 do begin
    H[N]:=S[61-N];
  end;
(***********  CONVOLVE INPUT AND IMPULSE RESPONSE  ***********)
  for N:=101 to 420 do begin
    for K:=1 to 100 do begin
      Y[N]:=Y[N]+H[K]*X[N-K];
    end;
  end;
(******  NORMALISE ARRAYS TO MAX VALUE OF UNITY FOR PLOT  ******)
  MX:=0;MY:=0;
    for N:=101 to 420 do begin
      if abs(X[N])>MX then MX:=abs(X[N]);
      if abs(Y[N])>MY then MY:=abs(Y[N]);
    end;
  for N:=1 to 420 do begin
    X[N]:=X[N]/MX;Y[N]:=Y[N]/MY;
  end;
(***********  PLOT INPUT AND OUTPUT SIGNAL ARRAYS  ***********)
  clearscreen;
    for N:=101 to 420 do begin
      draw(N-100,50,N-100,round(50-X[N]*22),1);
    end;
    for N:=101 to 420 do begin
      draw(N-100,150,N-100,round(150-Y[N]*22),1);
    end;
end.
(**************************************************************)
```

Appendix A2
Continuous-time Fourier Analysis

A2.1 SIGNALS, VECTORS, AND ORTHOGONAL FUNCTIONS

A fundamental notion of Fourier analysis is that signals may be broken down into sets of sinusoidal (or exponential) functions of appropriate frequency, amplitude, and phase. Conversely, we may in principle build up, or *synthesize*, a signal waveform by adding together a number of such basic functions. In theory, it may be necessary to add an infinite number of them in order to synthesize the signal perfectly. In practice, of course, we must work with a finite number, producing a waveform which is an *approximation* to the actual shape.

If a continuous signal is strictly repetitive, or *periodic*, then the sinusoidal waves needed to synthesize it are harmonically related. In other words, their frequencies bear a simple integer relationship to one another. Let us take an example. Suppose we wish to build up the periodic sawtooth signal shown in Figure A2.1. Using formulae which we develop in the next section, this may be represented by the following *Fourier Series*

$$x(t) = \frac{2}{\pi} \sin\omega_0 t - \frac{1}{\pi} \sin 2\omega_0 t + \frac{2}{3\pi} \sin 3\omega_0 t - \frac{1}{2\pi} \sin 4\omega_0 t$$

$$+ \frac{2}{5\pi} \sin 5\omega_0 t - \ldots \tag{A2.1}$$

The sawtooth therefore contains the angular frequencies ω_0 (known as the *fundamental* component), $2\omega_0$ (the *second harmonic*), $3\omega_0$ (the *third harmonic*), and so on, with amplitudes which diminish with frequency. Note that the fundamental component $(2/\pi)\sin\omega_0 t$ has the same period $(T_0 = 2\pi/\omega_0)$ as the sawtooth itself. Theoretically, the series contains an infinite number of terms. If we sum a limited number of them — say just the first four, which are drawn in Figure A2.1(b) — we obtain an approximation to the true signal. This approximation is shown in part (a) of the figure. The approximation is poorest in those regions where the sawtooth undergoes sharp changes or transitions. This is not surprising, because sudden changes are rich in the higher-order harmonics which we have omitted.

The Fourier Series for the sawtooth is summarized graphically by the *frequency spectrum* of Figure A2.1(c), which indicates the amplitudes of the various

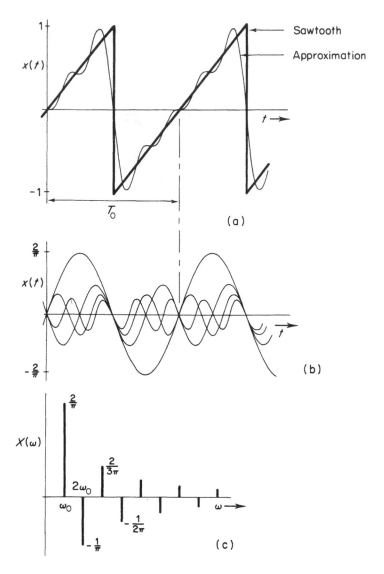

Figure A2.1 (a) A periodic sawtooth signal, (b) the first four terms of its Fourier Series, and (c) its frequency spectrum. Also shown in part (a) is an approximation of the signal formed by summing the four sinusoids in part (b).

components in the signal. A spectrum of this type is called a *line* spectrum because it contains a number of distinct frequency components which are drawn as a set of vertical lines. Note that we use the upper-case symbol $X(\omega)$ for the spectrum, to denote a frequency-domain function. In this particular case the complete spectral information is represented by a single diagram, because the phase relationships between the various components are very simple. Successive terms are merely inverted (equivalent to 180° phase-shift). But in most cases the phase relationships are more complicated, and must be indicated separately.

An important advantage of representing a practical signal as a set of sine and cosine (or exponential) functions is that these functions possess the property of *orthogonality*. This property is fundamental to Fourier analysis, and has a direct bearing on the question of approximation. In order to explain clearly what it means, it is helpful to consider briefly some analogies between signals and vectors.

Most readers will be familiar with simple vector quantities such as velocity or force, which are specified by a direction as well as a magnitude. Suppose we have two such vectors V_1 and V_2, as shown in Figure A2.2. We may define the component of vector V_1 along vector V_2 by constructing the perpendicular from the end of V_1 on to V_2, and write

$$V_1 = C_{12}V_2 + V_e \qquad (A2.2)$$

where C_{12} is a scalar multiplier, or coefficient. Thus if we are trying to approximate V_1 by a vector in the direction of V_2, the error in the approximation is the vector V_e. The best approximation is obtained when C_{12} is chosen to make the error vector as short as possible (as in the figure). We then say that the component of V_1 along V_2 equals $C_{12}V_2$. If C_{12} is zero, then one vector has no component along the other and the vectors are *orthogonal*. Conversely if C_{12} = 1 and V_e is zero, V_1 and V_2 must be identical in both magnitude and direction.

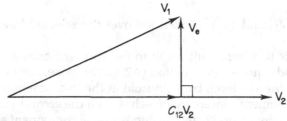

Figure A2.2 Vectors and orthogonality.

Similar ideas apply to signals. Suppose we wish to approximate a signal $x_1(t)$ over a certain interval $t_1 < t < t_2$ by another signal $x_2(t)$. In the present context we may think of $x_2(t)$ as a sinusoidal or exponential waveform. We therefore write

$$x_1(t) = C_{12}x_2(t) + x_e(t), \qquad t_1 < t < t_2 \qquad (A2.3)$$

where $x_e(t)$ is a third signal representing the approximation error. Now $x_e(t)$ will, in general, vary over the specified time interval, and we need to consider carefully how it may be 'minimized' by adjusting the value of C_{12}.

At first sight, it might appear sensible to minimize the *average* value of $x_e(t)$ over the interval $t_1 < t < t_2$. However, the disadvantage of such an *error criterion* is that positive and negative errors occurring at different instants would tend to cancel each other out. This difficulty is avoided if we adjust C_{12} to minimize the average, or mean, *squared* error (if we think of $x_e(t)$ as an electrical voltage or current this is equivalent to minimizing the *error power*; it is also the same as minimizing the *root mean square, or r.m.s. error*). The mean square error over the chosen time interval is given by

$$\overline{x_e^2(t)} = \frac{1}{(t_2 - t_1)} \int_{t_1}^{t_2} x_e^2(t) dt$$

$$= \frac{1}{(t_2 - t_1)} \int_{t_1}^{t_2} \{x_1(t) - C_{12}x_2(t)\}^2 \, dt \qquad \text{(A2.4)}$$

The value of C_{12} which minimizes this function may be found by differentiating with respect to C_{12} and equating to zero. Thus

$$\frac{d}{dC_{12}} \left[\frac{1}{(t_2 - t_1)} \int_{t_1}^{t_2} \{x_1(t) - C_{12}x_2(t)\}^2 \, dt \right] = 0 \qquad \text{(A2.5)}$$

This yields

$$C_{12} = \frac{\int_{t_1}^{t_2} x_1(t)x_2(t) \, dt}{\int_{t_1}^{t_2} x_2^2(t) \, dt} \qquad \text{(A2.6)}$$

By direct analogy with the vectors shown in Figure A2.2, if C_{12} is zero we say that $x_1(t)$ contains no component of $x_2(t)$, and that the signals are orthogonal in the interval $t_1 < t < t_2$. In this case it is clear that

$$\int_{t_1}^{t_2} x_1(t)x_2(t) = 0 \qquad \text{(A2.7)}$$

Conversely if $x_1(t)$ and $x_2(t)$ are identical over the selected interval, C_{12} must equal unity.

We now relate this important result to the Fourier Series for the sawtooth waveform already quoted in equation (A2.1). Suppose, for example, we try to approximate the sawtooth by a sinusoid at the fundamental frequency ω_0, without for the moment concerning ourselves with the second and higher-order harmonics. Since the sawtooth and the fundamental component are both strictly periodic with the same period T_0, it is clear that an approximation over one complete period must be equally valid for all other periods of the waveforms — and therefore for all time. Denoting the sawtooth signal in Figure A2.1(a) by $x_1(t)$, one period is

$$x_1(t) = \frac{2t}{T_0}, \qquad 0 < t < \frac{T_0}{2}$$

$$= \frac{2t}{T_0} - 2, \qquad \frac{T_0}{2} < t < T_0$$

Alternatively, if we write $T_0 = 2\pi/\omega_0$, then

$$x_1(t) = \frac{\omega_0 t}{\pi}, \qquad 0 < t < \frac{\pi}{\omega_0}$$

$$= \frac{\omega_0 t}{\pi} - 2, \qquad \frac{\pi}{\omega_0} < t < \frac{2\pi}{\omega_0} \qquad \text{(A2.8)}$$

Let us now approximate it over the interval $0 < t < 2\pi/\omega_0$ by the sinusoid

$$x_2(t) = \sin\omega_0 t$$

Equation (A2.6) becomes

$$C_{12} = \frac{\displaystyle\int_0^{\pi/\omega_0} \frac{\omega_0 t}{\pi} \sin(\omega_0 t)dt + \int_{\pi/\omega_0}^{2\pi/\omega_0} \left(\frac{\omega_0 t}{\pi} - 2\right) \sin(\omega_0 t) \, dt}{\displaystyle\int_0^{2\pi/\omega_0} \sin^2(\omega_0 t) \, dt} \tag{A2.9}$$

from which it may be shown that

$$C_{12} = \frac{2}{\pi} \tag{A2.10}$$

Therefore the 'amount' of the signal $x_2(t) = \sin\omega_0 t$ present in the sawtooth is $(2/\pi)\sin\omega_0 t$. Any other amount would give a larger mean-square error over a complete period.

Referring back to equation (A2.1), we see that the fundamental component of the sawtooth's Fourier Series is also $(2/\pi)\sin\omega_0 t$. This is because the Fourier method is also based upon a minimum mean-square (or 'least-square') error criterion. When we meet the formulae for calculating the amplitudes of the various harmonics in the next section, their similarity with equation (A2.6) will be apparent.

A little earlier we mentioned that an advantage of sine and cosine (or exponential) functions for signal representation is their mutual orthogonality. Thus the set of functions

$$\text{and} \quad \begin{matrix} \sin n\omega_0 t \\ \cos n\omega_0 t \end{matrix} \quad n = 0, 1, 2, 3, \ldots$$

are orthogonal over any time interval equal to one complete period of the fundamental frequency ω_0. When $n = 0$, $\sin n\omega_0 t = 0$ and $\cos n\omega_0 t = 1$, so that the complete set of orthogonal functions is

$$1, \sin\omega_0 t, \cos\omega_0 t, \sin2\omega_0 t, \cos2\omega_0 t, \ldots$$

The first member of the set represents a steady level or DC component, also referred to as a zero-frequency component.

The advantage of orthogonality for signal approximation may be summarized as follows. Suppose we have approximated a periodic signal — such as the sawtooth already discussed — by its fundamental component, and we now wish to improve the approximation by incorporating a second harmonic component. It is obviously a nuisance if, having found the second harmonic, we have to go back and recalculate the fundamental. But it may be shown that, since the various harmonics are mutually orthogonal, the value of any one component is unaffected by incorporating further harmonics in the approximation. The approximation is always the best (in the least-square error sense) that can be achieved with a given number of harmonic terms. This is a valuable feature of Fourier analysis.

The foregoing discussion of signal approximation and orthogonality leads naturally to the continuous-time Fourier Series, which will now be presented in more detail.

A2.2   THE FOURIER SERIES

A2.2.1   ANALYSIS AND SYNTHESIS OF PERIODIC SIGNALS

The basis of the Fourier Series is that a complicated periodic waveform may be analyzed into, or synthesized from, a number of harmonically related sine and cosine functions constituting an orthogonal set. If we have a continuous periodic signal $x(t)$ with a period T_0, then $x(t)$ may be represented by the series

$$x(t) = A_0 + \sum_{k=1}^{\infty} B_k \cos k\omega_0 t + \sum_{k=1}^{\infty} C_k \sin k\omega_0 t \qquad (A2.11)$$

where $\omega_0 = 2\pi/T_0$. Certain restrictions, known as the *Dirichlet conditions*, must be placed on $x(t)$ for the series to be valid. The integral of the magnitude of $x(t)$ over a complete period must be finite, and the signal can have only a finite number of discontinuities in any finite interval. Fortunately, these conditions do not exclude signals of practical interest.

Let us consider how the coefficients of the Fourier Series — A_0, B_k, and C_k — are determined. Recalling the minimum mean-square error criterion described in the previous section, and using equation (A2.6), we obtain

$$A_0 = \frac{\int_0^{T_0} x(t)\,(1)\mathrm{d}t}{\int_0^{T_0} (1)\mathrm{d}t} = \frac{1}{T_0} \int_0^{T_0} x(t)\mathrm{d}t = \frac{\omega_0}{2\pi} \int_0^{2\pi/\omega_0} x(t)\mathrm{d}t \qquad (A2.12)$$

A_0 is simply the average, or mean, value of the signal $x(t)$. It is also called its DC level, or zero-frequency component. The cosine coefficients B_k are given by

$$B_k = \frac{\int_0^{T_0} x(t)\,\cos(k\omega_0 t)\mathrm{d}t}{\int_0^{T_0} \cos^2(k\omega_0 t)\mathrm{d}t} \qquad (A2.13)$$

Now the r.m.s. value of a cosine (or sine) wave is $1/\sqrt{2}$, and its mean-square value is $1/2$. The latter may be found by integrating the square of the waveform over one (or more) complete periods, and dividing by the baseline, giving

$$\frac{1}{T_0} \int_0^{T_0} \cos^2(k\omega_0 t)\mathrm{d}t, \qquad k = 1, 2, 3$$

Since this equals $1/2$, the denominator of equation (A2.13) must equal $T_0/2$ for any integer value of k. Thus

$$B_k = \frac{2}{T_0} \int_0^{T_0} x(t)\,\cos(k\omega_0 t)\mathrm{d}t = \frac{\omega_0}{\pi} \int_0^{2\pi/\omega_0} x(t)\,\cos(k\omega_0 t)\mathrm{d}t \qquad (A2.14)$$

and similarly

$$C_k = \frac{2}{T_0} \int_0^{T_0} x(t)\,\sin(k\omega_0 t)\mathrm{d}t = \frac{\omega_0}{\pi} \int_0^{2\pi/\omega_0} x(t)\,\sin(k\omega_0 t)\mathrm{d}t \qquad (A2.15)$$

In other words, we may find the 'amount' of any sine or cosine harmonic in a periodic signal $x(t)$ by multiplying the signal by that sine or cosine, and integrating over a complete period. Although the integration limits are given as $t = 0$ to $t = 2\pi/\omega_0$ in the above formulae, we may in fact integrate over any complete period. For example, it is often convenient to use the limits $t = -\pi/\omega_0$ to $+\pi/\omega_0$.

If a signal is even, such that $x(t) = x(-t)$, then it can only be built up using functions which are themselves even. An odd signal, for which $x(t) = -x(-t)$, can only be built up from other odd functions. Now all cosines are even, and all sines are odd. So if $x(t)$ is even, all its Fourier Series coefficients C_k must be zero, and if it is odd, all coefficients B_k must be zero. The sawtooth signal in Figure A2.1 is a good example of this. Since it is an odd time function, its Fourier Series contains only sine components.

Let us now evaluate the Fourier Series coefficients for the sawtooth waveform and hence confirm the validity of equation (A2.1). We first note that all its cosine coefficients B_k must be zero. Furthermore, since the waveform clearly has equal areas above and below the axis, its average value, or DC component, is also zero. Thus $A_0 = 0$. In calculating the sine coefficients C_k it is convenient to integrate over the interval $t = \pm\pi/\omega_0$, since this avoids the discontinuity in the signal. Using equation (A2.15) we obtain

$$C_k = \frac{\omega_0}{\pi} \int_{-\pi/\omega_0}^{\pi/\omega_0} x(t)\,\sin(k\omega_0 t)\mathrm{d}t$$

$$= \frac{\omega_0}{\pi} \int_{-\pi/\omega_0}^{\pi/\omega_0} \frac{\omega_0 t}{\pi}\,\sin(k\omega_0 t)\mathrm{d}t$$

$$= \left(\frac{\omega_0}{\pi}\right)^2 \int_{-\pi/\omega_0}^{\pi/\omega_0} t\,\sin(k\omega_0 t)\mathrm{d}t \qquad (A2.16)$$

Integration by parts gives

$$C_k = \left(\frac{\omega_0}{\pi}\right)^2 \left\{ \left[\frac{-t}{k\omega_0}\cos(k\omega_0 t)\right]_{-\pi/\omega_0}^{\pi/\omega_0} + \int_{-\pi/\omega_0}^{\pi/\omega_0} \frac{\cos(k\omega_0 t)}{k\omega_0}\,\mathrm{d}t \right\}$$

$$= \left(\frac{\omega_0}{\pi}\right)^2 \left\{ \frac{-\pi}{k\omega_0^2}\{\cos(k\pi) + \cos(-k\pi)\} + \left[\frac{\sin k\omega_0 t}{k^2\omega_0^2}\right]_{-\pi/\omega_0}^{\pi/\omega_0} \right\} \qquad (A2.17)$$

Now $\cos k\pi = \cos(-k\pi)$, since a cosine function is even, and $\sin k\pi = 0 = \sin(-k\pi)$ for all values of the integer k. Hence

$$C_k = \left(\frac{\omega_0}{\pi}\right)^2 \left(\frac{-\pi}{k\omega_0^2}\right) 2\cos k\pi = -\frac{2}{k\pi}\cos k\pi \qquad (A2.18)$$

If k is 1, 3, 5 ... then $\cos k\pi = -1$; whereas if k is 2, 4, 6 ... then $\cos k\pi = 1$. Successive coefficients C_k are therefore as follows

$$C_1 = -\frac{2}{\pi}(-1) = \frac{2}{\pi}; \qquad C_2 = -\frac{2}{2\pi}(1) = -\frac{1}{\pi}$$

$$C_3 = -\frac{2}{3\pi}(-1) = \frac{2}{3\pi}; \qquad C_4 = -\frac{2}{4\pi}(1) = -\frac{1}{2\pi}$$

and so on. These results confirm the Fourier Series for the sawtooth signal given in equation (A2.1).

We have already noted that the Fourier Series is simplified in the case of an even or odd function, by losing either its sine or its cosine terms. A different type of simplification occurs if a signal possesses what is known as *half-wave symmetry*. In mathematical terms, half-wave symmetry exists when

$$x(t) = -x\{t + (T_0/2)\} = -x\{t + (\pi/\omega_0)\} \qquad (A2.19)$$

In other words, any two values of the signal separated by half a period are equal in magnitude but opposite in sign. Figure A2.3 shows that a sine wave of period T_0 and its third harmonic both have this property, whereas the second harmonic does not. Generalizing, only odd-order harmonics exhibit half-wave symmetry. Therefore any periodic signal $x(t)$ with half-wave symmetry cannot contain even-order harmonic components.

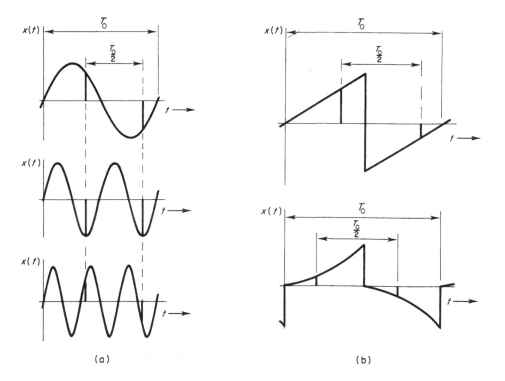

(a)   (b)

Figure A2.3 Illustrations of half-wave symmetry. All the signals shown exhibit this property, except the second-harmonic sinusoid in part (a) and the upper signal in part (b).

An important example of a waveform with half-wave symmetry is the square-wave shown in Figure A2.4. In this case $x(t)$ is an even time function with zero mean value. Hence all the sine coefficients C_k must be zero, and $A_0 = 0$. Since there is half-wave symmetry, we must expect the cosine coefficients B_k to be zero for $k = 2, 4, 6 \ldots$ Equation (A2.14) gives

$$B_k = \frac{\omega_0}{\pi} \int_0^{2\pi/\omega_0} x(t) \cos(k\omega_0 t) dt$$

However, in this case it is slightly more convenient to integrate over the period between $t = -\pi/2\omega_0$ and $t = 3\pi/2\omega_0$, giving

$$B_k = \frac{\omega_0}{\pi} \left\{ \int_{-\pi/2\omega_0}^{\pi/2\omega_0} (1)\cos(k\omega_0 t) dt + \int_{\pi/2\omega_0}^{3\pi/2\omega_0} (-1)\cos(k\omega_0 t) dt \right\} \qquad (A2.20)$$

This reduces to

$$B_k = \frac{\omega_0}{\pi} \frac{1}{k\omega_0} \left\{ \sin\left(\frac{k\pi}{2}\right) - \sin\left(\frac{-k\pi}{2}\right) - \sin\left(\frac{3k\pi}{2}\right) + \sin\left(\frac{k\pi}{2}\right) \right\}$$

$$(A2.21)$$

Since sine functions are odd

$$\sin \frac{(-k\pi)}{2} = -\sin \frac{k\pi}{2}$$

hence

$$B_k = \frac{1}{\pi k} \left\{ 3 \sin\left(\frac{k\pi}{2}\right) - \sin\left(\frac{3k\pi}{2}\right) \right\} \qquad (A2.22)$$

Now if k is an even number

$$\sin\left(\frac{k\pi}{2}\right) = 0 = \sin\left(\frac{3k\pi}{2}\right)$$

giving $B_k = 0$. This confirms that, as expected, the signal contains no even-order harmonics. If $k = 1, 5, 9 \ldots$ then

$$3 \sin\left(\frac{k\pi}{2}\right) - \sin\left(\frac{3k\pi}{2}\right) = 3 - (-1) = 4$$

and if $k = 3, 7, 11 \ldots$ then

$$3 \sin\left(\frac{k\pi}{2}\right) - \sin\left(\frac{3k\pi}{2}\right) = -3 - (1) = -4$$

Successive cosine coefficients are therefore as follows

$$B_1 = \frac{4}{\pi}; \quad B_2 = 0; \quad B_3 = -\frac{4}{3\pi}; \quad B_4 = 0; \quad B_5 = \frac{4}{5\pi} \text{ and so on}$$

The Fourier Series representation of the signal is therefore

$$x(t) = \frac{4}{\pi} \left(\cos \omega_0 t - \frac{1}{3} \cos 3\omega_0 t + \frac{1}{5} \cos 5\omega_0 t - \frac{1}{7} \cos 7\omega_0 t + \ldots \right)$$
(A2.23)

In the foregoing examples, we have always integrated over a complete period to derive the coefficients. However, in the case of an odd or even signal it is sufficient, and often simpler, to integrate over only one-half of a period and to multiply the result by 2. Furthermore, if the signal is not only even or odd, but also displays half-wave symmetry, it is adequate to integrate over a quarter of a period, and multiply by 4. Such closer limits are possible because the function being integrated repeats twice within one period when the signal is either even or odd, and four times within one period when it also exhibits half-wave symmetry.

The amount of work involved in calculating the Fourier Series coefficients of a signal is therefore reduced if the waveform is either even or odd, and this may sometimes be arranged by a judicious choice of time origin. For example, Figure A2.5 shows three versions of a periodic square wave which differ only in their time origin. Part (a) shows an even signal, symmetrial about $t = 0$, and as we have already seen its fundamental component is $(4/\pi)\cos\omega_0 t$. The square wave in part (b) is identical except that it is odd, and has a fundamental equal to $(4/\pi)\sin\omega_0 t$. The shift of time origin has therefore merely had the effect of converting a Fourier Series containing only cosines (a *cosine series*) into one containing only sines (a *sine series*). But the amplitude of a component at any given frequency is, as we would expect, unaltered. The situation in part

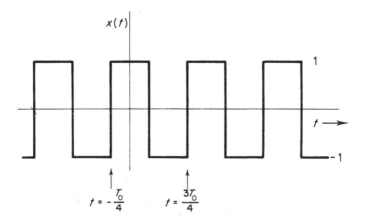

Figure A2.4   A periodic square-wave signal.

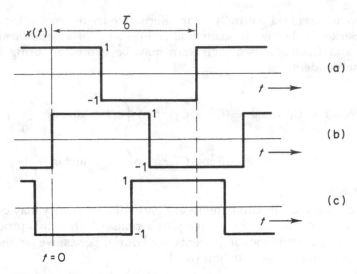

Figure A2.5 Three versions of a periodic square wave.

(c) of the figure is, however, rather more complicated because the square wave is neither even nor odd, and must be expected to include both sine and cosine terms in its Fourier Series.

The clue to the relationship between the values of the various coefficients in case (c) and those in cases (a) and (b) lies in the *average power* of the waveform. (This is the same as its mean-square value if we regard the signal as a voltage applied across, or a current flowing through, a resistor of value 1 ohm). Suppose we analyzed the signal in part (c) of Figure A2.5, and found that it contained fundamental components $B_1\cos\omega_0 t$ and $C_1\sin\omega_0 t$. Now the mean-square value of a unit-amplitude sine or cosine is 1/2. The total power represented by our two fundamental components is therefore

$$\frac{B_1^2}{2} + \frac{C_1^2}{2}$$

However, this value must be the same for all three signals shown in Figure A2.5, since the average power of a signal cannot be altered by a mere shift of time origin. Since in case (a) we have already found that $B_1 = 4/\pi$ and $C_1 = 0$, and we know that in case (b) $C_1 = 4/\pi$ and $B_1 = 0$, we conclude that for a waveform such as that shown in part (c) of the figure

$$\frac{B_1^2}{2} + \frac{C_1^2}{2} = \frac{1}{2}\left(\frac{4}{\pi}\right)^2, \quad \text{or } B_1^2 + C_1^2 = \left(\frac{4}{\pi}\right)^2 \qquad \text{(A2.24)}$$

Hence as the time-origin of a signal is shifted, the various sine and cosine coefficients of its Fourier Series change, but the *sum of the squares* of any pair of coefficients B_k and C_k remains constant.

The above ideas lead naturally to an alternative trigonometric form for the Fourier Series. If the two fundamental components of a periodic signal are $B_1\cos\omega_0 t$ and $C_1\sin\omega_0 t$, then their sum may be expressed using standard trigonometric identities

$$B_1 \cos\omega_0 t + C_1 \sin\omega_0 t = (B_1^2 + C_1^2)^{1/2} \cos\left(\omega_0 t - \arctan \frac{C_1}{B_1}\right)$$

$$= (B_1^2 + C_1^2)^{1/2} \sin\left(\omega_0 t + \arctan \frac{B_1}{C_1}\right) \quad (A2.25)$$

Thus the cosine and sine components at a particular frequency may be expressed as a single cosine, or sine, wave plus a phase-shift. If this procedure is applied to all the harmonic components of a Fourier Series, we get the following alternative forms for equation (A2.11)

$$x(t) = A_0 + \sum_{k=1}^{\infty} (B_k^2 + C_k^2)^{1/2} \cos(k\omega_0 t - \phi_k) \quad (A2.26)$$

or

$$x(t) = A_0 + \sum_{k=1}^{\infty} (B_k^2 + C_k^2)^{1/2} \sin(k\omega_0 t + \theta_k) \quad (A2.27)$$

where

$$\phi_k = \arctan^{-1} \frac{C_1}{B_1} \quad \text{and} \quad \theta_k = \arctan \frac{B_1}{C_1}$$

This is a good moment to consider the *total power* in the signal. The average power represented by any one frequency component is, as argued above, given by

$$\frac{1}{2} (B_k^2 + C_k^2)$$

The power represented by the signal's average value A_0 is simply A_0^2. Therefore the total mean power is given by

$$A_0^2 + \frac{1}{2} \sum_{k=1}^{\infty} (B_k^2 + C_k^2)$$

However, the total mean power must also equal the average value of $\{x(t)\}^2$ taken over one complete period, so that

$$A_0^2 + \frac{1}{2} \sum_{k=1}^{\infty} (B_k^2 + C_k^2) = \frac{1}{T_0} \int_{-T_0/2}^{T_0/2} \{x(t)\}^2 \, dt \qquad (A2.28)$$

This is a version of *Parseval's Theorem*, and expresses an interesting tie-up between the frequency-domain and the time-domain. The total power in any periodic signal may be found *either* by adding together the powers represented by the individual frequency components in its Fourier Series, *or* as the mean-square value of its time-domain waveform.

A2.2.2   THE EXPONENTIAL FORM OF THE SERIES

Our work on the continuous-time Fourier Series has so far concentrated on sines and cosines, because this is the approach which most people find easiest to visualize. However, these sines and cosines may be expressed as pairs of imaginary exponential signals, giving an equivalent *exponential form*. This is very important, because it leads naturally into the Fourier Transform which we develop and use in the next section.

The general trigonometric form of the Fourier Series has already been expressed in equation (A2.11)

$$x(t) = A_0 + \sum_{k=1}^{\infty} B_k \cos k\omega_0 t + \sum_{k=1}^{\infty} C_k \sin k\omega_0 t$$

It may be recast into the following exponential form

$$x(t) = \cdots + a_{-2} \exp(-2j\omega_0 t) + a_{-1} \exp(-j\omega_0 t)$$
$$+ a_0 + a_1 \exp(j\omega_0 t) + a_2 \exp(2j\omega_0 t) + \cdots$$

$$= \sum_{k=-\infty}^{\infty} a_k \exp(jk\omega_0 t) \qquad (A2.29)$$

These two forms look rather different, but they are in fact exact equivalents. This may be shown using the identities

$$\cos\theta = \frac{1}{2}\{\exp(j\theta) + \exp(-j\theta)\} \quad \text{and} \quad \sin\theta = \frac{-j}{2}\{\exp(j\theta) - \exp(-j\theta)\}$$

with $\theta = k\omega_0 t$. Substitution into equation (A2.11) yields equation (A2.29) without difficulty. The coefficients of the two forms are related as follows

$$a_0 = A_0; \quad a_k = \frac{1}{2}(B_k - jC_k), \quad k > 0; \quad a_k = \frac{1}{2}(B_k + jC_k), \quad k < 0$$
$$(A2.30)$$

We therefore see that the coefficients a_k of the exponential form are generally complex, and occur in complex conjugate pairs. The real part of a pair of complex coefficients denotes the magnitude of the cosine component at the relevant harmonic frequency, whereas the imaginary part denotes the magnitude of the sine component. If the signal $x(t)$ is even, it has a cosine series, and all the coefficients a_k in the exponential Fourier Series are real. If $x(t)$ is odd, it has a sine series, and all the coefficients a_k are imaginary.

The exponential form of the Fourier Series involves the concept of *negative frequency*, which stems directly from the identities for $\cos\theta$ and $\sin\theta$ quoted above. For example, if we put $\theta = k\omega_0 t$, a cosine signal of amplitude B_k is written as

$$B_k \cos k\omega_0 t = \frac{B_k}{2} \exp(j\{k\omega_0\}t) + \frac{B_k}{2} \exp(j\{-k\omega_0\}t) \qquad (A2.31)$$

If we plot its *exponential* components on a spectral diagram, the cosine is represented by two spectral lines of height $B_k/2$, one at frequency $(+k\omega_0)$ and the other at frequency $(-k\omega_0)$. The frequency scale is therefore formally extended to include negative as well as positive frequencies, with each cosine component providing *two* spectral lines. Similarly, each sine component is represented by a spectral line at frequency $(+k\omega_0)$ and another at $(-k\omega_0)$. Since these have imaginary coefficients they cannot be plotted on the same diagram as the real coefficients representing the cosines. We therefore see that negative frequencies are a consequence of using pairs of exponential signals to represent sines and cosines.

Although the coefficients a_k of the exponential Fourier Series may be found from the trigonometric coefficients B_k and C_k, this is not normally very economical. It is better to estimate the values of a_k directly. Since, as expression (A2.30) shows

$$a_k = \frac{1}{2}(B_k - jC_k), \qquad k > 0$$

we may use the expressions for B_k and C_k given earlier by equations (A2.14) and (A2.15) to yield

$$a_k = \frac{1}{2}\frac{\omega_0}{\pi} \left\{ \int_{-\pi/\omega_0}^{\pi/\omega_0} x(t) \cos(k\omega_0 t)dt - j \int_{-\pi/\omega_0}^{\pi/\omega_0} x(t) \sin(k\omega_0 t)dt \right\}$$

giving

$$a_k = \frac{\omega_0}{2\pi} \int_{-\pi/\omega_0}^{\pi/\omega_0} x(t) \exp(-jk\omega_0 t)dt \qquad (A2.32)$$

This result may also be written in terms of the fundamental period T_0 as

$$a_k = \frac{1}{T_0} \int_{-T_0/2}^{T_0/2} x(t) \exp(-2\pi jkt/T_0)dt \qquad (A2.33)$$

It is straightforward to show that these last two equations also hold good when $k < 0$, and when $k = 0$. They are therefore valid for all integer values of k, positive, negative, and zero. Just as equation (A2.29) shows how to *synthesize* any signal $x(t)$ by adding together, or superimposing, a whole set of weighted imaginary exponentials, so equations (A2.32) and (A2.33) show how to *analyze* $x(t)$ by finding out 'how much' of each exponential is present in it. These are key results. As we have already hinted, they lead on directly to the extremely important Fourier Transform covered in the following section.

Before ending our discussion on the exponential Fourier Series representation of continuous periodic signals, we should note that integration over half a period is admissible if $x(t)$ exhibits half-wave symmetry, just as it is when using the trigonometric series. On the other hand, the integration interval may *not* be shortened on account of $x(t)$ being either even or odd. This was possible with the trigonometric form because cosine and sine waveforms are themselves even and odd respectively, but $\exp(-jk\omega_0 t)$ is neither even nor odd, and such a simplification is not allowed.

We now illustrate the exponential Fourier Series by finding the coefficients a_k of the periodic triangular signal shown in Figure A2.6. Note that it exhibits half-wave symmetry, so the coefficients a_k must be zero for k even. We may integrate over the half-period between $t = 0$ and π/ω_0, and multiply the result by 2. Over this half-period the signal is given by

$$x(t) = \frac{2\omega_0 t}{\pi} - 1 \tag{A2.34}$$

and using equation (A2.32) we obtain

$$a_k = \frac{\omega_0}{\pi} \int_0^{\pi/\omega_0} \left(\frac{2\omega_0 t}{\pi} - 1 \right) \exp(-jk\omega_0 t)\, dt, \qquad k = 1, 3, 5 \ldots$$

$$\therefore a_k = \frac{2\omega_0^2}{\pi^2} \int_0^{\pi/\omega_0} t \exp(-jk\omega_0 t)\, dt - \frac{\omega_0}{\pi} \int_0^{\pi/\omega_0} \exp(-jk\omega_0 t)\, dt \tag{A2.35}$$

The first expression may be integrated by parts. Finally we obtain

$$a_k = \frac{2}{\pi^2} \left(\frac{\pi \exp(-jk\pi)}{-jk} + \frac{\exp(-jk\pi)}{k^2} - \frac{1}{k^2} \right) + \frac{1}{jk\pi} \{ \exp(-jk\pi) - 1 \} \tag{A2.36}$$

Now if k is odd, $\exp(-jk\pi) = -1$, and therefore

$$a_k = \frac{2}{\pi^2} \left(\frac{\pi}{jk} - \frac{2}{k^2} \right) - \frac{2}{jk\pi} = -\frac{4}{\pi^2 k^2} \qquad \text{for } k = 1, 3, 5, \ldots \tag{A2.37}$$

As we would expect, all the exponential coefficients are real, corresponding to a cosine series. Also, since a_k is a function of k^2, $a_k = a_{-k}$, and the spectral function $X\omega)$ is even. It is illustrated in part (b) of Figure A2.6. Note that since

$X(\omega)$ is purely real in this case it is possible to represent it entirely by a *single* diagram. The fact that all coefficients are negative may appear at first sight a little surprising. However, if we look carefully at the signal $x(t)$, we realize that it could only be synthesized by adding together cosines which are all negative at $t=0$. In other words, they are all inverted.

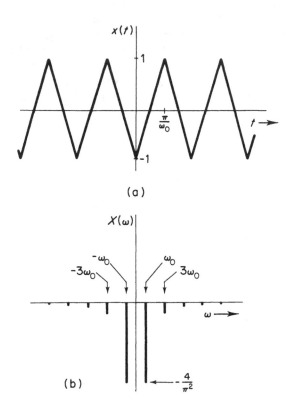

Figure A2.6 A periodic triangular signal.

A2.3 THE FOURIER TRANSFORM

A2.3.1 ANALYSIS AND SYNTHESIS OF APERIODIC SIGNALS

In the previous section we have seen how a periodic signal may be expressed as the sum of a set of sinusoidal or exponential functions which are harmonically related. The spectrum of such a signal contains a number of discrete frequencies. Although the analysis of strictly periodic signals gives results which can be of considerable practical value, in fact the great majority of signals are not of this type. First, even those signals which do repeat themselves a large number of times — such as the sinusoidal mains supply — are generally turned on and off. So they may not be assumed to exist for all time, past, present and future. It may be important to assess the effects which such *time-limitation* has upon their spectra, and upon systems which transmit or process them. Such transient effects are not covered by the Fourier Series. Secondly, and quite apart from any question of time-limitation, the majority of practical signals are simply not periodic in nature. For example, speech and TV signals tend to be very nonrepetitive.

However, the above comments are not meant to detract from the value or importance of the Fourier Series. Indeed, many of its central features form a very good starting point for discussing the spectra of nonrepetitive, or *aperiodic*, signals. These spectra are derived using the Fourier Transform, which may be thought of as a limiting case of the Fourier Series. It was one of Fourier's major achievements to show that an aperiodic signal may, in principle, be built up as an infinite sum — or integral — of sinusoidal or exponential functions which are *not* harmonically related.

A convenient and instructive way of introducing the Fourier Transform is to consider the recurrent pulse waveform shown in Figure A2.7(a). Note that the period of the signal is m times the pulse duration, and that the relative amount of time spent at level 1 (the 'high' state) and level 0 (the 'low' state) is therefore set by the parameter m. Our approach is as follows. We first derive the exponential Fourier Series for this signal, considered as a strictly periodic function. Then by allowing the value of m to increase, we may in the limit derive the spectrum of an isolated rectangular pulse at $t=0$ whose 'neighbors' have moved away to either side towards $t = \pm \infty$.

The signal is clearly even so we expect the Fourier coefficients a_k to be purely real, representing a cosine series. Equation (A2.32) gives

$$a_k = \frac{\omega_0}{2\pi} \int_{-\pi/\omega_0}^{\pi/\omega_0} x(t)\, \exp(-jk\omega_0 t)\mathrm{d}t$$

which in this case reduces to

$$a_k = \frac{\omega_0}{2\pi} \int_{-\pi/m\omega_0}^{\pi/m\omega_0} (1)\, \exp(-jk\omega_0 t)\mathrm{d}t = \frac{\omega_0}{2\pi} \left[\frac{\exp(-jk\omega_0 t)}{-jk\omega_0} \right]_{-\pi/m\omega_0}^{\pi/m\omega_0}$$

$$\therefore\ a_k = \frac{1}{2\pi jk} \{\exp(jk\pi/m) - \exp(-jk\pi/m)\} = \frac{1}{k\pi}\, \sin(k\pi/m)$$

Hence

$$a_k = \frac{1}{m} \left\{ \frac{\sin(k\pi/m)}{(k\pi/m)} \right\} \tag{A2.38}$$

As expected, all the coefficients are real. Their magnitudes follow an envelope of $(\sin x)/x$, or *sinc function*, form where $x = k\pi/m$. This function is commonly encountered in the theory of signals and linear systems, and takes the form of a decaying oscillation to either side of $x = 0$. It has unit value at $x = 0$, and passes through zero whenever $x = \pm n\pi$, for $n = 1, 2, 3, \ldots$ Parts (b) and (c) of the figure illustrate equation (A2.38) for two typical cases, $m = 3$ and $m = 5$ respectively. The sinc function envelope is shown dotted in each case. Although each diagram is labeled as a line spectrum, we may also consider it as a plot of a_k against k.

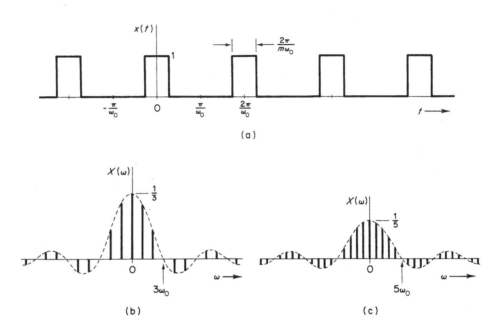

(a)

(b)   (c)

Figure A2.7 Waveform and spectral diagrams for a recurrent-pulse signal.

Two aspects of these spectral diagrams deserve particular attention. First, as m increases from 3 to 5 the fundamental frequency ω_0 reduces, and the coefficients a_k become more closely spaced. Since a reduction in ω_0 corresponds to an increasing period T_0 of the signal, our diagrams illustrate the effects of keeping the pulse duration constant, while altering the spacing between adjacent pulses. The second point to note is that as m increases, and more and more Fourier coefficients bunch together within the sinc function envelope, the magnitude of each coefficient reduces in proportion. Clearly, if m becomes large we obtain a line spectrum with a large number of closely bunched harmonics, all of small amplitude. In the limit as $m \rightarrow \infty$, the spectrum represents a single, 'isolated' pulse. Its neighbors have moved away on either side towards $t = \pm \infty$. The spectral lines are now extremely closely spaced, with vanishingly small amplitudes. The Fourier Series has become a Fourier Transform.

Mathematically, this situation may be expressed by modification to the exponential form of the Fourier Series given by equation (A2.29)

$$x(t) = \sum_{k=-\infty}^{\infty} a_k \exp(jk\omega_0 t)$$

where the coefficients a_k may be found using equation (A2.32), or the equivalent equation (A2.33)

$$a_k = \frac{1}{T_0} \int_{-T_0/2}^{T_0/2} x(t) \exp(-2\pi jkt/T_0)dt$$

As we let the period T_0 tend to infinity and each coefficient becomes vanishingly small, it might seem that the above equations are no longer useful. However, the product $a_k T_0$ does not vanish as $T_0 \to \infty$, so we now choose to write this as a variable X. Furthermore, as $T_0 \to \infty$, $\omega_0 \to 0$, and the term $k\omega_0$ tends to a continuous rather than a discrete variable. We will denote this as ω. Since X is a functiion of this continuous frequency variable ω, we now rewrite the second of the above equations as

$$X(\omega) = a_k T_0 = \int_{-\infty}^{\infty} x(t) \exp(-jk\omega_0 t)dt$$

$$\therefore X(\omega) = \int_{-\infty}^{\infty} x(t) \exp(-j\omega t)\, dt \qquad \text{(A2.39)}$$

Returning to the first equation which expresses $x(t)$ as the sum of an infinite set of harmonic components, we have

$$x(t) = \sum_{k=-\infty}^{\infty} \frac{X(\omega)}{T_0} \exp(jk\omega_0 t) = \sum_{k=-\infty}^{\infty} X(\omega) \frac{\omega_0}{2\pi} \exp(jk\omega_0 t)$$

Once again the term $k\omega_0$ is replaced by the continuous variable ω. The fundamental frequency ω_0 (which is now vanishingly small) is written as $d\omega$. The summation becomes an integration in the limit, so that

$$x(t) = \frac{1}{2\pi} \int_{-\infty}^{\infty} X(\omega) \exp(j\omega t)d\omega \qquad \text{(A2.40)}$$

The two equations (A2.39) and (A2.40) are of central importance in the theory of continuous signals and systems, and are known as the *Fourier Transform pair*. $X(\omega)$ is the Fourier Transform of $x(t)$, and $x(t)$ is the inverse Fourier Transform of $X(\omega)$. It is very important to grasp the significance of these equations. The first tells us how the spectral energy of the signal $x(t)$ is continuously distributed in the frequency domain. The second shows how, in effect, an aperiodic signal may be synthesized from an infinite set of exponential functions of the form $\exp(j\omega t)$, each weighted by the relevant value of $X(\omega)$.

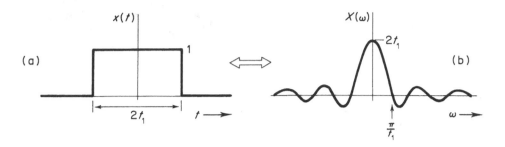

Figure A2.8 Fourier transformation of two pulse signals.

Let us straightaway illustrate the Fourier Transform by a simple example. Consider the single, isolated, rectangular pulse shown in Figure A2.8(a). The frequency spectrum of this signal may be found using equation (A2.39)

$$X(\omega) = \int_{-\infty}^{\infty} x(t) \exp(-j\omega t)\mathrm{d}t = \int_{-t_1}^{t_1} (1) \exp(-j\omega t)\mathrm{d}t$$

$$\therefore X(\omega) = -\frac{1}{j\omega} [\exp(-j\omega t)]_{-t_1}^{t_1} = \frac{1}{j\omega} \{\exp(j\omega t) - \exp(-j\omega t)\}$$

$$\therefore X(\omega) = \frac{2}{\omega} \sin \omega t_1 = 2t_1 \left\{ \frac{\sin \omega t_1}{\omega t_1} \right\} \qquad (A2.41)$$

As would be expected from the above discussion, this function is of sinc form and is illustrated in Figure A2.8(b). Note that a double-ended arrow is drawn between $x(t)$ and $X(\omega)$, to show that they are a Fourier Transform pair. The spectrum passes through zero whenever $\sin \omega t_1 = 0$, which occurs when ω is an integer multiple of π/t_1. It may seem strange that the pulse contains no energy at certain frequencies, but this is not in fact hard to demonstrate. Consider, for example, the frequency $\omega = \pi/t_1$, or $f = 1/2t_1$ Hz. If we wish to find out 'how much' of this frequency is contained in the pulse, the rule is to multiply the pulse by a sinusoid at the relevant frequency, and integrate over the interval of interest (as explained in our discussion of signal orthogonality in Section A2.1). It is clear that the result must be zero because in this case the product is simply equal to the sinusoid itself, and the integral of the sinusoid over any interval equal to $2t_1$ is zero. Therefore the pulse contains no energy at $1/2t_1$ Hz.

The spectrum of the isolated rectangular pulse illustrates two important general ideas. First, equation (A2.41) and Figure A2.8(b) show that the frequency at which the spectrum first crosses zero ($\omega = \pi/t_1$) is inversely proportioned to the pulse duration ($2t_1$). Thus a signal which is very *time-limited* occupies a wide spectral bandwidth, but a signal which is very *band-limited* has a wide spread in the time-domain — in other words it tends to be of long duration.

The second idea concerns an essential *symmetry* between the time and frequency domains, as shown by the Fourier Transform pair

$$X(\omega) = \int_{-\infty}^{\infty} x(t)\, \exp(-j\omega t)\,dt \qquad (A2.42)$$

and

$$x(t) = \frac{1}{2\pi} \int_{-\infty}^{\infty} X(\omega)\, \exp(j\omega t)\,d\omega \qquad (A2.43)$$

Apart from the $1/2\pi$ multiplier in the second equation (which arises from the use of angular frequency ω rather than frequency expressed in Hz), and the change of sign in the exponential index, the equations are symmetrical in form. The symmetry becomes perfect for even functions, such as the rectangular pulse shown in part (a) of Figure A2.8. Just as an even rectangular pulse in the time-domain has a spectrum of sinc form, so a sinc function 'pulse' in the time-domain *must* have a rectangular distribution of spectral energy. Such equivalence between time and frequency domains is often referred to as the *duality property* of the Fourier Transform.

Before we evaluate more Fourier Transforms, it is perhaps helpful to consider rather more carefully the physical meaning of a continuous spectrum $X(\omega)$. There is no doubt that many people find the idea of a signal being composed of an infinite set of exponentials, all of vanishingly small amplitudes, rather hard to envisage. The more familiar situations illustrated by Figure A2.9 may help. Part (a) shows a simply-supported beam loaded at several points. In part (b) of the figure the beam is continuously loaded along its length by, say, gravel or sand. In the first case it is clear that the loading is applied only at discrete points, just as a periodic signal contains only discrete frequencies. However, if one is asked what the load on the continuously loaded beam is at a point such as P, the answer must be that at that point (or any other) the applied load is vanishingly small. The sensible approach is to ask what the *average* loading is over a small distance such as Q–R, and to give the answer as a loading *density* in kilograms per metre. In an analogous way, a continuous frequency spectrum $X(\omega)$ implies that the component at any point-frequency is vanishingly small, and that we should talk instead about the energy contained

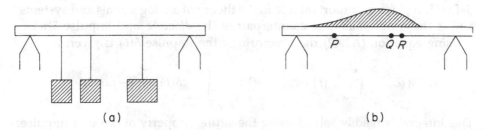

(a)   (b)

Figure A2.9 Two types of beam-loading; (a) discrete, and (b) continuous.

over a small band of frequencies centered around that point. $X(\omega)$ is therefore best thought of as a *frequency density* function.

This is a good point to reconsider Parseval's Theorem, quoted in Section A2.2.1 for the case of periodic signals and the Fourier Series. In essence, the theorem states that the power or energy in a signal may be found either in terms of its time-domain waveform, or in terms of its frequency spectrum. When dealing with a strictly periodic signal which (at least in theory) continues forever, it is appropriate to consider the signal's *average power*. However, in the case of a time-limited aperiodic signal the average power over all time tends to zero, so Parseval's relation is written in terms of total energy

$$\int_{-\infty}^{\infty} |x(t)|^2 \, dt = \frac{1}{2\pi} \int_{-\infty}^{\infty} |X(\omega)|^2 \, d\omega \qquad (A2.44)$$

This equation may be interpreted as follows. We consider $x(t)$ to represent a voltage across, or current through, a 1 ohm resistor. The total energy dissipated then equals the instantaneous power $x^2(t)$, integrated over all time (note that $x^2(t)$ may also be written as $|x(t)|^2$, thereby emphasizing the symmetry between time and frequency in the above equation). The left-hand side therefore gives the signal's total energy, evaluated in the time-domain. This must equal the sum of energies in all the signal's frequency components. Now the energy in a frequency component depends only upon the square of its amplitude, not on its phase. Since $X(\omega)$ is a frequency density function, $|X(\omega)|^2/2\pi$ is effectively *spectral energy density*, representing the amount of energy per unit frequency. If this function is integrated over all frequencies, as on the right-hand side of the above equation, we again obtain the signal's total energy.

So far, we have used the Fourier Transform equation to find the spectrum of only one type of aperiodic signal — an isolated rectangular pulse. To give practice in Fourier Transformation and to illustrate some important aspects of continuous spectra, we now consider some further examples. At the end of this appendix, our 'library'' of Fourier Transforms is supplemented by including a table of continuous signals and their transforms. Such tables are widely used in engineering practice, and save the trouble of evaluating transforms of commonly used functions.

We first find the Fourier Transforms of the continuous unit impulse function $\delta(t)$, and its shifted version $\delta(t - t_0)$. These signals are shown in Figure A2.10. The reader is probably aware that $\delta(t)$ is a very narrow, high, pulse or 'spike' centered at $t = 0$. Its area is unity, but its precise waveshape is not defined. It plays an important role in the theory of analog signals and systems, and is the continuous-time counterpart of the discrete unit impulse $\delta[n]$.

Using equation (A2.42) the spectrum of the impulse $\delta(t)$ is given by

$$X(\omega) = \int_{-\infty}^{\infty} x(t) \exp(-j\omega t)dt = \int_{-\infty}^{\infty} \delta(t) \exp(-j\omega t)dt$$

This integral is readily solved using the sifting property of the unit impulse. When the impulse is multiplied by another function, and the product integrated

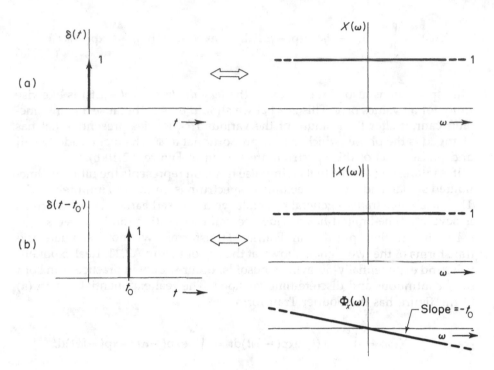

Figure A2.10 Fourier Transforms of continuous impulse signals.

between $\pm \infty$, the result is to *sift out* the value of the second function where the impulse occurs. Since $\delta(t)$ occurs at $t = 0$, $X(\omega)$ must simply equal the value of $\exp(-j\omega t)$ at $t = 0$. Thus

$$X(\omega) = \exp(-j\omega t)|_{t=0} = \exp(0) = 1 \qquad (A2.45)$$

This important result shows that a unit impulse centered at $t = 0$ theoretically contains an equal 'amount' of all frequencies. Since $X(\omega)$ is purely real, $\delta(t)$ could be synthesized by adding together an infinite set of cosines, with all frequencies equally represented. This may be visualized as follows. At $t = 0$ every infinitesimal cosine, regardless of its frequency, has its peak value. But at any other instant, some are positive and others negative, and their sum averages out to zero. So the synthesized signal has a large peak at $t = 0$, and is zero elsewhere. In practice, of course, we cannot expect *infinitely* high frequencies to be present. But the notion that a very narrow pulse contains an equal amount of a wide range of frequencies remains a valid, and very valuable, one. The spectrum $X(\omega)$ is illustrated in Figure A2.10(a). It is said to be *white*, because all spectral components are equally represented (just as white light contains an equal mixture of all the colors of the rainbow).

The spectrum of the shifted unit impulse shown in part (b) of Figure A2.10 may be found in similar fashion. We have

$$X(\omega) = \int_{-\infty}^{\infty} \delta(t - t_0)\, \exp(-j\omega t)\mathrm{d}t = \exp(-j\omega t)|_{t=t_0} = \exp(-j\omega t_0)$$

$$(A2.46)$$

This spectrum is also white, because the *magnitude* of $\exp(-j\omega t_0)$ is likewise unity for all values of ω. This is what we should expect, because a mere time-shift cannot alter the amount of the various frequencies present. What has changed is the phase, which is now proportional to ω. The magnitude $|X(\omega)|$ and phase $\phi_x(\omega)$ of this spectrum are shown in Figure A2.10(b).

It is interesting to note that an impulse function represents the ultimate time-limited signal, and we now see that its spectrum is, in theory, infinitely wide. This nicely illustrates a general principle we discussed earlier. A signal which is severely time-limited has a wide spectral distribution, and vice versa.

To give further practice in Fourier Transforms, we now evaluate the transforms of the two signals shown at the top of Figure A2.11. Real exponentials, and exponentially-decaying sinusoids, occur widely in practice — in both their continuous and discrete-time versions. The real exponential in part (a) of the figure has the Fourier Transform

$$X(\omega) = \int_{-\infty}^{\infty} x(t)\, \exp(-j\omega t)\mathrm{d}t = \int_{0}^{\infty} \exp(-\alpha t)\, \exp(-j\omega t)\mathrm{d}t$$

$$\therefore\; X(\omega) = \int_{0}^{\infty} \exp(-\{\alpha + j\omega\}t)\mathrm{d}t = \left[\frac{\exp(-\{\alpha + j\omega\}t)}{-(\alpha + j\omega)} \right]_{0}^{\infty}$$

$$\therefore\; X(\omega) = 0 - \frac{\exp(0)}{-(\alpha + j\omega)} = \frac{1}{(\alpha + j\omega)} \qquad (A2.47)$$

This spectrum is complex, implying that both sine and cosine waves would be needed to synthesize the decaying exponential signal. $X(\omega)$ cannot be represented on a single diagram, and the most usual approach is to plot its magnitude and phase separately. The magnitude is given by

$$|X(\omega)| = \frac{1}{|\alpha + j\omega|} = \frac{1}{(\alpha^2 + \omega^2)^{1/2}} \qquad (A2.48)$$

The phase angle in radians represented by the term $(\alpha + j\omega)$ is $\arctan(\omega/\alpha)$. Since this appears in the *denominator*, and the numerator is real, the phase of $X(\omega)$ is simply

$$\Phi_x(\omega) = -\arctan(\omega/\alpha) \qquad (A2.49)$$

Magnitude and phase are separately shown in the figure. Although these spectral diagrams do extend to negative frequencies, we generally show only the positive frequency range. In fact, the spectral magnitude function of any real signal or time function is always even, and the phase function is always odd.

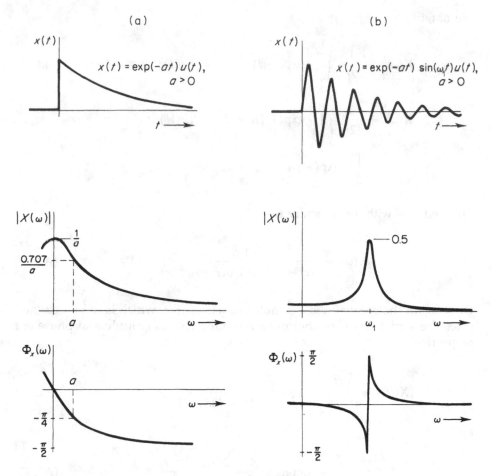

Figure A2.11 Two more examples of continuous signals and their spectra.

In this particular case it is clear that the signal is relatively rich in low-frequency components. When $\omega = \alpha$, the magnitude function has already fallen to $1/\sqrt{2}$ of its peak value at $\omega = 0$, and the phase is $-\pi/4$. As $\omega \to \infty$, the phase asymptotically approaches $-\pi/2$ (or $-90°$).

The exponentially-decaying sinusoid shown in part (b) of the figure has the transform

$$X(\omega) = \int_{-\infty}^{\infty} x(t) \exp(-j\omega t)dt \int_{0}^{\infty} \exp(-\alpha t) \sin \omega_1 t \exp(-j\omega t)dt$$

Using the identity

$$\sin \omega_1 t = \frac{1}{2j} \{\exp(j\omega_1 t) - \exp(-j\omega_1 t)\}$$

we obtain

$$X(\omega) = \frac{1}{2j} \int_0^\infty \{\exp(j\omega_1 t) - \exp(-j\omega_1 t)\} \exp(-\{\alpha + j\omega\}t)dt$$

$$\therefore X(\omega) = \frac{1}{2j} \left\{ \int_0^\infty \exp(-\{\alpha + j\omega - j\omega_1\}t)dt \right.$$

$$\left. - \int_0^\infty \exp(-\{\alpha + j\omega + j\omega_1 t)dt \right\}$$

This reduces without difficulty to

$$X(\omega) = \frac{1}{\alpha^2 + \omega_1^2 + 2j\alpha\omega - \omega^2} \tag{A2.50}$$

Once again the spectrum is a complex function of ω, which is to be expected since the signal $x(t)$ is neither even nor odd. The magnitude and phase are respectively

$$|X(\omega)| = \frac{\omega_1}{\{(\alpha^2 + \omega_1^2 - \omega^2)^2 + (2\alpha\omega)^2\}^{1/2}} \tag{A2.51}$$

and

$$\Phi_x(\omega) = - \arctan \frac{2\alpha\omega}{(\alpha^2 + \omega_1^2 - \omega^2)} \tag{A2.52}$$

It is difficult to sketch these functions as they stand, because their shapes depend markedly on the values of α and ω_1. Note, however, that the signal shown in the figure displays several oscillations of the sinusoid within one time constant of its exponential decay, so that $\omega_1 \gg \alpha$. If we take $\alpha = 1$ and $\omega_1 = 20$ as an example, then clearly $\omega_1^2 \gg \alpha^2$, and to a good approximation we may write

$$|X(\omega)| \approx \frac{\omega_1}{\{(\omega_1^2 - \omega^2)^2 + (2\alpha\omega)^2\}^{1/2}} = \frac{20}{\{(400 - \omega^2)^2 + 4\omega^2\}^{1/2}} \tag{A2.53}$$

and

$$\Phi_x(\omega) = - \arctan \frac{2\alpha\omega}{\omega_1^2 - \omega^2} = - \arctan \frac{2\omega}{400 - \omega^2} \tag{A2.54}$$

These functions are also shown in the figure. The most important point they reveal is that freqencies close to ω_1 are very strongly represented in $|X(\omega)|$. This is what we would expect, since the signal is essentially a sinusoid at this frequency. The spreading of spectral energy to either side of ω_1 is due to time-limitation of the sinewave — caused by 'switching it on' at $t = 0$, and by the subsequent exponential decay. The sharpness of the resonant peak is very dependent on the value of α. The plot of $\Phi_x(\omega)$ shows that there is a sudden phase change of π radians as resonance is passed. The phase of frequency components close to ω_1 is near $\pm \pi/2$, denoting sines rather than cosines. Again, this is what would be expected.

So far, our discussion may have suggested that the Fourier Transform is applicable to any continuous, aperiodic signal. But just as certain restrictions are placed on periodic signals if they are to be represented by a Fourier Series, so equivalent convergence criteria — again referred to as *Dirichlet conditions* — must be met in the case of a Fourier Transform.

First, the signal $x(t)$ to be transformed must be *absolutely integrable*. This means that

$$\int_{-\infty}^{\infty} |x(t)|\,dt < \infty \qquad (A2.55)$$

In other words, the total area under a 'rectified' version of $x(t)$ must be finite. Secondly, $x(t)$ can have only a finite number of maxima and minima, or of finite discontinuities, in any finite time interval. Although this second condition is invariably obeyed by practical signals, the first may not be. For example, continuous-time step functions and eternal sinusoids are not absolutely integrable. Fortunately there are ways round the problem, and it is possible to derive spectral representations of such signals in terms of frequency-domain 'impulses'. We need not go into further detail here, although the reader should be aware of the problem. Several spectra involving such frequency-domain impulses appear in the table of transforms at the end of this appendix.

A2.3.2 PROPERTIES OF THE TRANSFORM

The Fourier Transform, considered as a mathematical operation, possesses a number of important properties. Among the most valuable of these from the point of view of signals and systems are the following: linearity, time-shifting and scaling, differentiation and integration, convolution and modulation. These properties give further insight into the nature of Fourier transformation, and the relationships between the time and frequency domains.

Rather than use the Fourier Transform and inverse Fourier Transform equations each time we need to define or discuss a property, it is helpful to introduce the following shorthand notation

$\mathfrak{F}\{x(t)\}$ denotes the Fourier Transform of a signal $x(t)$
$\mathfrak{F}^{-1}\{X(\omega)\}$ denotes the inverse transform of a spectrum $X(\omega)$
$x(t) \leftrightarrow X(\omega)$ signifies that $x(t)$ and $X(\omega)$ are a Fourier Transform pair

The *linearity* of the Fourier Transform is an important property which stems directly from its mathematical definition. It may be summarized as follows

If $x_1(t) \leftrightarrow X_1(\omega)$ and $x_2(t) \leftrightarrow X_2(\omega)$

then $ax_1(t) + bx_2(t) \leftrightarrow aX_1(\omega) + bX_2(\omega)$ (A2.56)

Thus the Fourier Transform of a composite signal, formed by the addition of weighted individual signals $x_1(t)$, $x_2(t)$, ... equals the sum of their individual transforms, duly weighted. It must be emphasized that since spectra are generally complex functions of ω, a frequency-domain summation must take proper account of relative phase as well as magnitude.

The Fourier Transform also has an important *time-shifting* property, which may be expressed as follows

If $x(t) \leftrightarrow X(\omega)$

then $x(t - t_0) \leftrightarrow X(\omega) \exp(-j\omega t_0)$ (A2.57)

Since $\exp(-j\omega t_0)$ has unit magnitude for all values of ω, it follows that time-shifting does not affect the magnitude of a Fourier Transform. But it does affect the phase characteristic, introducing a phase-shift proportional to frequency. The time-shifting property can be very useful. If we already know the Fourier Transform of a signal, we may at once write down the transform of its time-shifted version without further analysis.

The *time-scaling* property of the transform means that if a signal $x(t)$ is expanded in time, then its spectrum $X(\omega)$ is compressed in proportion, and vice versa. An example of time-scaling, probably familiar to many readers, is the playback of a tape or disk recording at a speed other than the recording speed. For example, if the time-scale is *expanded* by reducing the playback speed, the frequency scale is shortened or *compressed* by the same factor, reducing the frequency (or pitch) of musical notes and speech. This property is summarized as

If $x(t) \leftrightarrow X(\omega)$

then $x(at) \leftrightarrow \dfrac{1}{a} X\left(\dfrac{\omega}{a}\right)$ (A2.58)

where the constant a is real and positive. Note that the spectrum undergoes a simple change of amplitude, as well as of frequency scale. This may be explained quite easily. Suppose, for example, $a = 0.5$, corresponding to a doubling of the signal's total duration and hence of its total energy. Since the spectrum is compressed by a factor of 2, $X(\omega/a)$ has half the total energy of $X(\omega)$. In order to obey Parseval's Theorem the energy in $X(\omega/a)$ must therefore be increased by a factor of 4. But energy is proportional to the square of amplitude, so the amplitude must be doubled, or multiplied by $(1/a)$.

When a signal is *differentiated* or *integrated*, the effect on its Fourier Transform is readily predictable. This is a particularly useful property of the transform. The performance of continuous LTI systems is often described in terms of differential or integral equations relating input and output signals. Therefore the frequency-domain effects of time-domain differentiation or integration are of special interest for the analysis of signal flow through such systems. The *differentiation property* of the Fourier Transform may be stated as follows

If $\quad x(t) \leftrightarrow X(\omega)$

then $\quad x(t) = \dfrac{1}{2\pi} \displaystyle\int_{-\infty}^{\infty} X(\omega) \exp(j\omega t)\, d\omega$

Since $X(\omega)$ is not a function of time, differentiation of both sides with respect to time gives

$$\frac{dx(t)}{dt} = \frac{1}{2\pi} \int_{-\infty}^{\infty} X(\omega) j\omega \exp(j\omega t)\, d\omega = \mathcal{F}^{-1}\{j\omega X(\omega)\}$$

Therefore

$$\mathcal{F}\left\{\frac{dx(t)}{dt}\right\} = j\omega X(\omega) \tag{A2.59}$$

This shows that the transform of a differentiated signal is just the transform of the signal itself, multiplied by $j\omega$. The relatively complicated process of *differentiation* in the time-domain is therefore equivalent to a much simpler *multiplication* in the frequency domain.

Since differentiation is equivalent to multiplication by $j\omega$, the reader may suspect that *integration* is equivalent to *division* by $j\omega$. This is so, but unfortunately it is not quite the whole story, because such a division does not take account of any DC level possessed by the signal. The mathematical argument is quite complicated, but the precise relationship may be shown to be

$$\int_{-\infty}^{t} x(\tau) d\tau \leftrightarrow \frac{X(\omega)}{j\omega} + X(0)\pi\delta(\omega) \tag{A2.60}$$

where the left-hand side denotes the running integral of $x(t)$, and $X(0)$ is the zero-frequency, or DC, value of $X(\omega)$.

We now turn to the very important *convolution property* of the Fourier Transform. As will become clear below, this property is responsible for the widespread use of frequency-domain methods to analyze and predict signal flow through LTI systems. Like other properties already mentioned, it is essentially a consequence of the mathematical definition of the Fourier Transform. We may introduce it in general terms by considering the convolution of two time functions or signals $x_1(t)$ and $x_2(t)$, producing a third time function $x_3(t)$. The form of the convolution integral, which is the counterpart of the convolution sum in discrete time, is

$$x_3(t) = \int_{-\infty}^{\infty} x_1(\tau) x_2(t - \tau) d\tau \tag{A2.61}$$

where τ is an auxiliary time variable. Now if

$$x_3(t) \leftrightarrow X_3(\omega)$$

then

$$X_3(\omega) = \int_{-\infty}^{\infty} \int_{-\infty}^{\infty} \{ x_1(\tau) x_2(t - \tau) d\tau \} \, \exp(-j\omega t) dt \tag{A2.62}$$

$x(\tau)$ is not a function of t, so by changing the order of integration we obtain

$$X_3(\omega) = \int_{-\infty}^{\infty} x_1(\tau) \left\{ \int_{-\infty}^{\infty} x_2(t - \tau) \, \exp(-j\omega t) dt \right\} d\tau$$

Using the shifting property of the Fourier Transform, if

$$x_2(t) \leftrightarrow X_2(\omega)$$

then

$$\int_{-\infty}^{\infty} x_2(t - \tau) \, \exp(-j\omega t) dt = \exp(-j\omega \tau) X_2(\omega)$$

and hence

$$X_3(\omega) = \int_{-\infty}^{\infty} x_1(\tau) \, \exp(-j\omega \tau) X_2(\omega) d\tau \tag{A2.63}$$

Finally, if

$$x_1(t) \leftrightarrow X_1(\omega)$$

then

$$X_3(\omega) = X_1(\omega) X_2(\omega) \tag{A2.64}$$

Equation (A2.64) shows that the spectrum of $x_3(t)$ equals the product of the individual spectra of $x_1(t)$ and $x_2(t)$. Since $x_1(t)$ and $x_2(t)$ have been convolved to produce $x_3(t)$, we conclude that *time-domain convolution is equivalent to frequency-domain multiplication*. In mathematical terms we may therefore write

$$x_3(t) = x_1(t) * x_2(t) \leftrightarrow X_3(\omega) = X_1(\omega) X_2(\omega) \tag{A2.65}$$

This is a perfectly general result which applies to any time and frequency functions, but it has a special relevance to work on signals and systems. This is because the output signal from an LTI system may be found by convolution of the input signal with the system's impulse response. Equation (A2.65) shows that this must be precisely equivalent to *multiplying* the corresponding spectral functions. The first of these is the spectrum of the input signal. The second is another frequency-domain function equal to the Fourier Transform of the system's impulse response. The latter function is the *frequency response* of the system.

The last property to be discussed is *modulation*. Closely related to the convolution property, it is of particular relevance to the field of electronic communications, where *modulation processes* are widely used for the transmission of signals by such means as cable, or optical fiber. Although it is quite straightforward to derive the modulation property mathematically (as we have done with the convolution property) we may instead infer it from the other features of the Fourier Transform already described. It may be stated as follows. If two time functions are multiplied together, then the Fourier Transform of their product may be found by convolving their individual transforms. In other words, just as time-domain convolution is equivalent to frequency-domain multiplication, so *time-domain multiplication is equivalent to frequency-domain convolution*. Thus

$$x_3(t) = x_1(t)x_2(t) \leftrightarrow X_3(\omega) = \frac{1}{2\pi} \{X_1(\omega) * X_2(\omega)\} \tag{A2.66}$$

In view of the symmetry of the Fourier Transform pair, and the duality between time and frequency domains, this result is surely what we should expect.

The reader who requires a convenient reference to the various properties of the Fourier Transform described in this section, will find them summarized in a table at the end of the appendix.

A2.4 FREQUENCY RESPONSES AND IMPULSE RESPONSES OF ANALOG LTI SYSTEMS

A closely-related application of Fourier analysis is the representation of analog LTI systems in the frequency-domain. Fortunately, much of the work we have done on signals is applicable to systems. The initial task is to show that, just as an LTI system is completely described in the time-domain by its impulse response, so an equivalent function is available in the frequency-domain. This function is called the *frequency response* of the system.

A diagrammatic summary of signal flow through an analog LTI system is given in Figure A2.12. In the time-domain, the input signal $x(t)$ is convolved with the system's impulse response $h(t)$ to produce the output signal $y(t)$. The alternative approach is to work in the frequency-domain. If we start with a known signal waveform $x(t)$, we must first find its spectrum $X(\omega)$ by Fourier transformation. We then *multiply* $X(\omega)$ by a function $H(\omega)$ which represents the system, to produce the spectrum $Y(\omega)$ of the output signal. Thus

$$Y(\omega) = X(\omega)H(\omega) \tag{A2.67}$$

or

$$H(\omega) = \frac{Y(\omega)}{X(\omega)} \tag{A2.68}$$

However, in this case the output signal $y(t)$ is by definition the system's impulse response $h(t)$. Therefore, $h(t)$ and $H(\omega)$ *must be related as a Fourier Transform pair*. The impulse response gives a complete description of the LTI system in

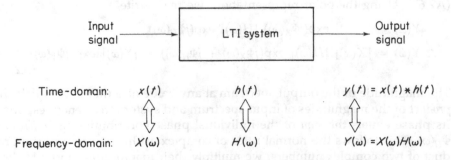

Figure A2.12 Signal flow through an analog LTI system.

the time-domain, and its frequency response $H(\omega)$ gives a complete description in the frequency-domain. This is a satisfying result, which ties in neatly with our earlier work on the Fourier Transform. For, as we have seen, all practical signals and time functions have their frequency-domain counterparts, and an impulse response is no exception. *Its* counterpart is $H(\omega)$, which describes the system's response in both magnitude and phase to all frequency components present in an input signal.

Before proceeding, we must be clear about the nature of frequency responses which describe LTI systems. The main point to note is that, whereas we generally concern ourselves with *real* time functions, their spectral counterparts are very often *complex* functions of ω. Thus, a function $H(\omega)$ may in general be expressed in terms of real and imaginary parts.

$$H(\omega) = H_r(\omega) + jH_i(\omega) \tag{A2.69}$$

where $H_r(\omega)$ is an even function and $H_i(\omega)$ is an odd function. The magnitude of $H(\omega)$ at any frequency is then given by

$$|H(\omega)| = \{H_r^2(\omega) + H_i^2(\omega)\}^{1/2} \tag{A2.70}$$

and its phase by

$$\Phi_H(\omega) = \arctan \frac{H_i(\omega)}{H_r(\omega)} \tag{A2.71}$$

A convenient alternative is to express $H(\omega)$ directly in terms of magnitude and phase using the so-called *polar representation*

$$H(\omega) = |H(\omega)| \exp(j\Phi_H(\omega)) \tag{A2.72}$$

Note that the exponential has unit magnitude, and represents just the phase of $H(\omega)$. The representation of a spectral function in terms of magnitude and phase is generally simpler to visualize than the alternative based on real and imaginary parts. Since in our present discussion $H(\omega)$ denotes the *frequency response of a system*, we interpret $|H(\omega)|$ as the factor by which the system multiplies the *magnitude* of any sinusoidal component present at its input. This factor is widely referred to as the *gain* of the system. $\Phi_H(\omega)$ represents the *phase-shift* imposed by the system as a function of the sinusoidal input frequency.

Let us now consider the multiplication of an input signal spectrum $X(\omega)$ by $H(\omega)$ to produce the output signal spectrum $Y(\omega)$, as specified by equation (A2.67). Using the polar representation, we may write

$$Y(\omega) = |X(\omega)| \exp(j\Phi_X(\omega)) |H(\omega)| \exp(j\Phi_H(\omega))$$

$$\therefore Y(\omega) = |X(\omega)| |H(\omega)| \exp(j\Phi_X(\omega) + j\Phi_H[\omega)) = |Y(\omega)| \exp(j\Phi_Y(\omega))$$
$$\tag{A2.73}$$

The magnitude of the output spectrum at any value of ω therefore equals the *product* of the magnitudes of input spectrum and system frequency response. Its phase equals the *sum* of the individual phase contribution of $\Phi_X(\omega)$ and $\Phi_H(\omega)$. This follows the normal rules of complex arithmetic: to find the product of two complex numbers, we multiply their magnitudes, and add their phase angles.

Table A2.1 The continuous-time Fourier Transform: pairs

Waveform	Signal $x(t)$	Spectrum $X(\omega)$
Unit impulse	$\delta(t)$	1
Shifted unit impulse	$\delta(t - t_0)$	$\exp(-j\omega t_0)$
Unit step	$u(t)$	$\pi\delta(\omega) + \dfrac{1}{j\omega}$
DC level	1	$2\pi\,\delta(\omega)$
Rectangular pulse	$u(t + T) - u(t - T)$	$2T\,\dfrac{\sin(\omega T)}{\omega T} = 2T\,\mathrm{sinc}(\omega T)$
Exponential	$\exp(-\alpha t)\,u(t)$	$\dfrac{1}{\alpha + j\omega}$
Eternal cosine	$\cos\,\omega_0 t$	$\pi\left[\delta(\omega - \omega_0) + \delta(\omega + \omega_0)\right]$
Eternal sine	$\sin\,\omega_0 t$	$\dfrac{\pi}{j}\left[\delta(\omega - \omega_0) - \delta(\omega + \omega_0)\right]$
Damped sine	$\exp(-\alpha t)\sin(\omega_0 t)\,u(t)$	$\dfrac{\omega_0}{(\alpha + j\omega)^2 + \omega_0^2}$

Table A2.2 The continuous-time Fourier Transform: properties

Property or operation	Aperiodic signal	Fourier Transform
Transformation	$x(t)$	$\int_{-\infty}^{\infty} x(t)\exp(-j\omega t)\,dt$
Inverse transformation	$\frac{1}{2\pi}\int_{-\infty}^{\infty} X(\omega)\exp(j\omega t)\,d\omega$	$X(\omega)$
Linearity	$a_1 x_1(t) + a_2 x_2(t)$	$a_1 X_1(\omega) + a_2 X_2(\omega)$
Time-reversal	$x(-t)$	$X(-\omega)$ $= X^{*}(\omega),\, x(t)\,\text{real}$
Time-shifting	$x(t-t_0)$	$\exp(-j\omega t_0)X(\omega)$
Time-scaling	$x(at)$	$\frac{1}{|a|}X\left(\frac{\omega}{a}\right)$
Time-differentiation	$\frac{d^n}{dt^n}x(t)$	$(j\omega)^n\, X(\omega)$
Time-integration	$\int_{-\infty}^{t} x(t)\,dt$	$\frac{1}{j\omega}X(\omega) + \pi X(0)\delta(\omega)$
Frequency-differentiation	$t^n x(t)$	$(j)^n \frac{d^n}{d\omega^n}X(\omega)$
Convolution	$x_1(t) * x_2(t)$	$X_1(\omega) X_2(\omega)$
Modulation	$x_1(t)\, x_2(t)$	$\frac{1}{2\pi}\left\{X_1(\omega) * X_2(\omega)\right\}$
Real time-function	$x(t)$	$X(\omega) = X^{*}(-\omega)$ $\mathrm{Re}\{X(\omega)\} = \mathrm{Re}\{X(-\omega)\}$ $\mathrm{Im}\{X(\omega)\} = -\mathrm{Im}\{X(-\omega)\}$ $|X(\omega)| = |X(-\omega)|$ $\Phi_x(\omega) = -\Phi_x(-\omega)$
Duality	if $g(t) \Longleftrightarrow$ then $f(t) \Longleftrightarrow$	$f(\omega)$ $2\pi g(-\omega)$

Answers to Selected Problems

CHAPTER 1

Q1.1   72 Mbit s^{-1}

Q1.2   96 Mbit s^{-1}

Q1.6   (a) $3u[n-1]$;   (b) $-2\delta[n+2]$;

       (c) $r[n+4] - r[n] - 2r[n-7] + 2r[n-9]$

Q1.8   (a) period = 18;   (d) period = 20

Q1.11   (a) is invertible

Q1.12   If $|\alpha| > 1$, system is unstable

CHAPTER 2

Q2.1   (a) $x[n] = \delta[n] + 2\delta[n-1] + \delta[n-2]$

       (b) $x[n] = 2\delta[n+2] + 2\delta[n] - 0.5\delta[n-1] + \delta[n-4]$

Q2.2   (a), (c), (d) are causal and stable

Q2.6   1.0, -0.0498, -0.0406, -0.0279, -0.0133,

       0.0019, 0.0162, 0.0281, 0.0367, 0.0414

Q2.9   (a) 2;   (b) 2/3

Q2.10   (a) values of $y[n]$, starting at $n = 0$, are:

       0, 0, 1, 2, 3, 3, 3, 3, 3, 2, 1, 0, 0 ...

       (b) values of $y[n]$, starting at $n = 0$, are:

       1, 1, -1.5, -1.0, -0.5, 0.8, 0.3, -0.1, 0, 0 ...

Q2.12 $y[0] = 5.29$; $y[6] = 11.29$; $y[20] = 13.43$

Q2.14   (a) $y[n] = x[n] + x[n-1] + \ldots x[n-6] = \displaystyle\sum_{k=0}^{6} x[n-k]$

       (b) $y[n] = x[n] + 0.9x[n-1] + 0.9^2 x[n-2] + \ldots$

           $= \displaystyle\sum_{k=0}^{\infty} (0.9)^k x[n-k]$

Q2.15   Overall $h[n]$ has sample values (starting at $n = 1$):
        1, 2.5, 4.333, 2.167, 1, 0, 0 ...
Q2.16   First few values of $h[n]$ are:
        2, 2.8, 3.64, 0.512, 0.4096 ...
Q2.19 $y[0]$ to $y[12]$: 1, -1, -0.5, 2, -1.75, -0.25, 2.125, -2,
        -0.0625, 2.0625, -2.0312, 0, 2.016
        $y_h[n]$, $-2 \le n \le 9$: 2, 0, -1, 1, -0.5, 0, 0.25, -0.25,
        0.12, 0, -0.06, 0.06

CHAPTER 3

Q3.1   (a) $a_0 = 5$, $a_1 = a_7 = 0.5$, $a_2 = -a_6 = -0.5j$,
        $a_3 = a_5 = 0$, $a_4 = 0$
        (b) $a_1 = (1-j)/2\sqrt{2}$, $a_3 = (1+j)/2\sqrt{2}$
        (c) $a_0 = 3$, $a_1 = -1+j$, $a_2 = -1$, $a_3 = -1-j$
Q3.2 $a_0 = 0$; $a_1 = 0.0107 + j0.5622$;
        $a_2 = 0.5671 + j0.4632$; $a_3 = 0.4223 + j1.4033$;
        $a_4 = 0.4223 - j1.4033$; $a_5 = 0.5671 - j0.4632$;
        $a_6 = 0.0107 - j0.5622$
Q3.5   All coefficients zero, except:
        $a_0 = 1$; $|a_1| = |a_{63}| = 0.5$ (phase = 0);
        $|a_8| = |a_{56}| = 0.5$ (phase = $\pm \pi/2$)
Q3.8   0.518, 75°
Q3.11   (a) $X(\Omega) = 1 + 2 \exp(-j\Omega) + \exp(-j2\Omega)$
        (b) $X(\Omega) = 2j \sin\Omega$
        (c) $X(\Omega) = 1 + 2 \cos\Omega + 2 \cos 2\Omega + 2 \cos 3\Omega$

Q3.13 $H(\Omega = \dfrac{0.1}{1 + 0.9 \exp(-j\Omega)}$;   (a) 0.0526; (b) 1.0

Q3.14 $|H(\Omega)| = \{(1-0.9 \cos\Omega + 0.8 \cos 2\Omega)^2$
        $\qquad\qquad\qquad + (0.9 \sin\Omega - 0.8 \sin 2\Omega)^2\}^{-1/2}$
Q3.15   (a) 1.021; (b) 0.0104

CHAPTER 4

Q4.1   (a) $\dfrac{3}{z^2(z-\alpha)}$; (b) $\dfrac{2(z^8-1)}{z^7(z-1)}$

Q4.3 $\dfrac{2(z^8-1)}{z^9(z-1)} - \dfrac{2}{z^7(z-0.5)}$

Q.4.5 $Y(z) = (1 + 3z^{-1} + 6z^{-2} + 6z^{-3} + 3z^{-4} + z^{-5} + z^{-7})$

Q4.6   (a) $x[n] = \dfrac{1}{\sqrt{2}}^{\,n} \sin \dfrac{n\pi}{4} u[n]$

        (b) $x[n] = \{2.5 - 1.5(0.8)^{n-2}\} u[n-2]$
Q4.8   (a) zeros at $z = 2$, $z = -1$
             poles at $z = 0.5$, $z = 0.8$

(b) zeros at $z = 0.5 \pm j0.866$
    poles at $z = \pm j$
    unstable

(c) zeros at $z = 1$, $z = \pm j$
    poles at $z = \pm 0.5$
    noncausal

(d) Nine zeros equally spaced around unit circle.
    pole at $z = 1$, eighth-order pole at origin.
    stable

Q4.10   (a) $x[n] = \{1-(-0.8)^n\}\, u[n]$

      (b) $x[n] = \left(\cos \dfrac{2\pi n}{3}\right) u[n]$

      (c) $x[n] = \left(0.8^n \sin \dfrac{\pi n}{2}\right) u[n]$

Q4.12   zeros at $z = 1$, $z = \pm 0.894j$
        pole at $z = -0.8$, second-order pole at origin

Q4.13 $H(z) = \dfrac{z^3 - 2z^2 + 2z - 1}{z(z^2 + 0.9z + 0.81)}$

      $y[n] = -0.9y[n-1] - 0.81y[n-2] + x[n]$
              $-2x[n-1] + 2x[n-2] - x[n-3]$
      first six values of $h[n]$ are:
      1, -2.9, 3.8, -2.071, -1.2141, 2.7702

Q4.15 $-2 < \alpha < 2$; $0 \leq \beta < 1$
      (a) $\alpha = 1$; (b) $\alpha = -1$

Q4.18   (a) $\dfrac{z^2 - z}{z^2 + 0.5z + 0.5}$

      (b) $\dfrac{z^2}{z^2 + 0.5z + 0.5}$; $y[-1] = -2$; $y[-2] = 4$

CHAPTER 5

Q5.3   Impulse response values $h[0]$ to $h[m]$ are:
      (a) 0.4, 0.30273, 0.09355, -0.06237, -0.07568 ...
      (b) 0.2, -0.1870, 0.15137, -0.10091, 0.04677 ...

Q5.7   Window values $w[0]$ to $w[m]$ are:
      (a) 1.0, 0.93301, 0.75, 0.5, 0.25, 0.06699
      (b) 1.0, 0.93837, 0.77, 0.54, 0.31, 0.14163, 0.08

Q5.10 (a) about -32 dB, -42 dB, -48 dB
      (b) about -40 dB, -39 dB, -40 dB

Q5.11 4.9899; 67

Q5.12 4.0910; 37

Q5.14 0.143 dB; 0.579 dB

Q5.15 $h[0] = 0$;   values $h[1]$ to $h[m]$ are:

$$h[n] = \frac{1}{\pi n^2} (-1)^{(n+1)/2}, \; n = 1,3,5,7 \ldots$$

$$= \frac{1}{2n} (-1)^{n/2}, \; n = 2,4,6,8 \ldots$$

CHAPTER 6

Q6.3   (a) 20;   (b) 0;   (c) -60;   (d) -68;   (e) $-\infty$

Q6.4 $y[n] = 1.1202 \, y[n-1] - 0.9080 \, y[n-2] + x[n]$
$\qquad\qquad\qquad + 1.6180 \, x[n-1] + x[n-2]$

Q6.5 $y[n] = 1.5164 \, y[n-1] - 0.8783 \, y[n-2] + x[n]$
$\qquad\qquad\qquad\qquad - 1.6180 \, x[n-1] + x[n-2]$

Q6.9   (a) polar coordinates of poles:
$\qquad$ $r = 0.32492, \quad \theta = 0$
$\qquad$ $r = 0.83360, \quad \theta = \pm 53.31°$
$\qquad$ $r = 0.57395, \quad \theta = \pm 47.10°$
$\qquad$ $r = 0.39599, \quad \theta = \pm 30.85°$
$\qquad$ (also 7th-order zero at $z = -1$)

Q6.10   6th-order

Q6.11   Polar coordinates of poles:
$\qquad$ $r = 0.90985, \quad \theta = 180°$
$\qquad$ $r = 0.95479, \quad \theta = \pm 163.71°$
$\qquad$ (also 3rd-order zero at $z = 1$)
$\qquad$ peak gain = 1126.2 (61.03 dB)
$\qquad$ cut-off at 0.8π -29 dB
$\qquad$ $y[n] = -2.7428 \, y[n-1] - 2.5793 \, y[n-2]$
$\qquad\qquad\quad -0.8294 \, y[n-3] + x[n] - 3x[n-1]$
$\qquad\qquad\quad +3x[n-2] - x[n-3]$

Q6.12 $w[n] = -0.90985 \, w[n-1] + 0.000\,888 \, \{x[n] - x[n-1]\}$
$\qquad$ $y[n] = -1.8329 \, y[n-1] - 0.9116 \, y[n-2] + w[n]$
$\qquad\qquad\qquad\qquad -2w[n-1] + w[n-2]$

Q6.13 $y[n] = 1.7236 \, y[n-1] - 0.74082 \, y[n-2]$
$\qquad\qquad\qquad +x[n] - 0.9909 \, x[n-1]$

Q6.14 $y[n] = 1.3196 y[n-1] - 0.4230 y[n-2] + x[n]$
$\qquad\qquad\qquad\qquad -0.9928 \, x[n-1]$

Q6.17 $w[n] = x[n] - 0.83512 x[n-90]$
$\qquad$ $p[n] = 1.5290 p[n-1] - 0.99600 p[n-2] + w[n]$
$\qquad$ $q[n] = 1.43580 \, q[n-1] - 0.99600 \, q[n-2] - w[n]$
$\qquad$ $r[n] = 1.33558 \, r[n-1] - 0.99600 \, r[n-2] + w[n]$
$\qquad$ $s[n] = 1.22886 \, s[n-1] - 0.99600 \, s[n-2] - w[n]$
$\qquad$ $t[n] = 1.11615 \, t[n-1] - 0.99600 \, t[n-2] + 0.5 \, w[n]$
$\qquad$ $u[n] = 0.99800 \, u[n-1] - 0.99600 \, u[n-2] - 0.5 \, w[n]$
$\qquad$ $y[n] = p[n] + q[n] + r[n] + s[n] + t[n] + u[n]$

Q6.19   (b) $y[n] = y[n-1] + x[n] - x[n-10]$
$\qquad\qquad$ $y[n] = -y[n-1] - y[n-2] + x[n] - x[n-12]$

Q6.20 $y[n] = -2y[n-1] - y[n-2] + x[n] - x[n-1] - 2x[n-12]$
$\qquad\qquad\qquad +2x[n-13] + x[n-24] - x[n-25]$

impulse response values:
$$1, -3, 5, -7, 9, -11 \ldots, 21, -23, 23, -21, 19 \ldots$$
$$-9, 7, -5, 3, -1$$

Q6.22 Compared with ideal integrator, responses at
$\Omega = 0.2\pi, 0.5\pi,$ and 0.9π are:

running sum: $+0.33, +2.10, +7.17$ dB
trapezoidal rule: $-0.67, -4.83, -29.9$ dB
Simpson's rule: $+0.02, +0.92, +23.26$ dB

CHAPTER 7

Q7.3 (a) $3\,X[k]$; (b) $X[k]W_N^{2k}$; (c) $X[k]\,\{2+W_N^{-k}\}$;

(d) $\dfrac{1}{N} \displaystyle\sum_{m=0}^{N-1} X[m]\,X[k-m]\,W_N^{k-m}$

Q7.6 (a) $X[0] = 0$; $X[1] = 2$
(b) 7; j; -3; $-j$
Q7.8 (a) $X[9]$ and $X[31]$; (b) $X[0] = X[39] = 0$
Q7.14 (a) 18.3; (b) 102.4; (c) 4096
Q7.16 0, 16, 8, 24, 4, 20, 12, 28, 2, 18, 10, 26, 6, 22, 14, 30,
1, 17, 9, 25, 5, 21, 13, 29, 3, 19, 11, 27, 7, 23, 15, 31,

CHAPTER 8

Q8.4 (a) 8th and 56th
(b) 7th and 57th; -6.55 dB
(c) 480 samples/second
Q8.5 (a) 42nd, 43rd, 44th; 0.286:1:0.667
(b) 40th, 41st, 42nd; 0.344:1:0.204
Q8.6 (a) 21.48 Hz
(b) 0.0491 radian
(c) 0.00586 Hz
Q8.8 0.99116
Q8.11 $N = 16$
Q8.12 (a) 15.625 Hz
(b) 1.953 Hz
Q8.14 One period of circular convolution has values:
4, 1, -3, 1, 1, 0
Linear convolution gives values:
1, 1, -1, 0, 0, 0, 3, 0, -2, 1, 1
$N \geq 11$

CHAPTER 9

Q9.1   (b) Mean $= 0$; mean-square $= 2/3$; variance $= 2/3$
Q9.3   (b) Mean $= 4$; mean-square $= 19$; variance $= 3$
Q9.9   Mean $= 0.4$; mean-square $= 0.8$; standard deviation $= 0.8$;
       3 sampling intervals

Q9.11 (a) $\phi_{xx}[m] = \dfrac{1}{4}\,\text{sinc}\left(\dfrac{\pi m}{4}\right)$

Q9.15 $1 + 2\exp(-j\Omega) + 2\exp(-j2\Omega) + \exp(-j3\Omega)$

CHAPTER 10

Q10.1   (a) 6
        (b) 0.5556
        (c) 2.857

Q10.2 $\dfrac{4}{1 - 2r\cos\theta + r^2}$

Q10.3   (a) 2
        (c) 10
        (d) $P_{yy}(\Omega) = 10 + 16\cos\Omega + 8\cos 2\Omega + 2\cos 3\Omega$

Q10.4   Variance $= 8$
Q10.5   (a) $H(\Omega) = 2 + \exp(-j\Omega) + \exp(-j2\Omega)$
Q10.8   (a) 10 dB
        (b) 17 dB
        (c) 16 dB
Q10.11 (a) 2.73 dB
        (c) 7.78 dB
Q10.13 (a) 100
        (b) 100,000

Bibliography

A. Books largely or exclusively concerned with Digital Signal Analysis and Processing:

1. Bellanger, M., *Digital Processing of Signals*, 2nd edn (Chichester: Wiley, 1989).
2. Bogner, R.E. and Constantinides, A.G., *Introduction to Digital Filtering* [Chichester: Wiley, 1975).
3. Bowen, B.A. and Brown, W.R., *VLSI Systems Design for Digital Signal Processing; Vol I; Signal Processing and Signal Processors* (Englewood Cliffs: Prentice-Hall, 1982).
4. Burrus, C.S. and Parks, T.W., *DFT/FFT and Convolution Algorithms* (New York: Wiley, 1985).
5. Hamming, R.W., *Digital Filters*, 2nd edn (Englewood Cliffs: Prentice-Hall, 1983).
6. Jury, E.I., *Theory and Application of the z-Transform Method* (New York: Wiley, 1964).
7. Kung, S.Y., Whitehouse, H.J. and Kailath, T., *VLSI and Modern Signal Processing* (Englewood Cliffs: Prentice-Hall, 1985).
8. Oppenheim, A.V. and Schafer, R.W., *Digital Signal Processing* (Englewood Cliffs: Prentice-Hall, 1975).
9. Parks, T.W. and Burrus, C.S., *Digital Filter Design* (New York: Wiley, 1987).
10. Rabiner, L.R. and Gold, B., *Theory and Application of Digital Signal Processing* (Englewood Cliffs: Prentice-Hall, 1975).
11. Ramirez, R.W., *The FFT-Fundamental and Concepts* (Englewood Cliffs: Prentice-Hall, 1985).
12. Smith, M.J.T. and Mersereau, R.M., *Introduction to Digital Signal Processing* (New York: Wiley, 1992).
13. Stanley, W.D., Dougherty, G.R. and Dougherty, R., *Digital Signal Processing*, 2nd edn (Reston, Virginia: Reston, 1984).
14. Taylor, F.J., *Digital Filter Design Handbook* (New York: Marcel Dekker, 1983).
15. Williams, C.S., *Designing Digital Filters* (Englewood Cliffs: Prentice-Hall, 1986).

B. Books covering analog and digital signals and systems:

16. Baher, H., *Analog and Digital Signal Processing* (Chichester: Wiley 1990).
17. Lynn, P.A., *Introduction to the Analysis and Processing of Signals*, 3rd edn (London: Macmillan, 1989).
18. Lynn, P.A., *Electronic Signals and Systems* (London: Macmillan, 1986)
19. McGillem, C.D. and Cooper, G.R., *Continuous and Discrete Signal and System Analysis* 2nd edn (New York: Holt, Rinehart and Winston, 1984).
20. Oppenheim, A.V., Willsky, A.S. and Young, I.T., *Signals and Systems* (Englewood Cliffs: Prentice-Hall, 1983).
21. Papoulis, A., *Circuits and Systems* (New York: Holt, Rinehart and Winston, 1980).
22. Ziemer, R.E., Tranter, W.H. and Fannin, D.R., *Signals and Systems* (New York: Macmillan, 1983).

C. More specific books and references for Chapters 5 to 8:

23. Bergland, G.D., 'A Guided Tour of the Fast Fourier Transform', *IEEE Spectrum*, **6**, pp 41–52, 1969.
24. Cooley, J.W., and Tukey, J.W., 'An Algorithm for the Machine Computation of Complex Fourier Series', *Math. Comp.*, **19**, pp 297–301, 1965.
25. Kaiser, J.F., 'Nonrecursive Digital Filter Design using the I_0-sinh Window Function', *Proc. 1974 IEEE Int. Symp. on Circuits and Syst.*, San Francisco, pp 20–23, 1974.
26. Niblack, W., *Digital Image Processing* (Englewood Cliffs: Prentice-Hall, 1986).
27. Rabiner, L.R. and Schafer, R.W., *Digital Processing of Speech* (Englewood Cliffs: Prentice-Hall, 1978).

D. Books covering probability, random signals and random DSP

28. Helstom, C.W., *Probability and Stochastic Processes for Engineers* (New York: Macmillan, 1991).
29. Leon Garcia, A., *Probability and Random Processes for Electrical Engineering* (Reading: Addison-Wesley, 1994).
30. Kay, S.M., *Fundamentals of Statistical Signal Processing & Estimation Theory* (Englewood Cliffs: Prentice-Hall, 1993).
31. Proakis, J.G., Rader, C.M., Ling, F. and Nikias, C.L., *Advanced Signal Processing* (New York: Macmillan, 1992).
32. Scharf, L.L., *Statistical Signal Processing: Detection, Estimation and Time Series Analysis* (Reading: Addison-Wesley, 1991).
33. Therrien, C.W., *Discrete Random Signals and Statistical Signal Processing* (Englewood Cliffs: Prentice-Hall, 1992).
34. Vaseghi, S.V., *Advanced Signal Processing and Digital Noise Reduction* (Chichester: Wiley-Teubner, 1996).

Index

Printed and bound in the UK by
CPI Antony Rowe, Eastbourne